Monte-Carlo Methods and Stochastic Processes
From Linear to Non-Linear

Monte-Carlo Methods and Stochastic Processes
From Linear to Non-Linear

EMMANUEL GOBET

ECOLE POLYTECHNIQUE - UNIVERSITY PARIS-SACLAY

CMAP, PALAISEAU CEDEX, FRANCE

CRC Press
Taylor & Francis Group
Boca Raton London New York

CRC Press is an imprint of the
Taylor & Francis Group, an **informa** business

A CHAPMAN & HALL BOOK

CRC Press
Taylor & Francis Group
6000 Broken Sound Parkway NW, Suite 300
Boca Raton, FL 33487-2742

© 2016 by Taylor & Francis Group, LLC
CRC Press is an imprint of Taylor & Francis Group, an Informa business

No claim to original U.S. Government works

International Standard Book Number-13: 978-1-4987-4622-9 (Hardback)

This book contains information obtained from authentic and highly regarded sources. Reasonable efforts have been made to publish reliable data and information, but the author and publisher cannot assume responsibility for the validity of all materials or the consequences of their use. The authors and publishers have attempted to trace the copyright holders of all material reproduced in this publication and apologize to copyright holders if permission to publish in this form has not been obtained. If any copyright material has not been acknowledged please write and let us know so we may rectify in any future reprint.

Except as permitted under U.S. Copyright Law, no part of this book may be reprinted, reproduced, transmitted, or utilized in any form by any electronic, mechanical, or other means, now known or hereafter invented, including photocopying, microfilming, and recording, or in any information storage or retrieval system, without written permission from the publishers.

For permission to photocopy or use material electronically from this work, please access www.copyright.com (http://www.copyright.com/) or contact the Copyright Clearance Center, Inc. (CCC), 222 Rosewood Drive, Danvers, MA 01923, 978-750-8400. CCC is a not-for-profit organization that provides licenses and registration for a variety of users. For organizations that have been granted a photocopy license by the CCC, a separate system of payment has been arranged.

Trademark Notice: Product or corporate names may be trademarks or registered trademarks, and are used only for identification and explanation without intent to infringe.

Visit the Taylor & Francis Web site at
http://www.taylorandfrancis.com

and the CRC Press Web site at
http://www.crcpress.com

To Sophie

and

To Alexandre, Marion, Maxime, *and* Antoine.

Contents

INTRODUCTION: BRIEF OVERVIEW OF MONTE-CARLO METHODS — 1

A LITTLE HISTORY: FROM THE BUFFON NEEDLE TO NEUTRON TRANSPORT — 3
THREE TYPICAL PROBLEMS IN RANDOM SIMULATION — 10
▷ PROBLEM 1 – NUMERICAL INTEGRATION: QUADRATURE, MONTE-CARLO AND QUASI MONTE-CARLO METHODS — 10
▷ PROBLEM 2 – SIMULATION OF COMPLEX DISTRIBUTIONS: METROPOLIS-HASTINGS ALGORITHM, GIBBS SAMPLER — 18
▷ PROBLEM 3 – STOCHASTIC OPTIMIZATION: SIMULATED ANNEALING AND THE ROBBINS-MONRO ALGORITHM — 23

PART A: TOOLBOX FOR STOCHASTIC SIMULATION — 29

CHAPTER 1 ▪ Generating random variables — 31

1.1 PSEUDORANDOM NUMBER GENERATOR — 31
1.2 GENERATION OF ONE-DIMENSIONAL RANDOM VARIABLES — 32

	1.2.1	Inversion method	32
	1.2.2	Gaussian variables	36
1.3	ACCEPTANCE-REJECTION METHODS		37
	1.3.1	Generation of conditional distribution	37
	1.3.2	Generation of (non-conditional) distributions by the acceptance-rejection method	38
	1.3.3	Ratio-of-uniforms method	40
1.4	OTHER TECHNIQUES FOR GENERATING A RANDOM VECTO		42
	1.4.1	The Gaussian vector	43
	1.4.2	Modeling of dependence using copulas	44
1.5	EXERCISES		47

CHAPTER 2 ▪ Convergences and error estimates 49

2.1	LAW OF LARGE NUMBERS		49
2.2	CENTRAL LIMIT THEOREM AND CONSEQUENCES		52
	2.2.1	Central limit theorem in dimension 1 and beyond	52
	2.2.2	Asymptotic confidence regions and intervals	54
	2.2.3	Application to the evaluation of a function of $\mathbb{E}(X)$	56
	2.2.4	Applications in the evaluation of sensitivity of expectations	61
2.3	OTHER ASYMPTOTIC CONTROLS		65
	2.3.1	Berry-Essen bounds and Edgeworth expansions	65
	2.3.2	Law of iterated logarithm	66
	2.3.3	"Almost sure" central limit theorem	66
2.4	NON-ASYMPTOTIC ESTIMATES		67
	2.4.1	About exponential inequalities	67
	2.4.2	Concentration inequalities in the case of bounded random variables	69
	2.4.3	Uniform concentration inequalities	70

	2.4.4	Concentration inequalities in the case of Gaussian noise	77
2.5	EXERCISES		85

CHAPTER 3 ▪ Variance reduction — 89

3.1	ANTITHETIC SAMPLING		89
3.2	CONDITIONING AND STRATIFICATION		92
	3.2.1	Conditioning technique	92
	3.2.2	Stratification technique	92
3.3	CONTROL VARIATES		94
	3.3.1	Concept	94
	3.3.2	Optimal choice	95
3.4	IMPORTANCE SAMPLING		97
	3.4.1	Changes of probability measure: basic notions and applications to Monte-Carlo methods	97
	3.4.2	Changes of probability measure by affine transformations	103
	3.4.3	Change of probability measure by Esscher transform	107
	3.4.4	Adaptive methods	110
3.5	EXERCISES		112

PART B: SIMULATION OF LINEAR PROCESSES — 115

CHAPTER 4 ▪ Stochastic differential equations and Feynman-Kac formulas — 117

4.1	BROWNIAN MOTION		119
	4.1.1	A brief history	119
	4.1.2	Definition	119
	4.1.3	Simulation	124
	4.1.4	Heat equation	128

	4.1.5	Quadratic variation	131
4.2	STOCHASTIC INTEGRAL AND ITÔ FORMULA		132
	4.2.1	Filtration and stopping times	133
	4.2.2	Stochastic integral and its properties	134
	4.2.3	Itô process and Itô formula	137
4.3	STOCHASTIC DIFFERENTIAL EQUATIONS		138
	4.3.1	Definition, existence, uniqueness	138
	4.3.2	Flow property and Markov property	139
	4.3.3	Examples	139
4.4	PROBABILISTIC REPRESENTATIONS OF PARTIAL DIFFERENTIAL EQUATIONS: FEYNMAN-KAC FORMULAS		142
	4.4.1	Infinitesimal generator	142
	4.4.2	Linear parabolic partial differential equation with Cauchy condition	144
	4.4.3	Linear elliptic partial differential equation	148
	4.4.4	Linear parabolic partial differential equation with Cauchy-Dirichlet condition	149
	4.4.5	Linear elliptic partial differential equation with Dirichlet condition	153
4.5	PROBABILISTIC FORMULAS FOR THE GRADIENTS		153
	4.5.1	Pathwise differentiation method	154
	4.5.2	Likelihood method	155
4.6	EXERCISES		156

CHAPTER 5 ▪ Euler scheme for stochastic differential equations — 163

5.1	DEFINITION AND SIMULATION		164
	5.1.1	Definition as an Itô process, quadratic moments	164
	5.1.2	Simulation	166

	5.1.3 Application to computation of diffusion expectation: discretization error and statistical error	168
5.2	STRONG CONVERGENCE	170
5.3	WEAK CONVERGENCE	173
	5.3.1 Convergence at order 1	173
	5.3.2 Extensions	176
5.4	SIMULATION OF STOPPED PROCESSES	178
	5.4.1 Discrete approximation of exit time	179
	5.4.2 Brownian bridge method	181
	5.4.3 Boundary shifting method	184
5.5	EXERCISES	186

CHAPTER 6 ▪ Statistical error in the simulation of stochastic differential equations — 191

6.1	ASYMPTOTIC ANALYSIS: NUMBER OF SIMULATIONS AND TIME STEP	191
6.2	NON-ASYMPTOTIC ANALYSIS OF THE STATISTICAL ERROR IN THE EULER SCHEME	194
6.3	MULTI-LEVEL METHOD	197
6.4	UNBIASED SIMULATION USING A RANDOMIZED MULTI-LEVEL METHOD	202
6.5	VARIANCE REDUCTION METHODS	206
	6.5.1 Control variates	206
	6.5.2 Importance sampling	207
6.6	EXERCISES	208

PART C: SIMULATION OF NON-LINEAR PROCESSES — 211

CHAPTER 7 ▪ Backward stochastic differential equations — 213

xii ■ Contents

7.1	EXAMPLES		214
	7.1.1	Examples coming from reaction-diffusion equations	214
	7.1.2	Examples coming from stochastic modeling	217
7.2	FEYNMAN-KAC FORMULAS		221
	7.2.1	A general result	221
	7.2.2	Toy model	224
7.3	TIME DISCRETIZATION AND DYNAMIC PROGRAMMING EQUATION		227
	7.3.1	Discretization of the problem	227
	7.3.2	Error analysis	228
7.4	OTHER DYNAMIC PROGRAMMING EQUATIONS		231
7.5	ANOTHER PROBABILISTIC REPRESENTATION VIA BRANCHING PROCESSES		233
7.6	EXERCISES		236

CHAPTER 8 ▪ Simulation by empirical regression			241
8.1	THE DIFFICULTIES OF A NAIVE APPROACH		241
8.2	APPROXIMATION OF CONDITIONAL EXPECTATIONS BY LEAST SQUARES METHODS		244
	8.2.1	Empirical regression	245
	8.2.2	SVD method	246
	8.2.3	Example of approximation space: the local polynomials	249
	8.2.4	Error estimations, robust with respect to the model	250
	8.2.5	Adjustment of the parameters in the case of local polynomials	252
	8.2.6	Proof of the error estimations	254
8.3	APPLICATION TO THE RESOLUTION OF THE DYNAMIC PROGRAMMING EQUATION BY EMPIRICAL REGRESSION		257
	8.3.1	Learning sample and approximation space	257

	8.3.2 Calculation of the empirical regression functions	258
	8.3.3 Equation of the error propagation	260
	8.3.4 Optimal adjustment of the convergence parameters in the case of local polynomials	266
8.4	EXERCISES	267

CHAPTER 9 ▪ Interacting particles and non-linear equations in the McKean sense — 273

9.1	HEURISTICS	273
	9.1.1 Macroscopic scale versus microscopic scale	273
	9.1.2 Examples and applications	275
9.2	EXISTENCE AND UNIQUENESS OF NON-LINEAR DIFFUSIONS	278
9.3	CONVERGENCE OF THE SYSTEM OF INTERACTING DIFFUSIONS, PROPAGATION OF CHAOS AND SIMULATION	279

APPENDIX A ▪ Reminders and complementary results — 285

A.1	ABOUT CONVERGENCES	285
	A.1.1 Convergence a.s., in probability and in L_1	285
	A.1.2 Convergence in distribution	286
A.2	SEVERAL USEFUL INEQUALITIES	287
	A.2.1 Inequalities for moments	287
	A.2.2 Inequalities in the deviation probabilities	290

Bibliography	293
Index	307

Preface

This book originates from a third-year course on Monte-Carlo methods at Ecole Polytechnique – University Paris Saclay. A version in French, with less material, is published by Editions de l'Ecole Polytechnique.

In fact, several Monte-Carlo methods exist. They all use random simulations, but they can be quite different in their techniques and objectives. It is not possible to present and analyze in detail all the existing methods in only one course. For this reason, I have made two choices.

1. The core of the book is related to the *simulation of stochastic processes in continuous time* and their link with partial differential equations. Although this link between Brownian motion and the heat equation goes back a century, the computational probabilistic aspects developed later. Progress in data-processing in the 1980s facilitated the task of researchers and engineers to test the algorithms, to improve them, to develop new ones, to strive to numerically simulate complex systems, and to increase the accuracy or the speed of simulation.

   The Monte-Carlo methods for the simulation of stochastic differential equations have many applications in biology (Wright-Fisher model), finance (valuation of options), geophysics (porous media), random mechanics (moving solid under random forces), fluid mechanics (Navier-Stokes equation for vorticity), etc. In Part **B**, we will study some related methods of simulation.

   Since the end of the 1990s, important progress has been made in the field of non-linear processes in connection with problems of control or modeling of interaction. This is connected to topical issues, in research and in applications (chemistry, ecology, economy, finance, neurosciences, material physics, etc); we devote Part **C** of this book to those non-linear processes. Part **A** gathers the basic tools of simulation and analysis of algorithm convergence.

2. In addition, because there are many Monte-Carlo methods, we wish to give to the unfamiliar reader a short overview (certainly incomplete though). This is the objective of the introductory chapter, which starts with a history and continues with the brief presentation of *three typical Monte-Carlo problems*: numerical integration and computation of expectation, simulation of complex distributions, and stochastic optimization. References are given progressively to enable readers to refer to published works handling one subject or another, according to their interest and needs. In this brief overview, we emphasize the ideas and not the rigorous supporting mathematics: Markov chains and discrete time martingales constitute the main probabilistic concepts for analyzing problems on simulation of complex distributions and stochastic optimization, see for instance [20], [32] and [26].

The problem of numerical integration and computation of expectation $\mathbb{E}X$ by the Monte-Carlo method is developed in this book thoroughly. A focus is made on the case where X is a path functional of a stochastic process in continuous time, possibly with non-linear dynamics.

Synopsis. The book is organized in three parts of progressive difficulty.

▷ **Part A: Toolbox for Stochastic Simulation.** The practice of random simulation requires the ability to simulate appropriate random variables. It is studied in Chapter 1. The convergence of the empirical mean of simulations towards the unknown expectation is studied in Chapter 2: we review the asymptotic tools (law of the large numbers, central limit theorem) and nonasymptotic ones (concentration inequalities, in a pointwise or uniform version). These tools are essential for suitably quantifying the statistical error of the Monte-Carlo method. Chapter 3 tackles the question of acceleration of convergence (variance reduction methods). We emphasize the methods of importance sampling, whose application to the evaluation of rare events is spectacular.

▷ **Part B: Simulation of Linear Processes.** Chapter 4 presents a very minimal background in stochastic calculus so that we can introduce and study stochastic differential equations and their simulation. The link with partial differential equations is then given, via Feynman-Kac formula, representing the solution to

PDE as an expectation of a functional of stochastic process. We also give representations of sensitivities as expectations. Chapter 5 tackles the question of simulating stochastic differential equation: exact simulation is not possible in general and we study the discretization by Euler scheme. We then extend these techniques to the problem of simulating exit time. In Chapter 6, we study the related statistical errors, the methods of variance reduction, and the multi-level Monte-Carlo methods.

▷ **Part C: Simulation of Non-Linear Processes.** The objective of the last part is to study non-linear dynamics, a field which is currently developing quickly, with a focus on their simulation. We present three generic non-linearities: backward equation and control, branching process, and mean-fields interaction. In Chapter 7, we define the solution to the backward stochastic differential equation. The link with semi-linear PDEs is established, extending the results of Part **B**. We present another interpretation of certain equations using branching diffusion. Then we study the time discretization of the backward process, which results in solving a dynamic programming equation. Chapter 8 is devoted to its resolution by the method of empirical regression (*supervised learning*), establishing the link with certain advanced statistical tools studied in Part **A**. The error analysis is fully carried out, displaying the adjustment of the convergence parameters. The resolution of the dynamic programming equation by empirical regression has many applications beyond the framework of backward stochastic differential equations, in the context of *machine learning, optimal stopping* [107], and *robust control*, with applications to non-linear finance [67], optimal decision and optimal investment for instance. Chapter 9 introduces the McKean-Vlasov stochastic equation whose non-linearity naturally translates an asymptotic interaction with a cloud of stochastic differential equations. A simulation algorithm results from this, which is different from the dynamic programming of Chapter 7.

At the end of each chapter, we provide some exercises of a theoretical or programming nature. Solutions and complementary material are available on the website
 http://montecarlo-polytechnique.blogspot.com

Throughout this book, we emphasize the main algorithms, the most important convergence phenomena, and the essential tools, ranging from the most basic to the most advanced. Certain aspects of Parts **A-B-C** are generally not taught at the level of a master's program, but we have made a quite significant pedagogical effort to demystify them and make them available (in a simplified but not denatured form) to master's students. Proofs of results are usually given, some only outlined, but we often choose the simplest presentation and we try to use the arguments requiring fewer mathematical prerequisites.

Nevertheless, this is a quite demanding textbook of applied mathematics, covering a broad spectrum of advanced and sometimes very modern tools of probabilities, statistics and partial differential equations, with systematic computational concerns regarding numerical efficiency. Moreover, we encourage readers to implement the algorithms in order to develop their own computational intuition: this is certainly an important skill for mastering and understanding the theory. Moreover, the reader should keep in mind that even if an algorithm converges *theoretically* quicker than another, it may be that its execution time is much longer, and that it is actually less efficient: thus, comparing a convergence speed or an error variance is not all that is required; computational time and memory requirements may be significant features, which can be assessed only by implementing the method on a computer.

Our presentation of algorithms assumes that the implementation is made sequentially on a machine with a single processor. It is clear that the implementation on parallel architecture could be performed alternatively, and this is also an active field of research.

Last, it is difficult to be very original on such a classic subject on Monte-Carlo method and this Ecole Polytechnique course took, as a starting point, that of my predecessors (in particular L. Elie, C. Graham, B. Lapeyre, D. Talay). I would like to thank my colleagues who have encouraged me to transform my lecture notes into a published book. I especially thank P. Del Moral, S. De Marco, M. Gubinelli, and B. Jourdain for their feedback on a first version of this book. Thank you also to U. Stazhynski for his assistance in the translation from the French version to the English one.

Emmanuel Gobet, Paris Saclay
December 2015

List of Figures

1	Koch snowflake set and randomly fallen beads	3
2	Petersen graph with differently colored neighbor vertices	4
3	Georges Louis Leclerc, count of Buffon (1707–1788)	5
4	Nicholas Constantine Metropolis (1915–1999)	6
5	19,000 tubes of the ENIAC	7
6	John von Neumann (1903–1957)	7
7	Stanislaw Marcin Ulam (1909–1984)	8
8	1000 independent random points uniformly distributed on the square	12
9	The first 1000 points of the two-dimensional sequence of the torus algorithm	13
10	Estimation of integral in dimension 3 with the random sequence and the three-dimensional sequence of the torus algorithm	15
11	Estimation of integral in dimension 3 with the random sequence and the one-dimensional sequence of the torus algorithm	16
1.1	Two samples of random variables in dimension 2, with Gaussian marginals, and Gaussian or Clayton copulas	46
1.2	Two samples of random variables in dimension 2, with Laplace marginals, and independent or Gaussian copulas	47
2.1	Several sets of 1000 simulations for the calculation of $\mathbb{E}(X)$ with uniformly distributed X	52
2.2	Monte-Carlo evaluation of $\mathbb{E}(e^{G/10})$ and $\mathbb{E}(e^{2G})$ with Gaussian random variable G	55
2.3	An example of ε-cover	72

2.4	Clipping $\mathcal{C}_L$	74
3.1	Monte-Carlo evaluation of $\mathbb{P}(|Y| \geq 6)$ by a naive method and by importance sampling method	106
4.1	The random walk after renormalization in time and space	119
4.2	A Brownian path and the two curves $f(t) = \pm 1.96\sqrt{t}$	121
4.3	Wiener approximation W^n for $n = 10, 20, 50$	122
4.4	Iterative construction of Brownian motion by the Brownian bridge technique	127
4.5	Brownian motion in dimension 2 and 3	129
4.6	Cauchy-Dirichlet condition and stopped process	150
5.1	Local approximation by half-space	182
5.2	Modified domain by locally shifting the boundary in the inward normal direction	186
7.1	Branching diffusion	234
8.1	Local polynomial function	250
8.2	Empirical regression error	252
9.1	Dyson Brownian motion	277

List of Algorithms

1	Metropolis-Hastings algorithm	20
2	Random Gibbs sampler	22
3	Simulated annealing algorithm	25
1.1	Generation of Gaussian random variables by the Box-Muller method	36
1.2	Acceptance-rejection method	39
1.3	Simulation of a Student distribution using the ratio-of-uniforms method	42
5.1	Computation by Monte-Carlo method of the expectation of diffusion process	169
5.2	Simulation of stopped process using discrete exit time	180
5.3	Simulation of stopped process by Brownian bridge	183
5.4	Simulation de stopped process by boundary shifting	185
8.1	Resolution of the dynamic programming equation by empirical regression	259

Frequently used notation

$|x|$ and $x \cdot y$: Euclidean norm and scalar product for $x, y \in \mathbb{R}^d$.

Id: Identity matrix.

A^T: the transpose of the matrix of A.

$\mathrm{Tr}(A)$ and $\det(A)$: trace and determinant of a square matrix A.

$:=$ defines the left term as equal to the right term.

$\mathcal{C}^k(\mathbb{R}^d, \mathbb{R}^q)$: the set of k-times continuously differentiable functions from $\mathbb{R}^d$ to $\mathbb{R}^q$.

$\mathcal{C}^k_b(\mathbb{R}^d, \mathbb{R}^q)$: the same set restricted to bounded functions with bounded derivatives.

∇f: gradient of $f \in \mathcal{C}^1(\mathbb{R}^d, \mathbb{R}^q)$ defined by $(\nabla f)_{i,j} = \partial_{x_j} f_i$ for $1 \leq i \leq q$ and $1 \leq j \leq d$ (note that when $q = 1$, ∇f is a row vector).

$\nabla^2 f$: Hessian of $f \in \mathcal{C}^1(\mathbb{R}^d, \mathbb{R})$ defined by $(\nabla^2 f)_{i,j} = \partial^2_{x_i, x_j} f$.

$\log(x)$: natural logarithm of x (in Neper basis, this is the only logarithm used).

$\mathbb{P}(B)$ and $\mathbb{E}(X)$: probability of B and expectation of X.

$\mathbb{P}(B|A)$ and $\mathbb{E}(X|A)$: the same but conditionally given A.

Some probability distributions:

- Bernoulli distribution $\mathcal{B}(p)$, see page 33.
- Exponential distribution $\mathcal{E}\mathrm{xp}(\lambda)$, see page 33.
- Gamma distribution $\Gamma(\alpha, \theta)$, see page 39.

- Gaussian distribution $\mathcal{N}(m, \sigma^2)$, see page 35.
- The cumulative distribution function of the standard Gaussian distribution is
$$\mathcal{N}(u) := \int_{-\infty}^{u} \frac{e^{-\frac{1}{2}x^2}}{\sqrt{2\pi}} \, \mathrm{d}x.$$
- Geometric distribution $\mathcal{G}(p)$, see page 34.
- Poisson distribution $\mathcal{P}(\theta)$, see page 33.
- Rademacher distribution, see page 119.
- Student distribution with k degrees of freedom, see 42.
- Uniform distribution on $[a,b]$, denoted by $\mathcal{U}([a,b])$.

$\stackrel{\mathrm{d}}{=}$ means the equality in distribution between two random variables or between two probability distributions or between a random variable and a probability distribution (by a slight abuse of notation).

L_2: quadratic norm (to simplify, we use very little the L_p-norm for $p \neq 2$).

Probabilistic convergences:

- $X_M \xrightarrow[M \to +\infty]{\mathrm{a.s.}} X$: almost sure convergence of X_M to X as $M \to +\infty$, i.e.,
$$\mathbb{P}(X_M \xrightarrow[M \to +\infty]{} X) = 1.$$

- $X_M \xrightarrow[M \to +\infty]{\mathrm{Prob.}} X$: convergence in probability of X_M to X as $M \to +\infty$, i.e., for any $\varepsilon > 0$
$$\mathbb{P}(|X_M - X| > \varepsilon) \xrightarrow[M \to +\infty]{} 0.$$

- $X_M \xrightarrow[M \to +\infty]{L_1} X$: convergence in L_1 of X_M to X as $M \to +\infty$, i.e.,
$$\mathbb{E}|X_M - X| \xrightarrow[M \to +\infty]{} 0.$$

- $X_M \underset{M \to +\infty}{\Longrightarrow} X$: convergence in distribution (i.e. weak convergence) of X_M to X as $M \to +\infty$, i.e., for any bounded continuous function φ with compact support
$$\mathbb{E}(\varphi(X_M)) \xrightarrow[M \to +\infty]{} \mathbb{E}(\varphi(X)).$$

For the most important basic notions of probability theory, we refer the reader to [80] for example. Some reminders and important results can be found in Appendix A.

INTRODUCTION: BRIEF OVERVIEW OF MONTE-CARLO METHODS

A LITTLE HISTORY: FROM THE BUFFON NEEDLE TO NEUTRON TRANSPORT

How can we measure complex state spaces by random exploration? Generally, Monte-Carlo methods refer to algorithms using random simulation. A simple example is the following: consider a set A on the plane, say, the interior of the square $[-1, 1]^2$ and let us randomly and repeatedly throw a bead of infinitesimal size on the square, so that the throws are *uniformly distributed* and *independent*. Denote by n_M^A

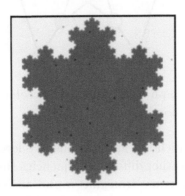

Figure 1 Set A (dark gray) that looks like a Koch snowflake and beads randomly fallen on the square (light gray).

the number of beads which fall on the set A after M throws: the larger the surface of A, the larger is the empirical frequency $f_M^A = \frac{n_M^A}{M}$ of the event "*a bead falls on A*". Precisely, as M goes to infinity, the empirical frequency converges to the ratio between the area of A — denoted by $|A|$ — and the area of the square $[-1, 1]^2$, i.e., with probability 1,

$$\lim_{M \to +\infty} f_M^A = \frac{|A|}{4}.$$

This result is simply the strong Law of Large Numbers, which validates the frequentist axiomatic of probability: the ratio $\frac{|A|}{4}$ is the probability for a uniform variable on the square to be in A. This simple experiment allows us to calculate — if only we are ready to repeat it infinitely many times — the area of A, even if A has complicated geometry. If A is the disk with center in the origin with radius 1, then we get a simple way for evaluating π.

Take an example of a graph with each vertex colored using one of 20 available colors. How can we determine the number of all colorings for which any two neighbors have different colors? This is a classic problem

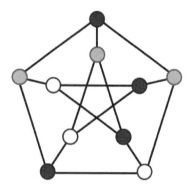

Figure 2 Petersen graph with differently colored neighbor vertices (among 3 possible colors).

of graph coloring, which has applications in telecommunications: when we want to allocate rather different frequencies (to avoid interferences), each vertex is a frequency transmitter and an edge specifies that the frequencies allocated to the ends must be different.

If the network has K vertices, the number of possible colored graphs is K^{20} and it quickly becomes impossible to enumerate all the graphs when K is about a hundred or more. Alternatively, rather than calculating the number of graphs having all the pairs of neighbors with different colors, we can evaluate their *proportion* p among the K^{20} possible graphs. For that, let us generate a sequence of random graphs, such that the color of each vertex is chosen uniformly among the 20 possible colors and independently of the others. After a large number M of draws, the proportion of admissible graphs becomes close to the proportion p, and it even converges to p as $M \to +\infty$.

These two examples sketch out how random simulation can provide a simple way for evaluating — certainly with error due to finiteness of the number of draws — deterministic and possibly very complicated quantities. One of the main questions concerns the quantification of the computational error of this probabilistic algorithm, and we will return to it later. In these two examples, if the area of A is small or if the number of admissible graphs is small, it is clear that the method will be less efficient due to the large number of wasted draws, and it may be no longer relevant to produce uniform draws. Instead we can try to properly give weights to the results of the experiment. This method is called *importance sampling* and it will also be studied in the sequel.

The advantage of Monte-Carlo methods relies on its *capacity to explore high-dimensional spaces of configurations* to get some information. Recently, they have been used to design software[1] for the *game of go*, which can be played with a human player and can even beat champions[2]; in this case, the space of possible configurations is approximately of the size 10^{600}!

The Buffon needle. Historically, one of the first examples of random simulation made for numerical computation, which was less trivial than the previous ones, is described by the experiment of the *Buffon needle*, proposed by the count with the same name in 1733. It consists

Figure 3   Georges Louis Leclerc, count of Buffon (1707–1788).

of randomly throwing a needle on a floor composed from parallel laths of width l, the needle having length $a \leq l$, and observing if the needle intersects one of the grooves of the parquet (i.e., if it touches two different laths or not). We can show that the probability of this event is $p = \frac{2a}{\pi l}$: so, independently repeating the experiment a large number of times, we get the convergence of the frequency of touching a groove to p, which again gives a way of calculating π.

---

[1] http://www.wired.com/2014/05/the-world-of-computer-go/
[2] the computer program AlphaGo developed by Google DeepMind beat the world champion Lee Sedol in March 2016.

Atomic bomb and the development of the computer. It was only in the middle of the 20th century when the numerical methods based on random simulation experienced an unprecedented development. From 1943, the USA, being dragged into the war, made efforts to develop a weapon of a new type, a nuclear weapon. The research was done secretly in the Los Alamos National Laboratory, in New Mexico, conducted by a large number of top-level scientists, including Noble Prize winners in physics. A physicist, Nicholas Metropolis, was among the researchers. The first nuclear testing in July 1945 in Alamogordo,

Figure 4 Nicholas Constantine Metropolis (1915–1999).

and later the devastating nuclear bombing of Hiroshima and Nagasaki in August 1945, hastened the end of the Second World War and gave the USA a certain confidence in mastery of nuclear weapons. The return to peace marks the start of massive atomic weapons production, largely supporting the activity in the Los Alamos laboratory.

Concurrent with this upheaval, another revolution was coming, the one in computer science. At the beginning of 1946, the first completely electronic computer was ready: this is the ENIAC, which weighted no less than 30 metric tons, and which was capable of performing 100,000 additions per second. It was first proposed to *Ballistics Research Laboratory* of Aberdeen (Maryland) for the computation of ballistics tables, then to Los Alamos through John von Neumann,[3] a

---

[3]Source of the photo: Los Alamos National Laboratory. *"Unless otherwise indicated, this information has been authored by an employee or employees of the University of California, operator of the Los Alamos National Laboratory under Contract No. W-7405-ENG-36 with the U.S. Department of Energy. The U.S. Government has rights to use, reproduce, and distribute this information. The public may copy and*

A little history: from the Buffon needle to neutron transport ■ 7

Figure 5   A technician changing one of the 19,000 tubes of the ENIAC (Electronic Numerical Integrator Analyzer and Computer).

mathematician who collaborated with both laboratories. The first com-

Figure 6   John von Neumann (1903–1957).

putational project on thermonuclear reactions was launched, gathering Metropolis, von Neumann and Enrico Fermi. A mathematician, Stanislaw Ulam[4], participating in one of the presentations of the project in 1946, is impressed by the speed of computations and the quality of

---

use this information without charge, provided that this Notice and any statement of authorship are reproduced on all copies. Neither the Government nor the University makes any warranty, express or implied, or assumes any liability or responsibility for the use of this information."

[4]Source of the photo: Los Alamos National Laboratory http://www.lanl.gov/history/wartime/staff.shtml. "Unless otherwise indicated, this information has been authored by an employee or employees of the

the available numerical results: he understands immediately that the appearance of modern computer will lead to a new era in experimental mathematics. Ulam has a great interest in randomness, both in

Figure 7   Stanislaw Marcin Ulam (1909–1984).

his hobbies, he devotes himself to games involving randomness (solitaire, poker), and in his research. He knows very well the methods of stochastic sampling, thus he proposes to von Neumann to use ENIAC for neutron transport computations based on random simulation. The studied model is a model of neutron diffusion in a fissile medium. Later, he realizes that Fermi already used the ideas of random sampling in the 1930s. One day, Metropolis proposes the name *Monte-Carlo* for this method, joking about the uncle of Ulam who borrowed money to go to Monte-Carlo (a neighborhood in Monaco known for its famous casino). The name *Monte-Carlo method* remained. This was followed by numerous studies on *Monte-Carlo methods* in Los Alamos, with a program code that still exists — *Monte-Carlo N-Particle* (MCNP) transport code — and the methods rapidly spread all over the world. The interested reader is referred to the following historical articles, summaries or accounts: [116, 140, 11, 72, 114, 35].

In 2000 [29], the Society for Industrial and Applied Mathematics

---

University of California, operator of the Los Alamos National Laboratory under Contract No. W-7405-ENG-36 with the U.S. Department of Energy. The U.S. Government has rights to use, reproduce, and distribute this information. The public may copy and use this information without charge, provided that this Notice and any statement of authorship are reproduced on all copies. Neither the Government nor the University makes any warranty, express or implied, or assumes any liability or responsibility for the use of this information."

(SIAM) selected the Monte-Carlo method among the 10 algorithms which have most influenced the development and practice of the engineering sciences during the 20th century.

THREE TYPICAL PROBLEMS IN RANDOM SIMULATION

▷ PROBLEM 1 – NUMERICAL INTEGRATION: QUADRATURE, MONTE-CARLO AND QUASI MONTE-CARLO METHODS

Consider a problem of numerical computation of the following integral

$$I = \int_{[0,1]^d} f(x)\mathrm{d}x \tag{1}$$

where $f : [0,1]^d \mapsto \mathbb{R}$ is an integrable function. If the integral is in $\mathbb{R}^d$, it is always possible to make, as a preliminary step, a change of variable such that the domain of integration becomes $[0,1]^d$.

The Monte-Carlo method. The evaluation of I by Monte-Carlo method is based on generating independent random variables $(U_1, \ldots, U_m, \ldots)$ uniformly distributed on $[0,1]^d$. Of course, the simulation of U_m can be made via the generation of d independent random variables uniformly distributed on $[0,1]$. Then the Monte-Carlo estimator is written as

$$I_M := \frac{1}{M} \sum_{m=1}^{M} f(U_m). \tag{2}$$

The law of large numbers (see Section 2.1) assures that with probability 1,

$$\lim_{M \to +\infty} I_M = \mathbb{E}(f(U_1)) = \int_{[0,1]^d} f(x)\mathrm{d}x. \tag{3}$$

Moreover, if we suppose that f is square integrable $\int_{[0,1]^d} f^2(x)\mathrm{d}x < +\infty$, then the central limit theorem (Theorem 2.2.1) gives a convergence rate (in distribution) equal to $\sqrt{M}$: indeed, denoting $\sigma^2 = \mathbb{V}\mathrm{ar}(f(U_1))$, we have

$$\sqrt{M}(I_M - I) \underset{M \to +\infty}{\Longrightarrow} \mathcal{N}(0, \sigma^2). \tag{4}$$

It is remarkable that the convergence rate does not depend on the dimension d, which becomes especially interesting if d is large (or even infinite). Moreover, $\sigma^2 = \mathbb{E}(f^2(U_1)) - (\mathbb{E}(f(U_1)))^2$ can also be written as an expectation and evaluated using the same draws: this allows us to construct a completely explicit error estimation. We will return to this later.

Deterministic discretization methods. The rectangle and trapezoid methods and their different versions consist of putting N points in each of the d directions, regularly placed with ad hoc weighting. If the function f is Lipschitz, the accuracy is of order N^{-1}: so, for the computational cost $\mathcal{C} = N^d$ (number of space points), the accuracy is $\mathcal{C}^{-1/d}$, showing significant sensitivity to the dimension d (the famous *curse of dimensionality*). With more regularity on f and using integration methods of higher order, the accuracy can be improved, lowering the effect of dimension, without removing it.

Quadrature formulas. Let us start by describing the case $d = 1$. The approximation is written as

$$\int_0^1 f(x)\mathrm{d}x \approx \sum_{n=1}^{N} w_n f(x_n)$$

where $(w_n, x_n)_{1 \leq n \leq N}$ are the weights/points of Gauss-Legendre quadrature of degree N. These $2N$ parameters are such that the calculation is exact for the polynomials of degree at most $2N - 1$. If the integral is written instead as $\int_{\mathbb{R}} f(x) \frac{1}{\sqrt{2\pi}} e^{-\frac{x^2}{2}} \mathrm{d}x$, we use the Gauss-Hermite quadrature, whereas we use the Gauss-Laguerre quadrature if the weights in the integral are $e^{-x} \mathbf{1}_{x \geq 0}$, etc. We refer the reader to [48, Chapters 3 and 6] or [15]. If the function f is very close to a polynomial (thus very regular), the approximation is generally excellent.

For a multi-dimensional computation, we tensorize the formula of dimension 1. Here the method is again sensitive to dimension.

It is worth mentioning that in general,[5] there does not exist a simple way to pass from a formula with N points to a formula with $N + 1$ points: thus, adding points to increase the expected accuracy requires that we restart the calculation from the beginning.

Quasi Monte-Carlo methods. The convergence equation (3) of the Monte-Carlo method tells us that the empirical distribution of the points $(U_m)_{1 \leq m \leq M}$ converges, with probability 1, to a uniform distribution on the cube $[0, 1]^d$. We could complain that this completely random method fills the cube not efficiently enough, leaving too many *holes*, although it obviously finishes with the complete filling of the cube; see Figure 8. Quasi Monte-Carlo methods contain nothing ran-

---

[5] An exception is related to the Clenshaw-Curtis quadrature method, which is based on Chebyshev polynomials and nodes.

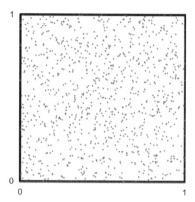

Figure 8 1000 independent random points uniformly distributed on the square $[0,1]^2$.

dom because they are based on a deterministic sequence of points, which fills the cube in a more regular manner. Unlike the methods of quadrature or discretization, mentioned earlier, *one point can be added without recalculation of the first M points.*

Discrepancy. Let us describe more quantitatively how a sequence $(x_m)_{m\geq 1}$ fills the cube $[0,1]^d$, comparing the proportion of points in $[0,y_1]\times\cdots\times[0,y_d]$ with respect to the uniform measure.

Definition 1 *The discrepancy of the sequence $(x_m)_{m\geq 1}$ is defined by*

$$D_M((x_m)_{m\geq 1}) = \sup_{y\in[0,1]^d} \left| \frac{1}{M}\sum_{m=1}^M 1_{x_m\in[0,y_1]\times\cdots\times[0,y_d]} - y_1\times\cdots\times y_d \right|.$$

The sequence is equidistributed if $\lim_{M\to+\infty} D_M((x_m)_{m\geq 1}) = 0$.

One can show that a sequence of independent random variables distributed uniformly on $[0,1]^d$ is equidistributed with probability 1, with discrepancy asymptotically bounded by $C\sqrt{\frac{\log(\log(M))}{M}}$. The rate cannot be improved (law of iterated logarithm, see Theorem 2.3.3).

We say that a sequence has a low discrepancy if its discrepancy is asymptotically smaller than that of a sequence of independent uniform random variables. However, the discrepancy of any sequence must fulfill

$$D_M((x_m)_{m\geq 1}) \geq C_d \frac{(\log M)^{\max(1,\frac{d}{2})}}{M}$$

for infinitely many values of M, giving a lower bound on what is possible in the best case. Let us give examples of a low-discrepancy sequence.

- **Irrational winding of a torus (torus algorithm).** These sequences are of the form

$$x_m = \bigl(\mathrm{Frac}(m\,\alpha_1),\ldots,\mathrm{Frac}(m\,\alpha_d)\bigr)$$

where $\mathrm{Frac}(y)$ is the decimal part of y and $(\alpha_i)_{1\leq i\leq d}$ is a free family in $\mathbb{Q}$. We can take for example $\alpha_i = \sqrt{p_i}$, where p_i is the i-th prime number. For this sequence, for any $\varepsilon > 0$ we have

$$D_M((x_m)_{m\geq 1}) \leq \frac{c_\varepsilon}{M^{1-\varepsilon}},$$

which improves the convergence of the random sequence and gets close the best possible bound. Figure 9 shows how the sequence regularly fills the square in dimension 2.

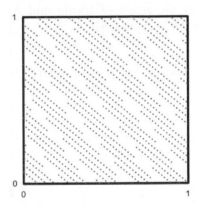

Figure 9 The first 1000 points of the sequence of the torus algorithm in dimension 2 ($\alpha_1 = \sqrt{2}$, $\alpha_2 = \sqrt{3}$).

- **Van Der Corput sequence.** Let p be an integer greater than 1 and write m in the form of its unique p-adic decomposition:

$$m = a_0 + \cdots + a_r p^r,$$

with $r \geq 0$, $a_i \in \{0,\ldots,p-1\}$ and $a_r > 0$. We set then

$$\phi_p(m) = \frac{a_0}{p} + \cdots + \frac{a_r}{p^{r+1}}.$$

- **Halton sequence.** This is a multi-dimensional generalization of the Van Der Corput sequence, taking for $(p_1, \ldots, p_d)$ the first d prime numbers, and setting

$$x_m = (\phi_{p_1}(m), \ldots, \phi_{p_d}(m)).$$

For this sequence, we have

$$D_M((x_m)_{m\geq 1}) \leq \frac{1}{M} \prod_{i=1}^{d} \frac{p_i \log(p_i M)}{\log(p_i)} \underset{M \to +\infty}{\sim} C \frac{(\log(M))^d}{M}. \quad (5)$$

There exist other low-discrepancy sequences, as those of Faure and Sobol, for which the construction is complicated; see [122]. The estimation of the discrepancy allows us to estimate the convergence rate for integrating f using an equidistributed sequence.

Proposition 2 (Koksma-Hlawka inequality) *Let f be a function of finite variation in the sense of Hardy and Krause, with variation $\mathbf{V}(f)$: then, for any $M \geq 1$, we have*

$$\left| \int_{[0,1]^d} f(x)\mathrm{d}x - \frac{1}{M} \sum_{m=1}^{M} f(x_m) \right| \leq \mathbf{V}(f) D_M((x_m)_{m \geq 1}).$$

We will not go into detail on the calculation of $\mathbf{V}(f)$, which is generally complicated: in dimension d, in the case when the function f is d-times continuously differentiable, we have

$$\mathbf{V}(f) = \sum_{k=1}^{d} \sum_{1 \leq i_1 < \cdots < i_k \leq d} \int_{[0,1]^d} \left| \frac{\partial^k f(x)}{\partial x_{i_1} \ldots \partial x_{i_k}} \right| \mathrm{d}x.$$

The Koksma-Hlawka inequality proves that we can achieve the convergence rate close to M for the numerical computation of I with a low-discrepancy sequence, while the rate is $\sqrt{M}$ if we use the random sequence (see the convergence (4)). In Figure 10, we check these results for the calculation of the integral of $f(x) = \exp(x_1 + x_2 + x_3)$. The exact value equals $I = (\exp(1) - 1)^3 \approx 5.073$. Clearly, the low-discrepancy sequence converges faster in this example.

The above expression of $\mathbf{V}(f)$ shows that to benefit from a low discrepancy of $(x_m)_{m \geq 1}$ in large dimension, the function f must be

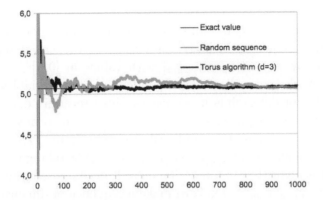

Figure 10 Estimation of $\int_{[0,1]^3} \exp(x_1 + x_2 + x_3) \mathrm{d}x$ with the random sequence and the sequence of the torus algorithm in dimension 3 ($\alpha_1 = \sqrt{2}, \alpha_2 = \sqrt{3}, \alpha_3 = \sqrt{5}$), as a function of the number M of points (from 1 to 1000).

more and more regular,[6] which limits the practical use of the above error bound. However, in view of the definition of discrepancy, the convergence of the error to 0 is guaranteed as soon as the function f is Riemann-integrable.

Finally, the low-discrepancy sequences offer the possibility of faster convergence than the random sequences, but without explicit error control. To learn more, see [24].

Behavior as the dimension increases. In the example of graph coloring (see page 4), the dimension d of the uniform random variable (i.e., the dimension of the variable of integration) can be very large (equal to the number of vertices of the graph) and may exceed several hundreds random variables. For simulating a process, as seen later we will also have very high dimensions of simulation. In fact, it is difficult to find efficient low-discrepancy sequences in high dimensions: we clearly see in the bound (5) that the impact of the dimension on the discrepancy is exponential. This heuristic is confirmed numerically for all known sequences.

How do we account for the high-dimensional situation? It is tempt-

---

[6]This is again a form of the curse of dimensionality: the convergence rate is not worsened by increasing dimension provided that the functions are more and more regular.

ing to try to gather several consecutive terms of a low-discrepancy sequence of dimension d_0 to build a sequence of higher dimension $d > d_0$: if the initial sequence in $[0,1]^{d_0}$ is $(x_m)_{m \geq 1}$, the new sequence can be defined by $\tilde{x}_m = [x_{2m-1}, x_{2m}]$ with values in $[0,1]^{2d_0}$. This construction works naturally for independent random variables, but unfortunately, for deterministic sequences, this procedure does not work because the new sequence $(\tilde{x}_m)_{m \geq 1}$ has no reason to have a low discrepancy and even to be equidistributed. In Figure 11, we consider again the calculation of $\int_{[0,1]^3} \exp(x_1 + x_2 + x_3) dx$ taking the sequence of the torus algorithm in dimension $d_0 = 1$, decomposed into groups of 3 ($\widehat{x}_m = [x_{3m-2}, x_{3m-1}, x_{3m}]$) to make a sequence of dimension $d = 3$. We observe a fast convergence, but to a wrong value.

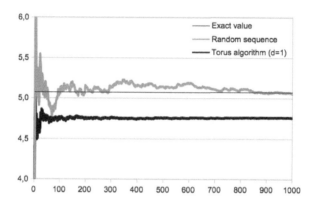

Figure 11 Estimation of $\int_{[0,1]^3} \exp(x_1 + x_2 + x_3) dx$ with the random sequence and the sequence $(\widehat{x}_m)_{m \geq 1}$ of the torus algorithm in dimension 1 extended to dimension 3.

Another way to get a sequence of dimension d is to complete a low-discrepancy sequence of dimension $d_0 < d$ with $d - d_0$ independent random variables having the uniform distribution on $[0,1]$. Unfortunately, this procedure generally leads to the loss of benefits from a low discrepancy of the first d_0 components and does not provide the desired accuracy. Even so, there are situations where this can work well, in particular, when certain directions of integration contain most of the variability of the function f. Instead of developing a theory, we consider a simple example illustrating this idea: for $d = 5$, consider $f(x) = \exp(x_1 + x_2 + x_3 + 0.001(x_4 + x_5))$. Then a good choice is to take $(x_m = [x_{1,m}, \ldots, x_{5,m}])_{m \geq 1}$, a low-discrepancy sequence in dimension 3, for the first three components $([x_{1,m}, x_{2,m}, x_{3,m}])_{m \geq 1}$, and then to take independent random variables to complete the last two coordinates.

Conclusion. For the numerical integration, it is commonly accepted that if the dimension is low, the discretization and quadrature formulas are the most efficient, and also that it is preferable to use a Quasi Monte-Carlo method for medium dimensions, while for high dimensions, the Monte-Carlo method has the advantage. The latter, in addition, offers the advantage of a priori error controls, with a rate $\sqrt{M}$, which is universal enough (robust to the regularity of the function or other specificities of the problem).

We can always represent an integral in the form of an expectation using a probabilistic interpretation of the Lebesgue measure dx, but an expectation cannot always be written in the form of an integral with an explicit density function. In the latter case, a Monte-Carlo method can be the only approach to the numerical integration.

Finally, we remark that a low-discrepancy sequence is biased while a Monte-Carlo method gives an unbiased estimator of the sought quantity: indeed, the empirical mean (2) fulfills $\mathbb{E}(I_M) = I$. To recover this property for a low-discrepancy sequence, we can *randomize* the sequence in the following way: if U is a random variable uniformly distributed on $[0,1]^d$, then the m-th term of the new sequence is

$$x_m^U = \big(\mathrm{Frac}(U_1 + x_{m,1}), \ldots, \mathrm{Frac}(U_d + x_{m,d})\big).$$

We can easily check that the property of low discrepancy is preserved. Moreover, the Quasi Monte-Carlo estimator is unbiased:

$$\mathbb{E}\big(\frac{1}{M}\sum_{m=1}^{M} f(x_m^U)\big)$$
$$= \frac{1}{M}\sum_{m=1}^{M} \int_{[0,1]^d} f(\mathrm{Frac}(u_1 + x_{m,1}), \ldots, \mathrm{Frac}(u_d + x_{m,d})) du$$
$$= \int_{[0,1]^d} f(u) du = I$$

using the translation invariance of the Lebesgue measure. See [102] for more details.

▷ **PROBLEM 2 – SIMULATION OF COMPLEX DISTRIBUTIONS: METROPOLIS-HASTINGS ALGORITHM, GIBBS SAMPLER**

The prototype of a complex distribution is the Gibbs distribution (called also the Boltzmann distribution). In Example 5 later, we give another illustration of a complex distribution coming from Bayesian statistics.

The Gibbs distribution describes the law of a physical system and is written in the form

$$\pi(x) = \frac{1}{\mathcal{Z}_T} e^{-\frac{1}{T} U(x)}$$

where $x \in \mathcal{X}$ is a configuration of the system, $U(x)$ is its energy (or potential), T represents temperature; the probability π appears naturally as the distribution of the thermodynamic equilibrium of the system with parameters U and T.

The constant $\mathcal{Z}_T$ is the renormalization constant for the distribution π. The problem is that in applications, the number of configurations (cardinal of $\mathcal{X}$) is often extremely large and, as a consequence, the calculation of $\mathcal{Z}_T$ is very difficult numerically, or even out of reach.

Example 3 (statistical physics, interactions, social networks)
The *Ising model* is a model in statistical physics, which is a simplified representation of a magnetic system following the Gibbs distribution. In dimension 2, for example, the particles are disposed on a regular grid of size $N \times N$, and have a spin -1 or $+1$ representing their orientation. The energy of the configuration $x \in \mathcal{X} = \{-1, +1\}^{N^2}$ has the form

$$U(x) = -\frac{1}{2} \sum_{i \sim j} J_{i,j} x_i x_j - \sum_i h_i x_i$$

where $(J_{i,j})$ is the interaction force, (h_i) is the exterior magnetic field, and $i \sim j$ means that i and j are two neighbor sites. If $J_{i,j} > 0$, the interaction is *ferromagnetic* and it prefers identical spin orientations. Note that the calculation of $\mathcal{Z}_T$ requires the evaluation of 2^{N^2} configurations, impossible in practice as soon as N become larger than ten.

The *Potts model* is similar: the number of possible values per site are greater than 2.

This type of model is also inspired by the modeling of social networks with the interactions between friends, friends of friends ... see [144].

Example 4 (Markov fields, images) The Gibbs distribution is also very common in image modeling by Markov fields, where the $N \times N$ magnetized particles are replaced by 256×256 pixels and the spins are replaced by the levels of gray, see [51]. This allows us to solve difficult problems of image restoration or segmentation.

For numerical experiments (for example, in the Ising model, in order to develop intuition on the phase transitions and on the critical temperature at which they appear), the capability to simulate configurations with the distribution π becomes a crucial question.

Acceptance-Rejection method. The sampling method by acceptance-rejection — seemingly first proposed by Von Neumann in 1947; see [35] — allows us to generate a random variable with a given distribution starting from the simulation of another random variable, that is easier to generate, and a uniformly distributed random variable. If π is the target distribution defined on a discrete space $\mathcal{X}$ (as in the Ising model, for example), the simplest version of the acceptance-rejection sampling method consists of the following steps:

1. Generate a configuration Y randomly and uniformly on $\mathcal{X}$ (with probability $\pi'(Y) = \frac{1}{\text{Card}(\mathcal{X})}$).

2. Then accept this simulation with probability $\frac{\pi(Y)}{c\pi'(Y)}$ where c is a constant greater than $\max_{x \in \mathcal{X}} \pi(x)/\pi'(x)$.

We iterate this procedure until the first acceptance and the random variable at the output has the distribution π; for the justification, see Section 1.3.2 and Proposition 1.3.2.

The implementation of this algorithm faces two problems: the first is the time of simulation, which is in expectation equal to c, a constant that can be very large if π is very different from the uniform distribution (for the Ising model, this undesirable case corresponds to a low temperature and strong interactions). The second problem is the major one: as noted before, in applications it is often possible to access the values of π only up to the constant $\mathcal{Z}_T$; and thus the acceptance-rejection method cannot be implemented.

Metropolis-Hastings algorithm. Metropolis and his co-authors [115] proposed an algorithm in 1953 that allows an approximate simulation of π, considering π as the stationary distribution of a Markov chain $(X_n)_n$ representing the successive steps of the algorithm. After waiting long enough, X_n has a distribution close to π. Later, Hastings [74] gave a generalization. The Metropolis-Hastings algorithm name remains. See [106, Chapter 5].

We take a transition kernel[7] $q(x,y) > 0$[8] of a Markov chain defined on $\mathcal{X}$, which is easy to simulate: this kernel will serve for proposing moves from one configuration to another. We start from an arbitrary initial configuration x_0. The algorithm is written as follows.

---
1 n, N: **int**;
2 $x_0, \ldots, x_N$: **configuration**;
3 y: **configuration**;
4 $x_0 \leftarrow$ initial configuration;
5 **for** $n = 1$ **to** N **do**
6 $\quad$ $y \leftarrow$ simulation according to the transition probability $q(x_{n-1}, \cdot)$;
7 $\quad$ **if** rand $\leq \frac{\pi(y) \times q(y, x_{n-1})}{\pi(x_{n-1}) \times q(x_{n-1}, y)}$ **then**
8 $\quad\quad$ $x_n \leftarrow y$; $\quad$ /* acceptance of the proposal y */
9 $\quad$ **else**
10 $\quad\quad$ $x_n \leftarrow x_{n-1}$; $\quad$ /* reject */
11 **Return** $x_1, \ldots, x_N$; $\quad$ /* the empirical measure of $(x_1, \ldots, x_N)$ is close to π */

Algorithm 1: Metropolis-Hastings algorithm.

The intermediary step is based on the acceptance-rejection method (see the previous paragraph), which allows us to simulate a random variable distributed with probabilities (for given x)

$$p(x, y) = q(x, y) \min\left(1, \frac{\pi(y) q(y, x)}{\pi(x) q(x, y)}\right), \quad \forall y \neq x.$$

The successive steps of the algorithm are thus described by the trajectory of a Markov chain $(X_n)_{1 \leq n \leq N}$ with transition $P = (p(x,y))_{x,y}$, starting from $X_0 = x_0$. If q is symmetric, then the test for acceptance and the expression of p can be simplified as $p(x,y) = q(x,y) \min(1, \frac{\pi(y)}{\pi(x)})$. In this case we avoid the problem of vanishing q.

---
[7] Probability of going from the state x to y.
[8] Strict positivity is assumed to simplify the presentation.

We check immediately that $\pi(x)p(x,y) = \pi(y)p(y,x)$ (the transition kernel P is reversible for π): π is thus a natural candidate for the stationary distribution of $(X_n)_n$. If we suppose irreducibility in addition, we can apply the ergodic theorem (see for example [138]) and show that, indeed, the empirical distribution

$$\pi_N := \frac{1}{N} \sum_{n=1}^{N} \delta_{X_n} \qquad (6)$$

converges to π as $N \to +\infty$. To get a random variable with distribution π_N, it is enough then to simulate I, uniformly distributed on $\{1, \ldots, N\}$, and to take X_I. With this procedure, for N large enough, we can consider that the random variable X_I has the required distribution π. If the chain is aperiodic, the convergence is reinforced and we have directly the convergence of X_N to π in distribution.

It is remarkable that the implementation of the algorithm requires only the knowledge of π via the ratio $\pi(y)/\pi(x)$, so the simulation of the Gibbs distribution is possible (without knowing the normalization constant $\mathcal{Z}_T$).

Gibbs sampler. The Metropolis-Hastings algorithm is conditioned by the choice of the transition distribution q proposing the modifications of configuration. In practice, the choices of q often provide local changes to the nearest neighbor in the space of configurations, as in the case of a simple random walk. It is sometimes argued that the propositions are too noisy regarding the target distribution π. The Gibbs sampler proposed by Geman and Geman [52] in 1984 is an alternative, exploiting conditional distributions constructed directly starting from π.

We describe hereafter the version of the *random Gibbs sampler*. For more details, ramifications, and references, we refer the reader to [106, Chapter 6]. We keep the notation inspired from the Ising model denoting $\mathcal{X}$ as the space of possible configurations that are supposed to have a tensored form $x = (x^1, \ldots, x^d)$ with $x^i \in \mathcal{E}$: in the Ising model in dimension 2, $d = N^2$ and $\mathcal{E} = \{-1, +1\}$. We define $x^{-i} = (x^1, \ldots, x^{i-1}, x^{i+1}, x^d)$, the configuration x for which the i-th component is removed, and by abuse of notation we write $x = (x^i, x^{-i})$. We set

$$\pi_{x^{-i}}^{(i)}(x^i) = \frac{\pi(x)}{\sum_{y^i} \pi(y^i, x^{-i})},$$

which is interpreted as the conditional probability that the i-th component equals x^i given that the others are x^{-i}. In the Ising model, note that the calculation of $\pi^{(i)}_{x^{-i}}(x^i)$ is immediate as the i-th component takes only two values and this does not require calculation of the normalization constant $\mathcal{Z}_T$. In this model and in many other applications, simulation with respect to the distribution $\pi^{(i)}_{x^{-i}}(\cdot)$ is easy.

So the final algorithm can be written, starting from an initial configuration $x_0 = (x_0^0, \ldots, x_0^d)$.

---

1 n, N: **int**;
2 $x_0, \ldots, x_N$: **configuration**;
3 y: element of $\mathcal{E}$;
4 i: **int**;
5 $x_0 \leftarrow$ initial configuration;
6 for $n = 1$ **to** N **do**
7 $\quad i \leftarrow$ simulation according to the uniform distribution on $\{1, \ldots, d\}$;
8 $\quad y \leftarrow$ simulation according to the distribution $\pi^{(i)}_{x_{n-1}^{-i}}(\cdot)$;
9 $\quad x_n \leftarrow (y, x_{n-1}^{-i})$;
10 Return $x_1, \ldots, x_N$; /* the empirical measure of $(x_1,\ldots,x_N)$ is close to π */

Algorithm 2: Random Gibbs sampler.

We then obtain a sample $(X_1, \ldots, X_N)$ (with strongly dependent elements) having the empirical distribution π_N (defined as in (6)) that converges to π, as $N \to +\infty$.

The two previous algorithms are particular examples of Markov chain Monte-Carlo methods — commonly called MCMC, see [5, Chapter 13] — which generally implement a simulation of an ergodic Markov chain $(X_n)_n$, whose stationary distribution is the target distribution π. An important problem that we have not discussed in this short presentation is to find an algorithm together with its tuning that gives the highest possible speed of convergence.

During recent years, MCMC methods have undergone a lot of developments and applications in statistics (see for instance [129], [26]), uncertainty quantification [87], simulation of rare events [130], etc. We finish by giving an example in statistics.

Example 5 (Bayesian statistics) Suppose that we have access to observations $X = (X_1, \ldots, X_d)$ coming from a parametric model depending on an unknown parameter θ (eventually multidimensional) to be estimated.

For a given parameter θ, the distribution of the observations is given by $p_\theta(X)$. A *classic approach* by likelihood maximization yields $\theta_X = \arg\max_\theta p_\theta(X)$ to estimate the unknown parameter. With a number of observations going to infinity and under certain conditions, the estimator is consistent: $\theta_X \to \theta$ in probability.

The *Bayesian approach* is different. It assumes an uncertainty on the parameter θ, becoming thus random and following some known *a priori distribution* $p(\theta)$. Hence, $p_\theta(X)$ becomes a conditional probability $p(X|\theta)$, which we suppose also to be known. The *a posteriori distribution* is written using the Bayes formula

$$p(\theta|X) = \frac{p(\theta, X)}{p(X)} = \frac{p(X|\theta)p(\theta)}{p(X)}. \tag{7}$$

Thus the goal is to produce simulations according to this distribution, for example, to calculate the Bayes estimator $\mathbb{E}(\theta|X)$ via an empirical mean.

Unfortunately, in very few situations is the distribution $p(\theta|X)$ simple, and most often sampling according to this distribution is very delicate. In fact, in the representation (7), we retrieve common features with the Gibbs distribution:

– The numerator $p(X|\theta)p(\theta)$ is explicit and known.

– As for the denominator, $p(X)$ is a normalization constant, and difficult to obtain in usual situations because (X, θ) lives in spaces of very high dimension.

The Gibbs sampler thus offers a method for simulating θ based on the observations X.

▷ PROBLEM 3 – STOCHASTIC OPTIMIZATION: SIMULATED ANNEALING AND THE ROBBINS-MONRO ALGORITHM

Here we present two algorithms, showing how stochastic simulation allows us to resolve optimization problems of either a deterministic or stochastic nature. We will speak about *stochastic algorithms* rather than Monte-Carlo methods. For general information, we refer the reader to [32] and [14]. For the first algorithm (simulated annealing), the space of the optimization variable is finite, extremely large in applications, and the minimization function has a general form. For the second algorithm (Robbins-Monro), the optimization variable is in $\mathbb{R}^d$ and the minimization function is continuous and locally convex near the minima.

Simulated annealing algorithm. Certain combinatorial optimization problems are known to be difficult to solve exactly; here are some standard examples.

- **Traveling salesman**: Which is the shortest closed path to connect N given points? This is the problem of a salesman who has to visit a list of towns and return to the initial point.

- **Graph coloring**: How do we color the vertices of a given graph with as few colors as possible so that the neighbor vertices have different colors? The application in telecommunications is mentioned at the beginning of the introduction.

More generally, for a function $U : \mathcal{X} \mapsto \mathbb{R}^+$ (or $\mathbb{R}^-$), consider the problem of minimization:

Find an element of $\mathcal{U}_{\min} = \{x \in \mathcal{X} : U(x) = \inf_{y \in \mathcal{X}} U(y)\}$.

For example, in the case of the traveling salesman, $\mathcal{X} = \{(x_1, \ldots, x_N) : x_i \neq x_j, i \neq j\}$ is a space of all possible paths of length N passing through different towns and $U(x) = \sum_{i=1}^{N} d(x_i, x_{i+1})$ is the length of the path (with the convention $x_{N+1} = x_1$).

We can easily check that the Gibbs distribution of temperature T given by

$$\pi(x) = \frac{1}{\mathcal{Z}_T} e^{-\frac{1}{T} U(x)}$$

converges to the uniform distribution on $\mathcal{U}_{\min}$ as the temperature T goes to 0. This heuristic works in the Metropolis-Hastings algorithm if we couple the simulation of the Markov chain $(X_n)_n$ with a scheme of decreasing temperatures $(T_n)_n$, giving rise to the simulated annealing algorithm,[9] see [71]. When the transition probabilities $(q(x,y))_{x,y}$ are symmetric, the algorithm becomes

---

[9] In reference to metallurgical techniques of slow cooling of metal to optimize its mechanical properties.

```
1  n, N: int;
2  x, y: configuration;
3  x ← initial configuration;
4  for n = 1 to N do
5  │  y ← simulation according to the transition probability q(x, ·);
6  │  if rand ≤ exp(− 1/T_n × [U(y) − U(x)]) then
7  │  └  x ← y ;            /* acceptance of the proposal y */
8  Return x
```

Algorithm 3: Simulated annealing algorithm.

Under certain hypotheses and taking logarithmic decreasing temperatures $T_n = c \log(n)$, we can show that the distribution of X_n converges to the uniform distribution on $\mathcal{U}_{\min}$, see [20, Chapter 7, Section 8]. The choice of the transition kernel q is made separately for each case and it requires a certain practice in order for the algorithm to be efficient.

Searching for level sets.

▷ *Simple version.* Consider the general problem of searching level sets of a continuous function $F : \theta \in \mathbb{R}^q \mapsto F(\theta) \in \mathbb{R}^q$, i.e., for a level $a \in \mathbb{R}^q$, the set
$$\mathcal{T}_a = \{\theta \in \mathbb{R}^q : F(\theta) = a\}.$$

If $F = D\Phi$ is the gradient of a function Φ and $a = 0$, this amounts to searching the minima/maxima of Φ.

The resolution by an algorithm of Newton-Raphson type is a classic iterative procedure which takes the form
$$\theta_{n+1} = \theta_n - \gamma_{n+1}(F(\theta_n) - a),$$

built from a sequence of steps $(\gamma_n)_{n \geq 1}$, for example, given by the inverse of the Jacobian $F'(\theta_n)$ for the Newton method or by finite differences in the case of the secant method.

Now suppose that $(F(\theta_n))_n$ is not evaluated exactly, but with noise, whose measure is modeled by a sequence of random variables $(Y_n)_n$ with zero mean. Then we need to replace $F(\theta_n)$ in the previous algorithm by its noisy version $F(\theta_n) + Y_n$. We get the simplest form of the Robbins-Monro algorithm, which provides a sequence $(\theta_n)_{n \geq 1}$ converging to $\theta^* \in \mathcal{T}$, with an initialization θ_0 arbitrary enough. Among the required hypotheses, we suppose that the steps are positive and fulfill the two conditions

$$\sum_{n\geq 1} \gamma_n = +\infty, \qquad \sum_{n\geq 1} \gamma_n^2 < +\infty, \tag{8}$$

leading to choices of steps of type $\gamma_n = n^{-\alpha}$ with $\alpha \in (\frac{1}{2}, 1]$.

▷ *Advanced version.* If F is written as an expectation $F(\theta) = \mathbb{E}(f(\theta, Y))$, then the application of the previous version for approaching $F(\theta)$ by an empirical mean $\frac{1}{M}\sum_{m=1}^{M} f(\theta, Y_m)$ over a large number M of independent simulations of the random variable Y, is not the simplest way. In fact, it is enough to replace $F(\theta_n)$ by $f(\theta_n, Y_n)$ (only one simulation!) where $(Y_n)_{n\geq 1}$ is a sequence of independent random simulations having the same distribution as Y. The convergence of the *Robbins-Monro algorithm* in its general version is based on the following classic result, which we present under improved hypotheses [104, Theorem 1].

Theorem 6 (Robbins-Monro algorithm) *Let $f : \mathbb{R}^q \times \mathbb{R}^d \mapsto \mathbb{R}^q$ be a measurable function, let Y be a d-dimensional random variable such that $f(\theta, Y)$ is integrable for any $\theta \in \mathbb{R}^q$ and define $F(\theta) = \mathbb{E}[f(\theta, Y)]$. Moreover, we suppose*

i) *for $a \in \mathbb{R}^q$ fixed, F separates the level set $\mathcal{T}_a = \{\theta \in \mathbb{R}^q : F(\theta) = a\}$, i.e., for any $\theta^* \in \mathcal{T}_a$ and $\theta \in \mathbb{R}^q$, we have $\langle \theta - \theta^*, F(\theta) - a \rangle > 0$*[10];

ii) *for a certain constant $C \geq 0$, we have $\mathbb{E}(f^2(\theta, Y)) < C(1 + |\theta|^2)$ for any $\theta \in \mathbb{R}^q$.*

Define

$$\theta_{n+1} = \theta_n - \gamma_{n+1}\Big(f(\theta_n, Y_{n+1}) - a\Big), \qquad n \geq 0$$

with θ_0 chosen arbitrarily.

Then for a sequence of steps $(\gamma_n)_n$ satisfying (8), there exists a random variable $\theta^* \in \mathcal{T}_a$ such that

$$\theta_n \xrightarrow{a.s.} \theta^*.$$

---

[10]In a problem of searching for extrema where $F = [\nabla \Phi]^{\mathsf{T}}$, the sense of the inequality is connected with a local convexity property of Φ and the algorithm provides (at the limit) the minima of Φ.

The proof uses martingale techniques. There are a lot of ramifications of the above result, and we will mention some of them.

- We can determine the convergence rate[11] of $\theta_n - \theta^*$ and study the convergence in distribution of the renormalized error: see [32, Chapter 2] for example.

- We can average the results $(\theta_n)_n$ (Polyak-Ruppert averaging procedure) to allow an acceleration of convergence while relaxing the constraints on decreasing the steps: see [98, Chapter 11] and [8].

- In the case $f(\theta, y) = [\nabla_\theta \varphi(\theta, y)]^\mathsf{T}$ and $a = 0$ (i.e., searching for minima of $\mathbb{E}(\varphi(\theta, Y)))$, the previous algorithm is also known as the stochastic gradient algorithm. Besides, we can replace the gradient by a finite difference of the function φ. This approach, which avoids the calculation of derivatives for φ, is known as the Kiefer-Wolfowitz algorithm; see [98, Chapter 1].

- Recently, variants of this algorithm have been designed to include a variance reduction effect; see [132].

---

[11] It happens to be $\sqrt{\gamma_n}$ under certain hypotheses.

PART A: TOOLBOX FOR STOCHASTIC SIMULATION

CHAPTER 1

Generating random variables

Computer generation of random variables, no matter how complex it can be, is based on the generation of very simple distributions and further applications of suitable transformations. The basic distribution is the uniform law on $[0, 1]$.

1.1 PSEUDORANDOM NUMBER GENERATOR

In practical implementation of algorithms, a random number generator (RNG) is represented by a deterministic sequence of real numbers $(u_1, \ldots, u_m, \ldots)$ between 0 and 1 in the double-precision format. According to a programming language and libraries used, one can call a random number generator using the command **drand48** in C, **random** in C++ and Java, **rand** in MATLAB®, **grand** in Scilab, and **numpy.random.rand** in Python, with different call parameters. In the following we denote it as simply **rand**.

On the first call of the generator we obtain the value u_1; on the second call, u_2. Indeed, the *seed* of the generator - i.e., the index of the first number returned - is equal to 1 by default. Here, there is nothing random because the sequence $(u_1, \ldots, u_m, \ldots)$ always remains the same. This is very convenient in the first stage of writing a simulation program, where generating truly random results makes finding coding errors difficult (one cannot easily reproduce errors). As a second step, it is recommended to change the seed in a *random* manner, using for example the time of the computer used (a priori, always different).

The generator is qualified to be random if the sequence

$(u_1, \ldots, u_m, \ldots)$ — despite being deterministic — has random behavior, similar to that of the sequence of uniformly distributed random variables. To verify this, one can use numerous *statistical tests* that allow rejection of the hypothesis of independence or the adequacy with respect to a given distribution. An interested reader can refer to [5, Chapter 2].

In practice, a generator is cyclic and after L calls it again returns the initial value, etc. Obviously it is important to ensure that the *period* L of the generator that is used is large enough compared with the number of calls to be made: in practice, most of the currently available generators satisfy this constraint.

A classic example of a generator is the *linear congruential generator*: for three parameters a, b and L, it is written as

$$x_{m+1} = ax_m + b \text{ modulo } L, \quad u_m = \frac{x_m}{L}$$

to achieve the maximal period equal to L. For a long time a popular choice was $a = 7^5$, $b = 0$ and $L = 2^{31} - 1 = 2147483647$.

The *Mersenne Twister*[1] generator is more recent, robust, and rapid, with a period equal to $2^{19937} - 1$, which is more than sufficient for applications.

1.2 GENERATION OF ONE-DIMENSIONAL RANDOM VARIABLES

To get acquainted with the myriad of simulation algorithms for a given random variable, the reader can refer to the encyclopedic work of Devroye [31]. Our presentation will rather follow [10], where the random variables are defined directly via the algorithms generating them from the uniform distribution.

1.2.1 Inversion method

The first approach is based on inversion of the cumulated distribution function (c.d.f.). This method was first proposed by Von Neumann in 1947 [35]. In the sequel, U stands for a random variable uniformly distributed on $[0, 1]$.

Proposition 1.2.1 *Let X be a real random variable with a c.d.f.*

---

[1] http://www.math.sci.hiroshima-u.ac.jp/ m-mat/MT/emt.html

$F(x) = \mathbb{P}(X \leq x)$, for which the generalized inverse (called the quantile) is defined by $F^{-1}(u) = \inf\{x : F(x) \geq u\}$. Then

$$F^{-1}(U) \stackrel{\mathrm{d}}{=} X.$$

Conversely, if F is continuous, then $F(X) \stackrel{\mathrm{d}}{=} \mathcal{U}([0,1])$.

PROOF:
A standard argument shows that the function F is increasing, continuous from the right, and with left limits. Similarly, the quantile F^{-1} is increasing, continuous from the left, and with right limits. Moreover, we have the following general properties (easy to prove):

a) $F(F^{-1}(u)) \geq u$ for all $u \in]0,1[$.
b) $F^{-1}(u) \leq x \iff u \leq F(x)$.
c) $F(F^{-1}(u)) = u$ if F continuous at $F^{-1}(u)$.

Then from b) we deduce that $\mathbb{P}(F^{-1}(U) \leq x) = \mathbb{P}(U \leq F(x)) = F(x)$ which justifies the first assertion. Under the continuity hypothesis for F, the application of c) gives $F(X) \stackrel{\mathrm{d}}{=} F(F^{-1}(U)) = U$, which proves the second statement. □

Definition and Proposition 1.2.2 (exponential distribution) *Let $\lambda > 0$. Then*

$$X = -\frac{1}{\lambda} \log(U)$$

has an exponential distribution with parameter λ (denoted by $\mathcal{E}\mathrm{xp}(\lambda)$) which has the density $\lambda e^{-\lambda x} \mathbf{1}_{x \geq 0}$.

Definition and Proposition 1.2.3 (discrete distribution) *Let $(p_n)_{n \geq 0}$ be a sequence of positive real numbers satisfying $\sum_{n \geq 0} p_n = 1$. Let $(x_n)_{n \geq 0}$ be a sequence of real numbers. Then*

$$X = \sum_{n \geq 0} x_n \mathbf{1}_{p_0 + \cdots + p_{n-1} \leq U < p_0 + \cdots + p_n}$$

is a discrete random variable such that $\mathbb{P}(X = x_n) = p_n$ for $n \geq 0$.

A few simple examples:

- The Bernoulli random variable $\mathcal{B}(p)$ corresponds to the case $(p_0, p_1) = (1-p, p)$ and $(x_0, x_1) = (0, 1)$.

- The binomial random variable $\mathcal{B}in(n,p)$ is written as a sum of n independent Bernoulli random variables $\mathcal{B}(p)$: $\mathbb{P}(X = k) = \binom{n}{k} p^k(1-p)^{n-k}$ for $1 \le k \le n$. Its generalization is the multinomial distribution.

- The Poisson distribution $\mathcal{P}(\theta)$ corresponds to $x_n = n$ and $p_n = e^{-\theta}\frac{\theta^n}{n!}$ for $n \ge 0$.

The geometric distribution is also discrete but can be generated in a simpler way.

Definition and Proposition 1.2.4 (geometric distribution) *Let $(X_m)_{m \ge 1}$ be a sequence of i.i.d. variables distributed according to $\mathcal{B}(p)$. The random variable $X = \inf\{m \ge 1 : X_m = 1\}$ follows the geometric distribution with parameter p (denoted $\mathcal{G}(p)$): $\mathbb{P}(X = n) = p(1-p)^{n-1}$ for $n \ge 1$. We also have*

$$X \stackrel{d}{=} 1 + \lfloor Y \rfloor \quad \text{where} \quad Y \stackrel{d}{=} \mathcal{E}xp(\lambda)$$

with $\lambda = -\log(1-p)$, and therefore

$$1 + \left\lfloor \frac{\log(U)}{\log(1-p)} \right\rfloor \stackrel{d}{=} \mathcal{G}(p).$$

Definition and Proposition 1.2.5 (Cauchy distribution) *Let $\sigma > 0$. Then*

$$X = \sigma \tan\left(\pi\left(U - \frac{1}{2}\right)\right)$$

is a Cauchy random variable with parameter σ, whose density is $\frac{\sigma}{\pi(x^2+\sigma^2)} 1_{x \in \mathbb{R}}$.

Definition and Proposition 1.2.6 (Rayleigh distribution) *Let $\sigma > 0$. Then*

$$X = \sigma\sqrt{-\log U}$$

is a Rayleigh random variable with parameter σ, whose density is $\frac{x}{\sigma^2} e^{-\frac{x^2}{2\sigma^2}} 1_{x \ge 0}$.

Definition and Proposition 1.2.7 (Pareto distribution) *Let $(a, b) \in\,]0, +\infty[^2$. Then*

$$X = \frac{b}{U^{\frac{1}{a}}}$$

is a Pareto random variable with parameters (a, b), whose density is $\frac{ab^a}{x^{a+1}} \mathbf{1}_{x \geq b}$.

Definition and Proposition 1.2.8 (Weibull distribution) *Let $(a, b) \in\,]0, +\infty[^2$. Then*

$$X = b(-\log U)^{\frac{1}{a}}$$

is a Weibull random variable with parameters (a, b), whose density is $\frac{a}{b^a} x^{a-1} e^{-(x/b)^a} \mathbf{1}_{x \geq 0}$.

As for the Gaussian distribution, there is no explicit expression for the c.d.f. and its inverse. However, there exist some excellent approximations for this function, which allow us to use the approximate inversion method.

Definition and Proposition 1.2.9 (Gaussian distribution, from [118]) *Let us define the function $u \in\,]0, 1[\mapsto \mathcal{N}^{-1}_{\text{Moro}}(u)$ by*

$$\mathcal{N}^{-1}_{\text{Moro}}(u) = \begin{cases} \dfrac{\sum_{n=0}^{3} a_n (u-0.5)^{2n+1}}{1 + \sum_{n=0}^{3} b_n (u-0.5)^{2n}}, & 0.5 \leq u \leq 0.92, \\ \displaystyle\sum_{n=0}^{8} c_n (\log(-\log(1-u)))^n, & 0.92 \leq u < 1, \\ -\mathcal{N}^{-1}_{\text{Moro}}(1-u), & 0 < u \leq 0.5 \end{cases}$$

with

$a_0 = 2.50662823884$, $a_1 = -18.61500062529$, $a_2 = 41.39119773534$,

$a_3 = -25.44106049637$, $b_0 = -8.47351093090$, $b_1 = 23.08336743743$,

$b_2 = -21.06224101826$, $b_3 = 3.13082909833$,

$c_0 = 0.3374754822726147$, $c_1 = 0.9761690190917186$,

$c_2 = 0.1607979714918209$, $c_3 = 0.0276438810333863$,

$c_4 = 0.0038405729373609$, $c_5 = 0.0003951896511919$,

$c_6 = 0.0000321767881768$, $c_7 = 0.0000002888167364$,

$c_8 = 0.0000003960315187$.

Then $\mathcal{N}_{\text{Moro}}^{-1}$ is an approximation[2] of the inverse function for $\mathcal{N}$, where $\mathcal{N}$ is the c.d.f. of the standard Gaussian distribution: $\mathcal{N}(u) = \int_{-\infty}^{u} \frac{1}{\sqrt{2\pi}} e^{-\frac{x^2}{2}} dx$.

Thus for $\mu \in \mathbb{R}$ and $\sigma \geq 0$,

$$X = \mu + \sigma \mathcal{N}_{\text{Moro}}^{-1}(U)$$

approximately follows the Gaussian distribution $\mathcal{N}(\mu, \sigma^2)$ with mean μ and variance σ^2.

1.2.2 Gaussian variables

It is possible to generate a Gaussian random variable directly, without approximation, with the help of the Box-Muller transform. It is based on the following result.

Proposition 1.2.10 *Let X and Y be two independent standard Gaussian random variables. Define (R, θ) as the polar coordinates of (X, Y):*

$$X = R\cos(\theta), \quad Y = R\sin(\theta)$$

with $R \geq 0$ and $\theta \in [0, 2\pi[$. Then R^2 and θ are two independent random variables ; the first one has the distribution of $\mathcal{E}xp(\frac{1}{2})$, the second one is uniformly distributed on $[0, 2\pi]$.

```
1  r, theta: double;
2  u, v: double;
3  x, y: double;
4  u ← rand;
5  v ← rand;
6  theta ← 2 × π × u;
7  r ← sqrt(−2 × log(v));
8  x ← r × cos(theta);
9  y ← r × sin(theta);
```

Algorithm 1.1: Generation of two independent standard Gaussian random variables by the Box-Muller method.

---

[2] The error is less than 3×10^{-9} up to 7 times the standard deviation (i.e. $\mathcal{N}(-7) \leq u \leq \mathcal{N}(7)$).

To get a random variable distributed as $\mathcal{N}(\mu, \sigma^2)$, it is enough to multiply x by the standard deviation σ and then to add the mean μ.

The Marsaglia algorithm is another version of the simulation algorithm that avoids the use of trigonometric functions (this part is considered to have longer running time); see [5].

1.3 ACCEPTANCE-REJECTION METHODS

1.3.1 Generation of conditional distribution

To obtain a simulation of Z given an event A, it is enough to repeatedly and independently generate (Z, A) and reject the results when A doesn't occur. In the next statement, Z can be multi-dimensional.

Proposition 1.3.1 *Let Z be a random variable and A be an event with non-zero probability. Consider $(Z_n, A_n)_{n \geq 1}$ the sequence of independent random elements having the same distribution as (Z, A). Denote $\nu = \inf\{n \geq 1 : A_n \text{ occurs}\}$: then, the r.v. Z_ν has the distribution of Z conditionally to A.*

PROOF:

For any Borel set B, we have

$$\begin{aligned} \mathbb{P}(Z_\nu \in B) &= \sum_{n \geq 1} \mathbb{P}(Z_n \in B; A_1^c; \cdots ; A_{n-1}^c; A_n) \\ &= \sum_{n \geq 1} (1 - \mathbb{P}(A))^{n-1} \mathbb{P}(Z_n \in B; A_n) \\ &= \frac{\mathbb{P}(Z \in B; A)}{\mathbb{P}(A)} = \mathbb{P}(Z \in B | A). \end{aligned}$$

$\square$

The previous algorithm has a random duration ν: this random variable is distributed as $\mathcal{G}(\mathbb{P}(A))$. Thus, the more likely is A, the faster the simulation is (with expected duration equal to $\frac{1}{\mathbb{P}(A)}$).

Let us give a simple example of the application of this result: to generate a random variable X that is uniformly distributed on a compact set $D \subset \mathbb{R}^d$, it is enough to generate the random variable Z uniformly on a cube containing D (easy to do), and retain the first simulation that falls into D. Indeed, it will be distributed as $Z|\{Z \in D\}$, whose density is $z \mapsto \frac{\mathbf{1}_{z \in \text{cube}}}{|\text{cube}|} \frac{\mathbf{1}_{z \in D}}{\mathbb{P}(Z \in D)} = \frac{\mathbf{1}_{z \in D}}{|\text{cube}| \mathbb{P}(Z \in D)} = \frac{\mathbf{1}_{z \in D}}{|D|}$, i.e., the same as that of X.

1.3.2 Generation of (non-conditional) distributions by the acceptance-rejection method

Here we suppose that the random variable of interest X (possibly multidimensional) has a known density f, but the direct generation of X is complicated. The principle of the method consists of generating another random variable Y with density g and to accept the result of Y as realization of X with a probability proportional to the ratio $f(Y)/g(Y)$. This idea was proposed by Von Neumann in 1947 [35]. The property can be stated precisely as follows.

Proposition 1.3.2 *Let X and Y be two random variables with values in $\mathbb{R}^d$, whose densities with respect to a reference measure μ are f and g respectively. Suppose that there exists a constant $c(\geq 1)$ satisfying*

$$c\,g(x) \geq f(x) \quad \mu - a.e. \tag{1.3.1}$$

Let U be a random variable uniformly distributed on $[0,1]$ and independent of Y. Then, the distribution of Y given $\{c\,U\,g(Y) < f(Y)\}$ is the distribution X.

PROOF:
Indeed, setting $A = \{c\,U\,g(Y) < f(Y)\}$ for any Borel set B in $\mathbb{R}^d$, we have

$$\mathbb{P}(Y \in B \mid A) = \frac{\mathbb{P}(Y \in B; A)}{\mathbb{P}(A)}$$

$$= \frac{1}{\mathbb{P}(A)} \int_{\{(y,u):y\in B, c\,u\,g(y)<f(y)\}} g(y)\,\mathbf{1}_{g(y)>0}\,\mathbf{1}_{[0,1]}(u)\,\mu(dy)\,du$$

$$= \frac{1}{\mathbb{P}(A)} \int_B g(y)\,\frac{f(y)}{c\,g(y)}\,\mathbf{1}_{g(y)>0}\,\mu(dy)$$

$$= \frac{1}{c\,\mathbb{P}(A)} \int_B f(y)\,\mathbf{1}_{g(y)>0}\,\mu(dy) = \frac{1}{c\,\mathbb{P}(A)} \int_B f(y)\,\mu(dy)$$

as μ – a.e., if $g(y) = 0$, then $f(y) = 0$. The choice $B = \mathbb{R}^d$ leads to $c\,\mathbb{P}(A) = 1$, and thus $\mathbb{P}(Y \in B \mid A) = \int_B f(y)\,\mu(dy)$ for all B. □

Next, for the effective simulation of the above conditional distribution we apply the Proposition 1.3.1, and this leads to the following algorithm.

```
1  c: double;
2  y: double;
3  u: double;
4  c ← upper bound for f/g;
5  repeat
6  |    y ← simulation according to density g;
7  |    u ← rand;
8  until c × u × g(y) ≤ f(y);
9  Return y ;                /* the output variable y has the
   distribution of X with density f */
```

Algorithm 1.2: Acceptance-rejection method.

While implementing the acceptance-rejection method, it is rather easy to find, for a given density f, another density g and a number c that verifies $c\,g(x) \geq f(x)$ for all x. However, the choice of g is sufficient if the constant c is small, such that the expected number of rejections remains small (on average, $c = \frac{1}{\mathbb{P}(A)}$ calls) and therefore the algorithm works fast.

Example 1.3.3 (simulation of Beta distribution) A random variable with the Beta distribution $\mathcal{B}(\alpha, \beta)$ (with $\alpha > 0$ and $\beta > 0$) has density

$$\frac{1}{B(\alpha, \beta)} x^{\alpha-1}(1-x)^{\beta-1} 1_{0<x<1}.$$

Suppose that $\alpha \geq 1$ and $\beta \geq 1$ so that this density is bounded. We can then use the acceptance-rejection method with $Y \stackrel{d}{=} \mathcal{U}([0,1])$. The acceptance-rejection constant is equal to

$$c_\alpha = \sup_{0<x<1} \frac{1}{B(\alpha, \beta)} x^{\alpha-1}(1-x)^{\beta-1} = \frac{1}{B(\alpha, \beta)} (x_{\alpha,\beta})^{\alpha-1}(1-x_{\alpha,\beta})^{\beta-1}$$

where $x_{\alpha,\beta} = \frac{\alpha-1}{\alpha-1+\beta-1}$.

Example 1.3.4 (simulation of Gamma distribution) A random variable with the Gamma distribution $\Gamma(\alpha, \theta)$ (with $\alpha > 0$ and $\theta > 0$) has density

$$\frac{1}{\Gamma(\alpha)} \theta^\alpha x^{\alpha-1} e^{-\theta x} 1_{x \geq 0},$$

where $\Gamma(\alpha) = \int_0^{+\infty} x^{\alpha-1} e^{-x} dx$ is the gamma function.

- If $\alpha = 1$, this coincides with the distribution of $\mathcal{E}\text{xp}(\theta)$.

- If α is a non-zero integer, a random variable following the $\Gamma(\alpha, \theta)$ distribution can be written as a sum of α independent random variables with the distribution $\mathcal{E}\mathrm{xp}(\theta)$: the simulation scheme follows immediately.

- If α is not an integer, it can be useful to use the acceptance-rejection method. Let us illustrate this without considering optimality. To simplify, suppose that $\theta = 1$ and $\alpha \in (n, n+1)$ with $n \geq 1$. Take for Y a random variable that follows $\Gamma(n, \frac{1}{2})$. We check then that the acceptance-rejection constant is equal to

$$\begin{aligned} c_\alpha = \sup_{x>0} \frac{f(x)}{g(x)} &= \sup_{x>0} \frac{\frac{1}{\Gamma(\alpha)} x^{\alpha-1} e^{-x}}{\frac{1}{\Gamma(n)} 2^{-n} x^{n-1} e^{-\frac{1}{2}x}} \\ &= \frac{\Gamma(n)}{\Gamma(\alpha)} 2^\alpha \sup_{y>0} y^{\alpha-n} e^{-y} \\ &= \frac{\Gamma(n)}{\Gamma(\alpha)} 2^\alpha (\alpha - n)^{\alpha-n} e^{-(\alpha-n)}. \end{aligned}$$

This constant increases rapidly as $\alpha \to +\infty$. For more efficient procedures, see [31, Chapter 9].

1.3.3 Ratio-of-uniforms method

The ratio-of-uniforms method, introduced by Kinderman and Monahan in [91] and further developed in [124], is aimed at generating a multi-dimensional random variable X simply as a ratio of random variables uniformly distributed on a set. The target distribution is assumed to have a density p w.r.t. the Lebesgue measure on $\mathbb{R}^d$. One feature of the method is that the target density p is required to be known only up to a constant, i.e.

$$p(x) = cf(x)$$

where f is a known integrable non-negative function and the constant $c = (\int_{\mathbb{R}^d} f(x) \mathrm{d}x)^{-1}$ is unknown or numerically costly to be computed.

Proposition 1.3.5 *Let $r > 0$ and define*

$$A_{f,r} := \left\{ (u, v_1, \ldots, v_d) \in \mathbb{R}^{d+1} : 0 < u \leq \left[f\left(\frac{v_1}{u^r}, \ldots, \frac{v_d}{u^r}\right) \right]^{1/(1+rd)} \right\}.$$

Then the Lebesgue measure of $A_{f,r}$ is finite and equal to $\frac{1}{c(rd+1)}$.

Moreover, let $(U, V_1, \ldots, V_d)$ be a random variable uniformly distributed on $A_{f,r}$, then the distribution of $(V_1/U^r, \ldots, V_d/U^r)$ has a density equal to p.

PROOF:

First, assume that the Lebesgue measure $|A_{f,r}|$ of $A_{f,r}$ is finite, so that the uniform density of $(U, V_1, \ldots, V_d)$ is well defined and given by $x \mapsto \mathbf{1}_{x \in A_{f,r}}/|A_{f,r}|$. Then for a measurable function $\varphi : \mathbb{R}^d \mapsto \mathbb{R}^+$, we have

$$\mathbb{E}(\varphi(V_1/U^r, \ldots, V_d/U^r))$$

$$= \int_{\mathbb{R}^{d+1}} \varphi(v_1/u^r, \ldots, v_d/u^r) \frac{1}{|A_{f,r}|} \mathbf{1}_{0 < u \le [f(\frac{v_1}{u^r}, \ldots, \frac{v_d}{u^r})]^{1/(1+rd)}} \, du \, dv_1 \ldots dv_d$$

$$= \int_{\mathbb{R}^{d+1}} \varphi(z_1, \ldots, z_d) \frac{1}{|A_{f,r}|} \mathbf{1}_{0 < u \le [f(z_1, \ldots, z_d)]^{1/(1+rd)}} (u^r)^d \, du \, dz_1 \ldots dz_d$$

(using the change of variables $z_i = v_i/u^r$)

$$= \int_{\mathbb{R}^d} \varphi(z_1, \ldots, z_d) \frac{1}{|A_{f,r}|(rd+1)} f(z_1, \ldots, z_d) dz_1 \ldots dz_d. \tag{1.3.2}$$

A similar computation shows that $|A_{f,r}|$ is finite: indeed,

$$|A_{f,r}| = \int_{\mathbb{R}^{d+1}} \mathbf{1}_{0 < u \le [f(\frac{v_1}{u^r}, \ldots, \frac{v_d}{u^r})]^{1/(1+rd)}} \, du \, dv_1 \ldots dv_d$$

$$= \int_{\mathbb{R}^d} \frac{1}{(rd+1)} f(z_1, \ldots, z_d) dz_1 \ldots dz_d = \frac{1}{c(rd+1)} < +\infty.$$

Plugging this into (1.3.2), we deduce

$$\mathbb{E}(\varphi(V_1/U^r, \ldots, V_d/U^r)) = \int_{\mathbb{R}^d} \varphi(z_1, \ldots, z_d) p(z_1, \ldots, z_d) dz_1 \ldots dz_d.$$

We are done. □

Illustration in dimension 1. We demonstrate the importance of the above method in the case $d = 1$, and the arguments are similar for $d > 1$. In Proposition 1.3.5, r is a free parameter to possibly take advantage of, here we simply set $r = 1$.

First we observe that $A_{f,r}$ is bounded under mild conditions on the bounds of f. Indeed, we have the following:

Lemma 1.3.6 *If $x \mapsto f(x)$ and $x \mapsto x^2 f(x)$ are bounded, then $A_{f,r}$ is bounded and*

$$A_{f,r} \in \widetilde{A_{f,r}} := \left[0, \sup_x \sqrt{f(x)}\right] \times \left[\inf_x x\sqrt{f(x)}, \sup_x x\sqrt{f(x)}\right].$$

42 ■ Generating random variables

The proof is left to the reader. It includes all bounded densities whose tails are dominated by Cauchy tails (i.e. all the usual practical examples).

For such densities, the nice consequence is that one can easily sample the uniform distribution on $A_{f,r}$ using the acceptance-rejection method, by taking the uniform distribution on $\widetilde{A_{f,r}}$ as a proposal.

Example 1.3.7 (Student distribution with $k > 0$ degrees of freedom) *This density is equal to*

$$p(x) = \frac{1}{\sqrt{k\pi}} \frac{\Gamma(\frac{k+1}{2})}{\Gamma(\frac{k}{2})} (1 + \frac{x^2}{k})^{-\frac{k+1}{2}} 1_{x \geq 0}.$$

With the ratio-of-uniforms method, there is no need to consider the normalization factor and we can just take

$$f(x) = (1 + \frac{x^2}{k})^{-\frac{k+1}{2}} 1_{x \geq 0}.$$

For such a choice and for $k > 1$, we easily check that the supremum of $x \mapsto x^2 f(x)$ on $\mathbb{R}^+$ is achieved at $x^\star := \sqrt{\frac{2k}{k-1}}$, so that $\widetilde{A_{f,r}} = [0,1] \times [-x^\star \sqrt{f(x^\star)}, x^\star \sqrt{f(x^\star)}]$. Therefore, the algorithm is written as follows.

---
1 u: **double**;
2 v: **double**;
3 **repeat**
4 $\quad u \leftarrow$ simulation according to $\mathcal{U}([0,1])$;
5 $\quad v \leftarrow$ simulation according to $\mathcal{U}([-x^\star \sqrt{f(x^\star)}, x^\star \sqrt{f(x^\star)}])$, independent of u;
6 **until** $u \leq (1 + (v/u)^2/k)^{-(k+1)/4}$;
7 **Return** v/u;

Algorithm 1.3: Simulation of a Student distribution (with $k > 1$ degrees of freedom) using the ratio-of-uniforms method.

1.4 OTHER TECHNIQUES FOR GENERATING A RANDOM VECTOR

When the components of a random vector are independent, the generation can be performed separately for each component. In the case of non-trivial dependence of the components, a deeper analysis is required.

1.4.1 The Gaussian vector

We recall that a vector $X = (X_1, \ldots, X_d)$ is Gaussian if any linear combination of its components $\sum_{i=1}^{d} a_i X_i$ (with $a_i \in \mathbb{R}$) has the Gaussian distribution. A Gaussian vector X is characterized by its mean m and its covariance matrix K, and we write $X \stackrel{d}{=} \mathcal{N}(m, K)$.

Generally, Gaussian vectors are generated by affine transforms of independent standard Gaussian random variables (i.e. distributed as $\mathcal{N}(0, \mathrm{Id})$).

Proposition 1.4.1 *Let d_0 and d be two non-zero integers, let X be a d_0-dimensional Gaussian vector with the distribution $\mathcal{N}(0, \mathrm{Id})$, and $m \in \mathbb{R}^d$ and L be a $d \times d_0$ matrix. Then*

$$m + LX \stackrel{d}{=} \mathcal{N}(m, LL^\mathsf{T}),$$

i.e., $m + LX$ is a d-dimensional Gaussian vector with mean m and covariance $K = LL^\mathsf{T}$.

We leave the proof to the reader. Conversely, a covariance matrix K — symmetric non-negative definite matrix of size d — can always be decomposed, non uniquely, in the form

$$K = LL^\mathsf{T},$$

which thus allows us to simulate any Gaussian vector using the previous case.

To calculate L, we can use the Choleski algorithm, which provides a lower triangular matrix (with $d_0 = d$). Its computational cost with respect to the dimension is proportional to d^3.

In the case of large dimension, we may wish to speed-up the procedure. For certain matrices it is possible. For example, if

$$K = \begin{pmatrix} 1 & \rho & \cdots & & \cdots & \rho \\ \rho & 1 & \rho & & \cdots & \rho \\ \vdots & & \ddots & 1 & & \ddots & \vdots \\ \rho & \cdots & & \rho & 1 & \rho \\ \rho & \cdots & & & \rho & 1 \end{pmatrix},$$

for $\rho \in [0, 1]$, we can take the following $d \times d_0$ matrix (with $d_0 = d+1$)

$$L = \begin{pmatrix} \sqrt{\rho} & \sqrt{1-\rho} & 0 & \cdots & \cdots & 0 \\ \sqrt{\rho} & 0 & \sqrt{1-\rho} & 0 & \cdots & \vdots \\ \vdots & \vdots & & \ddots & \sqrt{1-\rho} & 0 \\ \sqrt{\rho} & 0 & \cdots & \cdots & 0 & \sqrt{1-\rho} \end{pmatrix}.$$

In this case we use $d+1$ independent standard Gaussian random variables to generate the d Gaussian random variables with the required covariance. The computational cost is of order d instead of d^3 with the usual Choleski method, which produces an important improvement in large dimensions.

1.4.2 Modeling of dependence using copulas

When the variables have the Gaussian distribution, it is natural to model the dependence using the covariance matrix. However, the level sets of the Gaussian density are ellipsoids and they cannot account well for the dependencies in the extreme values. The modeling of the dependence is a delicate and complex question, and is fundamental for applications. It cannot be reduced to a correlation coefficient only, as in the case of a Gaussian vector.

In fact, the dependence can be modeled intrinsically without taking the marginal distributions into account, using the notion of the *copula*. This is a pure measure of dependence, whose foundation is based on the Sklar theorem [134].

Theorem 1.4.2 *Let us consider a random d-dimensional vector $X = (X_1, \cdots, X_d)$ with joint c.d.f. $F(x_1, \cdots, x_d) = \mathbb{P}(X_1 \leq x_1, \cdots, X_d \leq x_d)$. Then there exists a copula function $C : [0,1]^d \mapsto [0,1]$ such that:*

$$F(x_1; \cdots, x_d) = C(F_1(x_1); \cdots F_d(x_d)).$$

The copula C is unique if the marginal distributions are continuous.

We refer to [113, Chapter 5] for the detailed properties of copula functions. We can easily verify that copulas are invariant under strictly increasing transformations of the initial random vector X, confirming that this is an intrinsic measure of the dependence between X-components. This point of view splits the modeling of X into the modeling of each marginal distribution on the one hand, and the modeling of their dependence on the other hand.

Let us give some usual examples of copulas.

1. *Independence copula*: this is a copula of a vector with independent components, i.e., $C(u_1, \ldots, u_d) = u_1 \ldots u_d$.

2. *Co-monotone copula*: this copula corresponds to the case $X_i = \phi_i(Y)$ with ϕ_i increasing, which gives the copula $C^+(u_1, \cdots, u_d) = \min(u_1, \cdots, u_d)$.

3. *Fréchet-Hoeffding bounds*: the copulas are universally bounded as follows

$$(u_1 + \cdots + u_d - d + 1)_+ := C^-(u_1, \cdots, u_d)$$
$$\leq C(u_1, \cdots, u_d) \leq C^+(u_1, \cdots, u_d).$$

4. *Gaussian copula with invertible symmetric matrix K*: this is the copula of a Gaussian vector $\mathcal{N}(0, K)$, i.e.,

$$C(u_1, \cdots, u_d) = \int_{-\infty}^{\mathcal{N}^{-1}(u_1)} \cdots \int_{-\infty}^{\mathcal{N}^{-1}(u_d)} \frac{1}{(2\pi)^{d/2}\sqrt{\det(K)}}$$
$$\times \exp\left(-\frac{x \cdot K^{-1} x}{2}\right) dx.$$

5. *Archimedean copula*: this copula has the form

$$C(u_1, \cdots, u_d) = \phi^{-1}(\phi(u_1) + \cdots + \phi(u_d))$$

where ϕ^{-1} is the Laplace transform of a positive non-zero random variable Y, i.e., $\phi^{-1}(u) = \mathbb{E}(e^{-uY})$.

Simulations. We seek to generate a vector $(X_1, \cdots, X_d)$ with a copula C and given marginal distributions $F_1, \cdots, F_d$. It is enough to

1. generate random variables $(U_1, \cdots, U_d)$ with uniform marginal distributions and with copula C;

2. then calculate $X_i = F_i^{-1}(U_i)$.

To generate $(U_1, \cdots, U_d)$ we can

1. generate an r.v. $(Y_1, \cdots, Y_d)$ with arbitrary continuous marginal distributions and copula C;

2. then calculate $U_i = F_{Y_i}(Y_i)$.

The separation of dependence and marginal distributions permits us to generate random vectors having the dependence of the Gaussian copula type with marginal distributions that are exponential, Cauchy, etc. Figure 1.1 shows two samples of bi-dimensional vectors with standard Gaussian marginal distributions, with a Gaussian copula on the one hand and an Archimedean copula on the other hand (Y has exponential

46 ■ Generating random variables

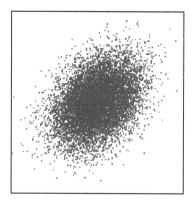 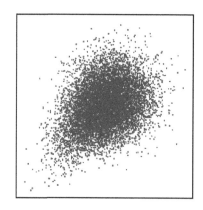

Figure 1.1 Two samples of random variables in dimension 2, for which the marginal r.v. has distribution $\mathcal{N}(0,1)$. On the left: Gaussian copula (Gaussian vector) with correlation 33%. On the right: Clayton copula (Archimedean copula with Y distributed as $\mathcal{E}\mathrm{xp}(1)$). Sample of size 10,000.

distribution): the correlation of the Gaussian copula is such that the two samples have the same empirical correlation (showing, however, different dependencies).

In Figure 1.2, each marginal distribution is exponential with parameter 1 and with random sign (i.e., obtained by εX where $\varepsilon = \pm 1$ with equal probability and $X \stackrel{\mathrm{d}}{=} \mathcal{E}\mathrm{xp}(1)$, also known as the Laplace distribution), with either independent components or a Gaussian copula with correlation 50%. These examples show the variety of possible distributions that we can generate.

Finally, let us mention that the Archimedean dependence admits an ad hoc simulation algorithm; see [109].

Proposition 1.4.3 (generation with Archimedean copula) *Let C be the Archimedean copula associated with the random variable Y (with the Laplace transform ϕ^{-1}), and suppose that $Y > 0$ a.s.. Let $(X_i)_{1 \leq i \leq d}$ be independent random variables with the uniform distribution $[0,1]$ and Y be a random variable independent of $(X_i)_i$. Define*

$$U_i = \phi^{-1}\left(-\frac{1}{Y}\log(X_i)\right).$$

Then the vector $(U_1, \ldots, U_d)$ has uniform marginal distributions and a copula C.

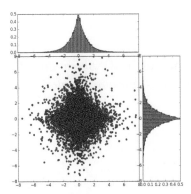 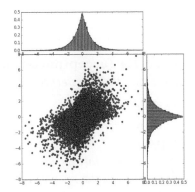

Figure 1.2 Two samples of random variables in dimension 2, for which the marginal distributions are exponential with random sign (Laplace distribution). On the left: independent components. On the right: Gaussian copula with correlation 50%. Sample of size 10,000.

To learn more, see [113].

1.5 EXERCISES

Exercise 1.1 (inversion method)

i) Prove the generation schemes of Propositions 1.2.2, 1.2.4, 1.2.5, 1.2.6, 1.2.7, 1.2.8.

ii) Show that $(1 - \sqrt{U})$ with $U \stackrel{d}{=} \mathcal{U}([0,1])$ has the triangular distribution on $[0,1]$ (with density $2(1-x)\mathbf{1}_{[0,1]}$).

Exercise 1.2 (Box-Muller method) *Prove Proposition 1.2.10.*

Exercise 1.3 (acceptance-rejection method) *We propose a variant of Proposition 1.3.2. Let $c > 0$. Prove the following statements.*

i) Let Y be a d-dimensional random variable with density g and let $U \stackrel{d}{=} \mathcal{U}([0,1])$ be independent of Y. Then, $(Y, cUg(Y))$ is a random vector uniformly distributed on

$$A_{cg} = \{(x,z) \in \mathbb{R}^d \times \mathbb{R} : 0 \leq z \leq cg(x)\}.$$

ii) Conversely, if (Y, Z) is uniformly distributed on A_{cg}, then the distribution of Y has a density equal to g.

48 ■ Generating random variables

From the above, deduce another proof of Proposition 1.3.2 when the reference measure μ is the Lebesgue measure.

Exercise 1.4 (acceptance-rejection method) *Show that the following algorithm generates a standard Gaussian random variable.*

```
1  x: double;
2  u: double;
3  repeat
4  |  x ← simulation according to Exp(1);
5  |  u ← simulation according to U([−1, 1]), independent of x;
6  until (x − 1)² ≤ −2 × log(|u|);
7  Return x if u > 0 and −x otherwise;
```

Exercise 1.5 (acceptance-rejection method) *What is the output distribution of the following algorithm?*

```
1  u: double;
2  v: double;
3  repeat
4  |  u ← simulation according to U([−1, 1]);
5  |  v ← simulation according to U([−1, 1]), independent of u;
6  until (1 + v²) × |u| ≤ 1;
7  Return v if u > 0 and 1/v otherwise;
```

Exercise 1.6 (ratio-of-uniforms method, Gamma distribution) *Using the ratio-of-uniforms method, design an algorithm for simulating the Gamma distribution $\Gamma(a, \theta)$ ($a \geq 1, \theta > 0$) whose density is*

$$p_{a,\theta}(z) = \frac{\theta^a z^{a-1}}{\Gamma(a)} e^{-\theta z} 1_{\{z>0\}}.$$

Hint: *first reduce to the case $\theta = 1$.*

Exercise 1.7 (ratio-of-uniforms method) *Generalize Lemma 1.3.6 to the multidimensional case. Make the algorithm explicit in the case of the two-dimensional density $p(x, y)$ proportional to $f(x, y) = (1 + x^2 + 2y^2)^{-4/3}$.*

Exercise 1.8 (Gaussian copula) *Write a simulation program for generating a bi-dimensional vector with Laplace marginals and Gaussian copula (like for Figure 1.2).*

Exercise 1.9 (Archimedean copula) *Prove Proposition 1.4.3.*

CHAPTER 2

Convergences and error estimates

In this chapter, we discuss the numerical evaluation of $\mathbb{E}(X)$ by the empirical mean

$$\overline{X}_M := \frac{1}{M}\sum_{m=1}^{M} X_m \qquad (2.0.1)$$

where $(X_m)_{m\geq 1}$ are independent simulations having the same distribution as X. This is the basic principle of Monte-Carlo methods for the calculation of expectation.

In particular, we study the convergence in the almost sure sense of the renormalized error $\sqrt{M}(\overline{X}_M - \mathbb{E}X)$, and also different related limit theorems and ramifications arising from them. The computation of the expectation sensitivity is discussed, too.

We also develop non-asymptotic estimates of the deviation of $\overline{X}_M$ with respect to $\mathbb{E}X$ (concentration inequalities). Some tools serve as preparation for the empirical regression methods in Chapter 8. The Gaussian case is analyzed in the light of the logarithmic Sobolev inequalities; these results are used in Chapter 6 to analyze the statistical error of the Euler scheme.

2.1 LAW OF LARGE NUMBERS

The almost sure convergence of $\overline{X}_M$ to $\mathbb{E}X$ is ensured by the strong law of large numbers.

Theorem 2.1.1 (strong law of large numbers) *Let X be a real-*

valued integrable random variable. Then, with probability 1,

$$\lim_{M \to +\infty} \overline{X}_M = \mathbb{E}X.$$

PROOF:

There exist several proofs, either supposing that X is of finite variance and carrying out the calculation of the second moment of $\overline{X}_M$ (see [80, Chapter 20]), or using a martingale convergence argument without assuming the extra condition on the variance (see [80, Chapter 27]). Here, we rather follow the historical proof of Kolmogoroff [93] supposing only that X is integrable.

We start with the general equivalence

$$\mathbb{E}|X| < +\infty \iff \sum_{m=1}^{M} \mathbb{P}(|X| > m) < +\infty, \qquad (2.1.1)$$

which easily follows from the inequality $\mathbb{E}|X| = \int_0^{+\infty} \mathbb{P}(|X| > x) \mathrm{d}x$. Now define

$$Y_m = X_m \mathbf{1}_{|X_m| \leq m}, \qquad Z_m = Y_m - \mathbb{E}(Y_m).$$

From equivalence (2.1.1), we note that

$$\sum_{m=1}^{+\infty} \mathbb{P}(X_m \neq Y_m) = \sum_{m=1}^{+\infty} \mathbb{P}(|X_m| > m) = \sum_{m=1}^{+\infty} \mathbb{P}(|X| > m) < +\infty,$$

which implies, by the Borel-Cantelli lemma (Theorem A.1.1 in Appendix), that a.s., the sequences $(X_m)_m$ and $(Y_m)_m$ have only a finite number of different elements. So with probability 1, if one of the limits $\lim_{M \to +\infty} \overline{X}_M$ or $\lim_{M \to +\infty} \frac{1}{M} \sum_{m=1}^{M} Y_m$ exist, and is finite, then the other limit also exists and the two limits are equal. By the dominated convergence theorem, we prove $\lim_{m \to +\infty} \mathbb{E}(Y_m) = \mathbb{E}(X)$ and hence $\lim_{M \to +\infty} \frac{1}{M} \sum_{m=1}^{M} \mathbb{E}(Y_m) = \mathbb{E}(X)$. Thus it is enough to show that $\lim_{M \to +\infty} \frac{1}{M} \sum_{m=1}^{M} Z_m = 0$ a.s. . In fact, we will prove that

$$S_M := \sum_{m=1}^{M} \frac{Z_m}{m} \text{ converges a.s. as } M \to +\infty.$$

Then, the Kronecker lemma (with $a_m = m$) completes the proof of the theorem.

Lemma 2.1.2 (Kronecker [125, Lemma 6.11]) *If $(a_m)_m$ is increasing to $+\infty$ and if $\sum_{m \geq 1} x_m$ is a convergent series, then $\frac{1}{a_M} \sum_{m=1}^{M} a_m x_m \xrightarrow[M \to +\infty]{} 0$.*

The idea is to get precise control of the variance of Z_m using only that X is integrable. We have $\mathrm{Var}(Z_m) = \mathrm{Var}(Y_m) \leq \mathbb{E}(Y_m^2) \leq \mathbb{E}(X^2 \mathbf{1}_{|X| \leq m})$ from which

$$\sum_{m \geq 1} \frac{\mathrm{Var}(Z_m)}{m^2} \underset{\text{Tonelli}}{=} \mathbb{E}\left(X^2 \sum_{m \geq 1 \vee |X|} \frac{1}{m^2}\right)$$
$$\leq \mathbb{E}\left(X^2 \frac{c}{1 \vee |X|}\right) \leq c\, \mathbb{E}|X| < +\infty. \qquad (2.1.2)$$

Now we control the fluctuations of $(S_m)_m$. Starting with the Cauchy-Schwarz inequality, we obtain for any $x > 0$ and any $n \geq 1$

$$\sqrt{\mathbb{E}(S_n^2)}\sqrt{\mathbb{P}(\max_{1 \leq j \leq n} S_j \geq x)} \geq \mathbb{E}(S_n \mathbf{1}_{\max_{1 \leq j \leq n} S_j \geq x})$$
$$= \sum_{k=1}^{n} \mathbb{E}(S_n \mathbf{1}_{\max_{1 \leq j \leq k-1} S_j < x, S_k \geq x})$$
$$= \sum_{k=1}^{n} \mathbb{E}(S_k \mathbf{1}_{\max_{1 \leq j \leq k-1} S_j < x, S_k \geq x})$$

$((Z_m)_m$ are independent and centered)

$$\geq \sum_{k=1}^{n} \mathbb{E}(x \mathbf{1}_{\max_{1 \leq j \leq k-1} S_j < x, S_k \geq x}) = x \mathbb{P}(\max_{1 \leq j \leq n} S_j \geq x).$$

Comparing the two sides of the inequality, we get $x^2 \mathbb{P}(\max_{1 \leq j \leq n} S_j \geq x) \leq \mathbb{E}(S_n^2) = \sum_{m=1}^{n} \frac{\mathrm{Var}(Z_m)}{m^2}$. Adapting these arguments for the numbers between k and $n = +\infty$, we get

$$\mathbb{P}\left(\max_{j \geq k} |S_j - S_k| \geq x\right) \leq \frac{2}{x^2} \sum_{m > k} \frac{\mathrm{Var}(Z_m)}{m^2} \xrightarrow[k \to +\infty]{} 0.$$

On the one hand, this shows that $\sup_{j \geq 1} |S_j|$ takes a.s. finite values. On the other hand, the two a.s. finite random variables $\underline{L} = \liminf_{k \to +\infty} S_k$ and $\overline{L} := \limsup_{k \to +\infty} S_k$ are equal (thus the sequence $(S_M)_M$ converges a.s.). Indeed, for all $x > 0$, and all $k \geq 0$,

$$\mathbb{P}(\overline{L} - \underline{L} \geq 2x) \leq \mathbb{P}(\max_{j,l \geq k} |S_j - S_l| \geq 2x) \leq 2\mathbb{P}(\max_{j \geq k} |S_j - S_k| \geq x).$$

Taking $k \to +\infty$, we deduce that $\mathbb{P}(\overline{L} - \underline{L} \geq 2x) = 0$. □

Let us finish with a remark on the assumption $\mathbb{E}|X| < +\infty$. It is not only sufficient but also necessary to have the strong law of large numbers: indeed, if $\overline{X}_M$ converges a.s., then $\frac{X_m}{m} = \overline{X}_m - \frac{m-1}{m}\overline{X}_{m-1} \to_{m\to+\infty} 0$ a.s.. In particular, $\mathbb{P}(|X_m| > m$ infinitely often$) = 0$ and the *independent* version of the Borel-Cantelli lemma (Theorem A.1.1) implies that $\sum_{m\geq 1} \mathbb{P}(|X_m| > m) = \sum_{m\geq 1} \mathbb{P}(|X| > m)$ is finite. From (2.1.1), X is integrable.

2.2 CENTRAL LIMIT THEOREM AND CONSEQUENCES

2.2.1 Central limit theorem in dimension 1 and beyond

If X is a real-valued square integrable random variable and $\sigma^2 = \mathrm{Var}(X)$, the variance of $\overline{X}_M$ is equal to $\frac{\sigma^2}{M}$ and it is natural to study the renormalized error $\sqrt{M}(\overline{X}_M - \mathbb{E}X)$, which has constant variance equal to σ^2. This renormalized error converges in distribution to the centered Gaussian distribution with variance σ^2. We give the statement directly in the multidimensional case.

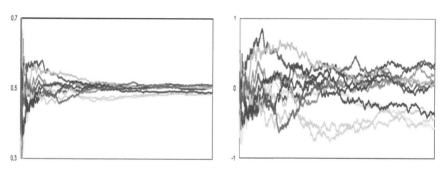

Figure 2.1 Several sets of 1000 simulations for the calculation of $\mathbb{E}(X)$ with X having the uniform distribution on $[0,1]$. On the left, $M \to \overline{X}_M$; on the right $M \to \sqrt{M}(\overline{X}_M - \frac{1}{2})$.

Theorem 2.2.1 (central limit theorem) *Let X be a d-dimensional square integrable random vector with the covariance matrix $K = (\mathrm{Cov}(X_i, X_j))_{i,j}$. Then*

$$\sqrt{M}(\overline{X}_M - \mathbb{E}(X)) \underset{M\to+\infty}{\Longrightarrow} \mathcal{N}(0, K).$$

PROOF:
    A fast proof is based on the Levy criterion (Theorem A.1.3). Suppose $\mathbb{E}(X) = 0$, otherwise we can recenter X. For $u \in \mathbb{R}^d$, define $Y = u \cdot X$: this is a square integrable variable with $\mathrm{Var}(Y) = u \cdot Ku$, its

characteristic function is $\Phi_Y(v) = \mathbb{E}(e^{ivY})$, and thus $\mathcal{C}^2$ with $\Phi'_Y(v) = \mathbb{E}(iYe^{ivY})$, $\Phi''_Y(v) = -\mathbb{E}(Y^2 e^{ivY})$. Using the independence and the zero mean property of $(X_m)_m$, we get

$$\mathbb{E}(e^{i\,u\cdot\sqrt{M}(\overline{X}_M - \mathbb{E}X)}) = \left(\mathbb{E}(e^{iu\cdot\sqrt{M}\frac{X}{M}})\right)^M = \left(\Phi_Y\left(\frac{1}{\sqrt{M}}\right)\right)^M$$

$$= \left(1 + \frac{1}{\sqrt{M}}\Phi'_Y(0) + \frac{1}{2M}\Phi''_Y(0) + o\left(\frac{1}{M}\right)\right)^M$$

$$= \left(1 - \frac{1}{2M}\mathrm{Var}(Y) + o\left(\frac{1}{M}\right)\right)^M$$

$$\xrightarrow[M \to +\infty]{} e^{-\frac{1}{2}\mathrm{Var}(Y)} = e^{-\frac{1}{2}u\cdot Ku},$$

which proves the announced convergence. □

In Figure 2.1, we see the normalized errors for several simulation sets: on the left, we observe the a.s. convergence of $\overline{X}_M$, on the right the renormalized error doesn't seem to converge a.s.. On the other hand, its distribution (not represented) is close to a Gaussian distribution.

This central limit theorem leads to several fundamental remarks.

1. If we want to increase the accuracy of $\mathbb{E}X$ by a factor 10, it turns out to be necessary to increase M - and thus the computational time - by a factor of $10^2 = 100$. This heuristic is valid in the sense of confidence intervals (for a fixed threshold) discussed later.

2. The error is random and it is hopeless to have a non-trivial control with probability 1. Besides, the error is asymptotically universally Gaussian: its characterization is thus simple via the determination of only one parameter, i.e. the *covariance matrix* K (a real number if $d = 1$). We will see how it can be estimated using the same sample. It is remarkable for a numerical method to yield an a posteriori error control.

3. The comparison of the calculation of $\mathbb{E}X$ by different Monte-Carlo methods can be done via their parameters K: the "larger" K is, the more numerically demanding the problem is, and the more significant the required efforts.

When K is invertible, we can rewrite the previous result to make a limit appear, independent of the model for X, emphasizing a certain universality and robustness - this will be useful to determine the confidence regions. For this we need to estimate the covariance matrix K using the sample $(X_m)_m$: we set

$$K_{i,j,M} = \frac{M}{M-1}\Big(\frac{1}{M}\sum_{m=1}^{M} X_{i,m}X_{j,m} - \underbrace{\Big(\frac{1}{M}\sum_{m=1}^{M} X_{i,m}\Big)\Big(\frac{1}{M}\sum_{m=1}^{M} X_{j,m}\Big)}_{:=\overline{X}_{i,M}}\Big)$$

$$= \frac{1}{M-1}\sum_{m=1}^{M}(X_{i,m} - \overline{X}_{i,M})(X_{j,m} - \overline{X}_{j,M}).$$

From the law of large numbers, $K_{i,j,M} \xrightarrow[M\to+\infty]{\text{a.s.}} \mathbb{E}(X_i X_j) - \mathbb{E}(X_i)\mathbb{E}(X_j) = \mathrm{Cov}(X_i, X_j)$. The factor $\frac{M}{M-1}$ is a minor improvement, because M is very large in practice, which gives a non-biased estimator K_M, i.e., $\mathbb{E}(K_M) = K$. If K is positive and invertible, so a.s. is K_M for M large enough, and we can take its square root, defined as the symmetric positive definite matrix $\sqrt{K_M}$ (abusive notation) such that $\sqrt{K_M}\sqrt{K_M} = K_M$. The Slutsky lemma (see Theorem A.1.5 in appendix) applied to Theorem 2.2.1 allows us to prove that

$$(\sqrt{K_M})^{-1}\sqrt{M}(\overline{X}_M - \mathbb{E}(X)) \underset{M\to+\infty}{\Longrightarrow} \mathcal{N}(0,\mathrm{Id}).$$

This convergence is interpreted as follows: for any continuous bounded function, we have

$$\lim_{M\to+\infty} \mathbb{E}\Big(f\big((\sqrt{K_M})^{-1}\sqrt{M}(\overline{X}_M - \mathbb{E}(X))\big)\Big) = \int_{\mathbb{R}^d} f(x)\frac{1}{(2\pi)^{d/2}}e^{-\frac{|x|^2}{2}}\,\mathrm{d}x.$$

This still holds if $f = \mathbf{1}_A$ with a measurable set A in $\mathbb{R}^d$ having a boundary of zero Lebesgue measure (a ball, for example): then

$$\lim_{M\to+\infty} \mathbb{P}\Big((\sqrt{K_M})^{-1}\sqrt{M}(\overline{X}_M - \mathbb{E}(X)) \in A\Big) = \int_A \frac{1}{(2\pi)^{d/2}}e^{-\frac{|x|^2}{2}}\,\mathrm{d}x.$$
(2.2.1)

2.2.2 Asymptotic confidence regions and intervals

As the normalized error converges in distribution, we cannot expect something better than having estimates in probability of $\sqrt{M}(\overline{X}_M - \mathbb{E}(X))$.

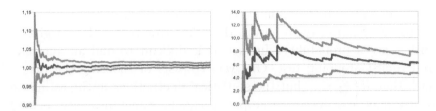

Figure 2.2 Monte-Carlo evaluation of $\mathbb{E}(e^{G/10}) = e^{\frac{1}{2}\frac{1}{10^2}} \approx 1.005$ on the left and $\mathbb{E}(e^{2G}) = e^{\frac{1}{2}2^2} \approx 7.389$ on the right, where G is a standard Gaussian random variable. The empirical mean and the confidence intervals of 95% are represented as a function of the number of simulations.

For example, the application of (2.2.1) with a centered ball of radius R ($A = B(0, R)$) gives

$$|(\sqrt{K_M})^{-1}\sqrt{M}(\overline{X}_M - \mathbb{E}(X))| \leq R \qquad (2.2.2)$$

with probability close to $p(R) := \int_{B(0,R)} \frac{1}{(2\pi)^{d/2}} e^{-\frac{|x|^2}{2}} dx$ as $M \to +\infty$. This defines a *confidence region* for the unknown value $\mathbb{E}(X)$, equal to the *ellipsoid* $\overline{X}_M + M^{-1/2}(\sqrt{K_M})B(0, R)$, at the approximate confidence level [1] $p(R)$.[2]

In the case $d = 1$ where $K_M := \sigma_M^2$ (with $\sigma_M > 0$), this is written as

$$\mathbb{E}(X) \in \left[\overline{X}_M - R\frac{\sigma_M}{\sqrt{M}}, \overline{X}_M + R\frac{\sigma_M}{\sqrt{M}}\right], \qquad (2.2.3)$$

where $p(R) = \int_{-R}^{R} \frac{1}{\sqrt{2\pi}} e^{-\frac{x^2}{2}} dx = 2\mathcal{N}(R) - 1$. The commonly used value is $R = 1.96$ giving $p(R) = 95\%$: we speak then about the confidence intervals (asymptotically symmetric) with the confidence level of 95%. The two given examples in Figure 2.2 show that the size of the confidence intervals may differ greatly from one case to another: therefore

> providing confidence intervals is essential and must be systematically done in evaluations of expectation by Monte-Carlo methods.

---

[1] We refer to probably approximatively correct bounds (*PAC bounds*).
[2] $p(R) = \mathbb{P}(\chi_d^2 \leq R^2)$, where χ_d^2 is the χ^2 random variable with d degrees of freedom (sum of d squares of independent standard Gaussian random variables).

When do we stop the simulations? In other words, is it possible to choose M adaptively to ensure that the error is less than a given ε, with a given probability? In fact, it is difficult to find an automatic and robust stopping test. One way consists of choosing the first M such that $R\frac{\sigma_M}{\sqrt{M}} \leq \varepsilon$, but the statistical fluctuations in the estimation of σ_M may lead to wrong stopping signals, sometimes early enough. For references and more detailed discussions, see [57].

In certain cases (X bounded or equal to a Lipschitz function of Gaussian random variables), non-asymptotic estimates may be available (see later in Section 2.4) and they can be used to construct a priori stopping tests.

2.2.3 Application to the evaluation of a function of $\mathbb{E}(X)$

Sometimes, the final purpose is not to calculate $\mathbb{E}(X)$, but to calculate a certain function f of $\mathbb{E}X$: it is natural to use the estimator $f(\overline{X}_M)$ (*substitution method*). For example, for the Buffon needle (see page 5), if $\overline{X}_M$ is the empirical frequency of the needles falling on a groove of the parquet, the limit is $\mathbb{E}X = \frac{2}{\pi}$, in the case where $a = l$, we want to approach π by $\frac{2}{\overline{X}_M}$.

Several questions appear: What is the bias of the new estimator? What is its convergence rate? What is its asymptotic variance?

▷ **Convergence rate.** If the function is regular, then the *delta method* also gives a central limit theorem for $f(\overline{X}_M)$.

Theorem 2.2.2 (substitution method) *Let X be a square integrable random vector in $\mathbb{R}^d$ (its covariance matrix is K) and $f : \mathbb{R}^d \mapsto \mathbb{R}$ be differentiable at the point $\mathbb{E}X$; then*

$$\sqrt{M}\big(f(\overline{X}_M) - f(\mathbb{E}X)\big) \underset{M \to +\infty}{\Longrightarrow} \mathcal{N}\big(0, \nabla f(\mathbb{E}X) \cdot K \, \nabla f(\mathbb{E}X)\big).$$

PROOF:
We start by assuming that f is of class C^2 with bounded derivatives. A Taylor formula gives

$$\sqrt{M}\big(f(\bar{X}_M) - f(\mathbb{E}X)\big) = \nabla f(\mathbb{E}X)\sqrt{M}(\bar{X}_M - \mathbb{E}X) \\ + O\Big(\sqrt{M}|\bar{X}_M - \mathbb{E}X|^2\Big).$$

As $\mathbb{E}(\sqrt{M}|\bar{X}_M - \mathbb{E}X|^2) = \dfrac{c}{\sqrt{M}}$, the last term tends to 0 in probability. Then the second one gives the stated limit.

If f is only differentiable, the above decomposition is modified to give a residual of the form $\sqrt{M}|\overline{X}_M - \mathbb{E}X|o(1)$, where $o(1)$ tends to 0 as $|\overline{X}_M - \mathbb{E}X| \to 0$: we then conclude that this residual term tends to 0 in probability because $\sqrt{M}(\overline{X}_M - \mathbb{E}X)$ is bounded in probability (as it is centered with constant variance) and $o(1) \xrightarrow[M \to +\infty]{\text{a.s.}} 0$ (due to $\overline{X}_M \xrightarrow[M \to +\infty]{\text{a.s.}} \mathbb{E}X$). □

Hence, as for $\overline{X}_M$, we have the central limit theorem for $f(\overline{X}_M)$ but the novelty is that the limit variance $\nabla f(\mathbb{E}X) \cdot K \, \nabla f(\mathbb{E}X)$ cannot be directly written as the limit of an empirical variance calculated simultaneously with the empirical mean.

How do we obtain a confidence interval?

- In certain cases, it is easy to estimate the limit variance $\nabla f(\mathbb{E}X) \cdot K \, \nabla f(\mathbb{E}X)$: for example, first estimate the matrix K on the sample $(X_m)_m$, then compute the derivatives of f at $\overline{X}_M \approx \mathbb{E}X$. In the case of the Buffon needle, $K = \frac{2}{\pi}(1 - \frac{2}{\pi})$ (variance of the Bernoulli random variable) and $f(x) = 2x^{-1}$, $f'(x) = -2x^{-2}$: the limit variance is then $\frac{2}{\pi}(1 - \frac{2}{\pi})4(\frac{2}{\pi})^{-4} = \pi^2(\pi/2 - 1)$.

- But we can also bypass these two stages using the *sectioning method*. This method is generally applied to probabilistic algorithms whose outputs are centered at μ (the quantity to calculate) and asymptotically have a Gaussian distribution with unknown (or difficult to calculate) variance σ^2.

    The principle consists of *sectioning* the M simulations in n subsamples of size M/n: to simplify the discussion, we suppose that M/n is an integer. Then the algorithm produces n independent outputs (approximately Gaussian), for which we calculate the mean and the variance: we are reduced to the classic statistical problem of parametric estimation of the unknown parameters of a Gaussian distribution. The confidence interval involves the Student distribution with $n-1$ degrees of freedom, see [143, Chapter 10]. The result is stated as follows.

Proposition 2.2.3 (sectioning) *Suppose that a probabilistic algorithm A performed with M simulations yields a one-dimensional estimator A_M of μ with an asymptotically Gaussian normalized error:*

$$\sqrt{M}(A_M - \mu) \underset{M \to +\infty}{\Longrightarrow} \mathcal{N}(0, \sigma^2)$$

for a certain parameter $\sigma^2 > 0$. Denote $(A_{i,M/n})_{1 \leq i \leq n}$ the n independent outputs of the algorithm, each one obtained with M/n simulations, and set

$$\overline{A}_{n,M/n} = \frac{1}{n} \sum_{i=1}^{n} A_{i,M/n}, \quad \overline{V}_{n,M/n} = \frac{1}{n-1} \sum_{i=1}^{n} (A_{i,M/n} - \overline{A}_{n,M/n})^2$$

for the empirical means and variances. Then, for fixed n, the random variable

$$\sqrt{n} \frac{(\overline{A}_{n,M/n} - \mu)}{\sqrt{\overline{V}_{n,M/n}}} \underset{M \to +\infty}{\Longrightarrow} \text{Student distribution with}$$

$n - 1$ degrees of freedom.

So, approximately, with probability 95%,

$$\mu \in \left[\overline{A}_{n,M/n} - \text{Stud.}^{-1}_{0.975, n-1} \sqrt{\frac{\overline{V}_{n,M/n}}{n}}, \right.$$

$$\left. \overline{A}_{n,M/n} + \text{Stud.}^{-1}_{0.975, n-1} \sqrt{\frac{\overline{V}_{n,M/n}}{n}} \right],$$

where $\text{Stud.}^{-1}(u, k)$ is the quantile at the level $u \in [0, 1]$ of the Student distribution with k degrees of freedom.[3]

Note that the effect of accuracy due to M is hidden in

$$\sqrt{\frac{\overline{V}_{n,M/n}}{n}} \underset{M/n \text{ and } n \text{ large}}{\approx} \sqrt{\frac{\frac{\sigma^2}{M/n}}{n}} = \frac{\sigma}{\sqrt{M}}.$$

Finally, we emphasize that σ^2 does not have to be estimated in the sectioning method. In practice it is enough to take $n = 10$.

Examples 2.2.4

– The application to Theorem 2.2.2 is made by taking

$$A_{i,M/n} = f\left(\frac{n}{M} \sum_{m=(i-1)\frac{M}{n}+1}^{i\frac{M}{n}} X_m \right).$$

---

[3] Recall that $\text{Stud.}^{-1}_{0.975, n-1} = 2.262$ is $n = 10$, 2.093 is $n = 20$ and converges to 1.96 if $n \to +\infty$.

- Consider the problem of finding the zero (supposed to be unique) of the function $\theta \mapsto f(\theta, \mathbb{E}(X))$: denote it by θ_0, i.e., $f(\theta_0, \mathbb{E}(X)) = 0$.

  If we denote θ_M, the zero of $\theta \mapsto f(\theta, \overline{X}_M)$ and if f is C^1, it is not difficult to show the central limit theorem for $\sqrt{M}(\theta_M - \theta_0)$. The sectioning method allows us then to construct an explicit (asymptotic) confidence interval.

▷ **Bias.** While $\overline{X}_M$ is an unbiased estimator of $\mathbb{E}X$ (i.e., $\mathbb{E}(\overline{X}_M) = \mathbb{E}X$), a bias may appear in the substitution method, i.e., $\mathbb{E}(f(\overline{X}_M)) \neq f(\mathbb{E}X)$; let us now analyze it.

Proposition 2.2.5 (bias of the substitution method)

i) *If f is continuous convex (resp. concave), the substitution method gives a over-estimation (resp. under-estimation) of the needed quantity.*

ii) *If f is C_b^4 and if X has finite moments of order $4 + \delta$ ($\delta \in]0, 1]$), then $\mathbb{E}(f(\overline{X}_M)) - f(\mathbb{E}X) = \dfrac{c_1}{M} + \dfrac{c_2}{M^2} + o(M^{-2})$ for two constants c_1, c_2 depending on the derivatives of f and moments of X.*

Note that the terms of order $M^{-1/2}$ and $M^{-3/2}$ disappear in this calculation of expectation, although the convergence in distribution of $\overline{X}_M$ to $\mathbb{E}X$ holds at rate $\sqrt{M}$.

PROOF:
The Jensen inequality (see (A.2.1)) gives (for convex f) $\mathbb{E}(f(\overline{X}_M)) \geq f(\mathbb{E}(\overline{X}_M)) = f(\mathbb{E}X)$. This shows the assertion i). The assertion ii) can be shown using Taylor expansion: to simplify the writing, suppose that X is one-dimensional and $\mathbb{E}X = 0$ (otherwise, we replace X by $X - \mathbb{E}X$). Then

$$f(\overline{X}_M) = f(0) + f^{(1)}(0)\overline{X}_M + f^{(2)}(0)\frac{\overline{X}_M^2}{2} + f^{(3)}(0)\frac{\overline{X}_M^3}{3!} + f^{(4)}(0)\frac{\overline{X}_M^4}{4!}$$
$$+ \overline{X}_M^4 \int_0^1 \frac{(1-u)^3}{3!}(f^{(4)}(u\overline{X}_M) - f^{(4)}(0))du. \qquad (2.2.4)$$

Now take the expectation in the above equality, the term with $f^{(1)}(0)$ disappears because $\mathbb{E}(\overline{X}_M) = 0$. The factor with $f^{(2)}(0)$ becomes $\dfrac{1}{2}\dfrac{\operatorname{Var}(X)}{M}$. As for the third term, observe that $\mathbb{E}(\overline{X}_M^3) = \dfrac{1}{M^3}\sum_{i,j,k} \mathbb{E}(X_i X_j X_k)$ includes only the M non-zero terms $i = j = k$

because of the independence and centering of $(X_m)_m$: thus this term equals γ_1/M^2 for a certain constant γ_1. A similar observation can be applied to $\mathbb{E}(\overline{X}_M^4)$ and shows that it is equal to $\gamma_2/M^2 + \gamma_3/M^3$ for two other constants γ_2 and γ_3. Hence, these first terms give the announced expansion $\frac{c_1}{M} + \frac{c_2}{M^2}$, plus a term $O(M^{-3})$. It remains to show that the expectation of the last term of (2.2.4) is $o(M^{-2})$: by the Hölder inequality, it is upper bounded by

$$(\mathbb{E}|\overline{X}_M|^{4+\delta})^{1/(1+\delta/4)}(\mathbb{E}|I_M|^{1+1/(\delta/4)})^{(\delta/4)/(1+\delta/4)}$$

where I_M stands for the integral term in (2.2.4). An application of the dominated convergence theorem shows that $\mathbb{E}|I_M|^{1+1/(\delta/4)} \xrightarrow[M \to +\infty]{} 0$. To estimate $\mathbb{E}|\overline{X}_M|^{4+\delta}$, we use the Rosenthal inequality (Theorem A.2.4) which provides a control of the moments for the sum of centered independent random variables:

$$\mathbb{E}|\overline{X}_M|^{4+\delta} \leq c_{4+\delta}\Big(\sum_{m=1}^{M}\mathbb{E}|\frac{X_m}{M}|^{4+\delta} + \big(\sum_{m=1}^{M}\mathbb{E}|\frac{X_m}{M}|^2\big)^{2+\delta/2}\Big)$$

$$= O\Big(\frac{1}{M^{2(1+\delta/4)}}\Big).$$

This finishes the proof. □

In the case of a smooth function, the bias converges to 0 at a speed faster than that for the central limit theorem, which makes it a priori negligible. However, it may happen that for medium values of M, the term c_1/M is of the same order as the statistical error of the central limit theorem: in this case, it is useful to decrease this bias. A possible approach is to use the *jackknife method* [36], or a simple version, the *Romberg extrapolation*, which takes advantage of clever combinations of the empirical means to erase the bias. Let us take for example an estimator

$$\widetilde{f}_M := 2f(\overline{X}_M) - f(\overline{X}_{M/2}) \qquad (2.2.5)$$

supposing to simplify that M is even. We check that its bias is of order M^{-2} (under the previous hypothesis):

$$\mathbb{E}(\widetilde{f}_M) = 2\Big[f(\mathbb{E}X) + \frac{c_1}{M} + \frac{c_2}{M^2} + o(\frac{1}{M^2})\Big]$$
$$- \Big[f(\mathbb{E}X) + \frac{c_1}{M/2} + \frac{c_2}{(M/2)^2} + o(\frac{1}{M^2})\Big]$$
$$= f(\mathbb{E}X) - 2\frac{c_2}{M^2} + o(\frac{1}{M^2}).$$

The variance of $\widetilde{f}_M$ decreases asymptotically again like $1/M$. One can consider other versions of (2.2.5) to optimize the method, regarding the bias and the limit variance.

For certain convex functions such that $f(x) = \max(x,a)^4$ for a given parameter a, in addition to $f(\overline{X}_M)$ it is possible to construct an estimator biased from below, that allows us to squeeze the sought value (at the level of the bias). For this, we divide the M-sample in two parts (M is still supposed to be even).

Proposition 2.2.6 (estimators biased from below and from above) *Let $\overline{X}_{1,M} = \frac{2}{M}\sum_{m=1}^{M/2} X_i$ and $\overline{X}_{2,M} = \frac{2}{M}\sum_{m=M/2+1}^{M} X_i$. Set*

$$\overline{f}_M = \max(\overline{X}_M, a), \qquad \underline{f}_M = 1_{\overline{X}_{1,M} \geq a} \overline{X}_{2,M} + 1_{\overline{X}_{1,M} < a} a.$$

Then $\overline{f}_M$ and $\underline{f}_M$ converge a.s. to $\max(\mathbb{E}X, a)$ as $M \to +\infty$, with

$$\mathbb{E}(\underline{f}_M) \leq \max(\mathbb{E}X, a) \leq \mathbb{E}(\overline{f}_M).$$

PROOF:
The a.s. convergence is almost immediate because $\overline{X}_{1,M}, \overline{X}_{2,M}, \overline{X}_M \xrightarrow[M\to+\infty]{a.s.} \mathbb{E}X$: we must be careful about $\underline{f}_M$ because of the discontinuous indicator function. If $\mathbb{E}X \neq a$, then $1_{\overline{X}_{1,M} \geq a} \xrightarrow[M\to+\infty]{a.s.} 1_{\mathbb{E}X \geq a}$ and $\underline{f}_M \xrightarrow[M\to+\infty]{a.s.} \mathbb{E}X 1_{\mathbb{E}X \geq a} + a 1_{\mathbb{E}X < a} = \max(\mathbb{E}X, a)$. If $\mathbb{E}X = a$, then $\underline{f}_M = \mathbb{E}X + 1_{\overline{X}_{1,M} \geq a}(\overline{X}_{2,M} - \mathbb{E}X) \xrightarrow[M\to+\infty]{a.s.} \mathbb{E}X = \max(\mathbb{E}X, a)$.

The upper bias of $\overline{f}_M$ was already shown in Proposition 2.2.5. For $\underline{f}_M$, we use the independence of the two sub-samples and write:

$$\mathbb{E}(\underline{f}_M) = \mathbb{E}(X)\mathbb{P}(\overline{X}_{1,M} \geq a) + a\mathbb{P}(\overline{X}_{1,M} < a) \leq \max(\mathbb{E}(X), a).$$

□

2.2.4 Applications in the evaluation of sensitivity of expectations

We study the Monte-Carlo evaluation of the derivative of $\mathbb{E}(X^\theta)$ with respect to a real parameter $\theta \in \Theta \subset \mathbb{R}$, where

- $X^\theta = f(Y^\theta)$ for a given function f, regular or not;

---

[4]These functions appear in certain problems of optimal stopping in financial engineering, namely for pricing *American options*.

62 ■ Convergences and error estimates

- Y^θ is a random variable depending on θ.

In other words, we want to numerically compute $\partial_\theta \mathbb{E}(f(Y^\theta))$, supposing that this derivative exists.

▷ **Resimulation method.** The derivative is approached using a centered finite difference method

$$\partial_\theta \mathbb{E}(f(Y^\theta)) \approx \frac{\mathbb{E}(f(Y^{\theta+\varepsilon})) - \mathbb{E}(f(Y^{\theta-\varepsilon}))}{2\varepsilon}$$

with a small ε. Then, each expectation is approximated by an empirical mean of the random variables $f(Y^{\theta+\varepsilon})$ and $f(Y^{\theta-\varepsilon})$, in the form of (2.0.1). However, it is important to use the same randomness[5] to generate $Y^{\theta+\varepsilon}$ and $Y^{\theta-\varepsilon}$: for example, if $Y^\theta = y(\theta, U)$ for a measurable function[6] $y(\theta, \cdot)$ and uniformly distributed random variables U, we calculate

$$\partial_\theta \mathbb{E}(f(Y^\theta)) \approx \frac{1}{M} \sum_{m=1}^{M} \frac{f(y(\theta+\varepsilon, U_m)) - f(y(\theta-\varepsilon, U_m))}{2\varepsilon}. \quad (2.2.6)$$

This little trick allows us to significantly reduce the global statistical error, in comparison with a method where independent simulations of U are used for the parameters $\theta + \varepsilon$ and $\theta - \varepsilon$.

This method gives a biased estimator of the derivative (because $\varepsilon \neq 0$) while the following techniques are without bias.

We also mention that it may happen (namely if one of the functions $y(\cdot, U)$ or $f(\cdot)$ is not regular) that the variance $\frac{1}{M}\mathrm{Var}\left(\frac{f(y(\theta+\varepsilon,U))-f(y(\theta-\varepsilon,U))}{2\varepsilon}\right)$ of the estimator (2.2.6) tends to infinity as $\varepsilon \to 0$. On the contrary, when $f(y(\cdot, U))$ is regular, the variance is approximately equal to $\frac{1}{M}\mathrm{Var}(\partial_\theta[f(y(\theta, U))])$ and thus depends little on ε.

Finally, the optimal adjustment of ε and M remains an important and delicate question. Unfortunately, there exists neither a unique answer, nor a robust approach, and the rules depend a lot on the regularity of f and of $y(\cdot, U)$. We will not go further into the details of this issue and refer the reader to [56] for example.

---

[5]That is we resimulate with the same random numbers (also known as *Common Random Numbers* [56]) modifying only the parameter.

[6]We can always write the measurable function in dimension 1 using the inversion of the c.d.f.

▷ **Pathwise differentiation method.** When the function f is regular and the random variable Y^θ is a.s. differentiable with respect to θ, we can simply differentiate under the expectation sign: we leave the proof to the reader.

Proposition 2.2.7 (pathwise differentiation method) *Suppose that*

i) *$\theta \in \Theta \mapsto Y^\theta \in \mathbb{R}^d$ are $\mathcal{C}^1$ a.s. with $|\partial_\theta Y^\theta| \leq \overline{Y}$ for all $\theta \in \Theta$ and $\overline{Y}$ is integrable;*

ii) *$f : \mathbb{R}^d \mapsto \mathbb{R}$ is a function $\mathcal{C}^1$ with bounded derivative.*

Then, $\partial_\theta \mathbb{E}(f(Y^\theta)) = \mathbb{E}(\nabla f(Y^\theta)\partial_\theta Y^\theta).$

As the derivative is again represented in the form of expectation, its evaluation by Monte-Carlo method is possible: this is why we are interested in such a formula.

Example 2.2.8 (Gaussian random variable) *Let G be a standard Gaussian random variable and set $Y^{m,\sigma} = m + \sigma G \stackrel{d}{=} \mathcal{N}(m, \sigma^2)$: then*

a) $\partial_m \mathbb{E}(f(Y^{m,\sigma})) = \mathbb{E}(f'(Y^{m,\sigma}));$

b) $\partial_\sigma \mathbb{E}(f(Y^{m,\sigma})) = \mathbb{E}\left(f'(Y^{m,\sigma})\frac{(Y^{m,\sigma}-m)}{\sigma}\right).$

When, in the centered finite difference method, the function $y(\cdot)$ is differentiable in θ, we can verify that the two methods coincide as $\varepsilon \to 0$.

▷ **Likelihood method.** In certain cases, the a.s. derivative of Y^θ does not exist: in this situation, we use the differentiability of the density of Y^θ with respect to θ.

Proposition 2.2.9 (likelihood method) *Suppose that*

i) *Y^θ is a d-dimensional random variable having a strictly positive density $p(\theta, .)$ with respect to a reference measure μ (independent of $\theta \in \Theta$);*

ii) *$(\theta, y) \in \Theta \times \mathbb{R}^d \mapsto p(\theta, y)$ is continuously differentiable with respect to θ and $|\partial_\theta p(\theta, y)| \leq \overline{p}(y)$ with $\int_{\mathbb{R}^d} \overline{p}(y)\mu(dy) < +\infty;$*

iii) *$f : \mathbb{R}^d \mapsto \mathbb{R}$ is a bounded measurable function.*

Then
$$\partial_\theta \mathbb{E}(f(Y^\theta)) = \mathbb{E}\Big(f(Y^\theta)\partial_\theta[\log(p(\theta,y))]|_{y=Y^\theta}\Big).$$

Note that no regularity on f is needed and that it is enough that the upper bound $\bar{p}$ is valid locally uniformly in θ.

PROOF:
Writing the expectation in the integral form, we obtain (the hypothesis allowing the differentiation)

$$\partial_\theta \mathbb{E}(f(Y^\theta)) = \partial_\theta \int_{\mathbb{R}^d} f(y)p(\theta,y)\mu(dy)$$
$$= \int_{\mathbb{R}^d} f(y)\partial_\theta p(\theta,y)\mu(dy) = \int_{\mathbb{R}^d} f(y)\frac{\partial_\theta p(\theta,y)}{p(\theta,y)}p(\theta,y)\mu(dy)$$
$$= \mathbb{E}\Big(f(Y^\theta)\frac{\partial_\theta p(\theta,y)}{p(\theta,y)}|_{y=Y^\theta}\Big).$$

□

Note that the weight $\frac{\partial_\theta p(\theta,y)}{p(\theta,y)}|_{y=Y^\theta}$ has zero expectation (take $f=1$). On the other hand, it is possible that it has infinite variance, which in this case prevents the calculation of the Gaussian confidence intervals for the resulting Monte-Carlo method.

Example 2.2.10 (Poisson random variable) *Let Y^θ be a random variable with the distribution $\mathcal{P}(\theta)$: in this case taking for μ the counting measure on $\mathbb{N}$, we have $p(\theta,y) = e^{-\theta}\frac{\theta^y}{y!}$ and $\partial_\theta \log(p(\theta,y)) = \frac{y}{\theta} - 1$. So*

$$\partial_\theta \mathbb{E}(f(Y^\theta)) = \mathbb{E}\Big(f(Y^\theta)(\frac{Y^\theta}{\theta} - 1)\Big).$$

Example 2.2.11 (Gaussian random variable (bis)) *Take again the Gaussian example in dimension 1 with $Y^{m,\sigma} \stackrel{d}{=} \mathcal{N}(m,\sigma^2)$. Suppose that $\sigma > 0$, so that $Y^{m,\sigma}$ has a density*

$$p((m,\sigma);y) = \frac{1}{\sigma\sqrt{2\pi}} e^{-\frac{(y-m)^2}{2\sigma^2}}$$

with respect to the Lebesgue measure. As its logarithm equals $\log(p((m,\sigma);y)) = \text{Cste} - \log(\sigma) - \frac{(y-m)^2}{2\sigma^2}$, we deduce the formulas

$$\partial_m \mathbb{E}(f(Y^{m,\sigma})) = \mathbb{E}\Big(f(Y^{m,\sigma})\frac{(Y^{m,\sigma}-m)}{\sigma^2}\Big),$$

$$\partial_\sigma \mathbb{E}(f(Y^{m,\sigma})) = \mathbb{E}\Big(f(Y^{m,\sigma}) \frac{1}{\sigma}\Big[\frac{(Y^{m,\sigma}-m)^2}{\sigma^2} - 1\Big]\Big).$$

▷ **Conclusion.** The interest in such formulas comes from the possibility to evaluate simultaneously $\mathbb{E}(f(Y^\theta))$ and its sensitivities by Monte-Carlo method. The pathwise differentiation method can be applied if the random variable is pathwise regular in the first parameter, while the likelihood method works without regularity on f but under the condition of having an explicit expression for the density.

In Part **B** (Section 4.5), we will see extensions to stochastic processes for which the distributions are not explicit.

2.3 OTHER ASYMPTOTIC CONTROLS

We complete our description of the picture on the asymptotic convergence results by some complementary and less common results, which can be skipped at first reading. We do not detail the proofs and refer the reader to the specialized works [125] [28].

2.3.1 Berry-Essen bounds and Edgeworth expansions

Theorem 2.3.1 (Berry-Essen 1941–42) *Let X be a real-valued random variable such that $\mathbb{E}|X|^3 < +\infty$. For a universal constant $A > 0$, we have*

$$\sup_{x \in \mathbb{R}} \Big|\mathbb{P}\Big(\sqrt{M}\frac{(\overline{X}_M - \mathbb{E}X)}{\sigma} \leq x\Big) - \mathcal{N}(x)\Big| \leq \frac{A}{\sqrt{M}} \frac{\mathbb{E}|X - \mathbb{E}(X)|^3}{\sigma^3}.$$

Now it is known that the best constant A is in the interval $[0.4097, 0.4748]$. Generally, the speed is optimal: it is attained if $X = \pm 1$ with probability $\frac{1}{2}$ (Rademacher variable).

When X has extra finite moments and has a density (implying $\limsup_{t \to +\infty} |\mathbb{E}(e^{itX})| < 1$), we can even get an asymptotic expansion in powers of $\frac{1}{\sqrt{M}}$, known as the Edgeworth expansion.

Theorem 2.3.2 (Edgeworth expansion) *Let X be a real-valued random variable such that $\mathbb{E}|X|^r < +\infty$ for an integer $r \geq 3$, and*

$\limsup_{t\to+\infty} |\mathbb{E}(e^{itX})| < 1$. *Then there exist polynomials $P_i(.)$ such that*

$$\sup_{x\in\mathbb{R}}(1+|x|)^k \left| \mathbb{P}\left(\sqrt{M}\frac{(\overline{X}_M - \mathbb{E}X)}{\sigma} \leq x\right) - \mathcal{N}(x) - \frac{1}{\sqrt{2\pi}}e^{-\frac{x^2}{2}} \sum_{i=1}^{r-2} \frac{P_i(x)}{M^{i/2}} \right|$$
$$= o\left(M^{-\frac{r-2}{2}}\right).$$

2.3.2 Law of iterated logarithm

Theorem 2.3.3 (Hartman-Wintner, 1941) *Let X be a real-valued square integrable random variable. Then with probability 1,*

$$\limsup_{M\to+\infty} \frac{\sqrt{M}}{\sqrt{2\log(\log(M))}} (\overline{X}_M - \mathbb{E}X) = \sigma,$$

$$\liminf_{M\to+\infty} \frac{\sqrt{M}}{\sqrt{2\log(\log(M))}} (\overline{X}_M - \mathbb{E}X) = -\sigma.$$

It is surprising to note that the factor $\sqrt{2\log(\log(M))}$ for passing from the CLT deviation to the a.s. deviation is approximately equal to 1,96 for $M = 1000$, and only 2,10 for $M = 10000$. The law of iterated logarithm (valid of course for $M \to +\infty$) finally gives the bounds that are close enough to the confidence intervals given by the central limit theorem (also valid as $M \to +\infty$), when M is several thousand.

2.3.3 "Almost sure" central limit theorem

The convergence in distribution given in the central limit theorem 2.2.1 can be interpreted as the following convergence (supposing here that the covariance matrix K is invertible)

$$\mathbb{E}\left(f(K^{-\frac{1}{2}}\sqrt{M}(\overline{X}_M - \mathbb{E}X))\right) = \int_{\mathbb{R}^d} f(x)\frac{1}{(2\pi)^{d/2}}e^{-\frac{|x|^2}{2}}\,\mathrm{d}x,$$

for any continuous bounded function f. We can ask whether the expectation of the left-hand side can be replaced by a mean on the realizations of $\sqrt{M}(\overline{X}_M - \mathbb{E}X) := Y_M$. Indeed, we can see Y as an auto-regressive process with the steps and the noise variables:

$$Y_M = Y_{M-1}\sqrt{1 - \frac{1}{M}} + \frac{X_M - \mathbb{E}X}{\sqrt{M}}$$

whose limit distribution is the Gaussian distribution $\mathcal{N}(0, K)$. In this situation, an appropriate version of the ergodic theorem can be justified and allows us to replace the expectation by a *time-average* on the path of $(Y_M)_{M \geq 1}$. So we obtain a so-called "almost sure" central limit theorem.

Theorem 2.3.4 *Let X be a square integrable random vector. For any continuous bounded function f we have*

$$\lim_{n \to +\infty} \frac{1}{\ln(n)} \sum_{M=1}^{n} \frac{1}{M} f\left(K^{-\frac{1}{2}} \sqrt{M}(\overline{X}_M - \mathbb{E}X)\right)$$

$$= \int_{\mathbb{R}^d} f(x) \frac{1}{(2\pi)^{d/2}} e^{-\frac{|x|^2}{2}} dx$$

with probability 1.

We refer the interested reader to [99].

2.4 NON-ASYMPTOTIC ESTIMATES

So far we have investigated statistical error estimates, which are valid asymptotically as $M \to +\infty$. With this approach, the choice of sufficiently large M remains a delicate problem.

An alternative viewpoint consists of looking for non-asymptotic error bounds, that are valid for each value of M. For a recent reference, see [18]. This is the purpose of this section. We need to reinforce the hypotheses on the variable X. Later in this chapter, all the random variables X are real-valued.

2.4.1 About exponential inequalities

A concentration inequality of exponential type is a bound of the form (for $\varepsilon \geq 0$)

$$\mathbb{P}\left(\frac{1}{M}\sum_{m=1}^{M}(X_m - \mathbb{E}(X_m)) > \varepsilon\right) \leq \exp\left(-\frac{M\varepsilon^2}{c(M,\varepsilon)}\right), \quad (2.4.1)$$

or $\quad \mathbb{P}\left(\left|\frac{1}{M}\sum_{m=1}^{M}(X_m - \mathbb{E}(X_m))\right| > \varepsilon\right) \leq 2\exp\left(-\frac{M\varepsilon^2}{c(M,\varepsilon)}\right), \quad (2.4.2)$

for a constant $c(M, \varepsilon) > 0$, which depends on M, ε and on the distribution of the random variables $(X_m)_m$. In this formulation, the random variables X_m are independent, real-valued, integrable, but they do not necessarily have the same distribution. We refer to (2.4.1)–(2.4.2) as *concentration inequalities* since they describe how the random variables concentrate around their expectations.

Such inequalities can be used in several ways.

1. They allow us to identify a level ε ensuring that the empirical mean $\frac{1}{M}\sum_{m=1}^{M} X_m$ is concentrated around $\frac{1}{M}\sum_{m=1}^{M} \mathbb{E}(X_m)$ with fluctuations smaller than ε, with high probability. This leads to a *non-asymptotic confidence interval*, of a form similar to that of the central limit theorem.

   Proposition 2.4.1 (non-asymptotic confidence interval) *Suppose that $(X_m)_{1 \leq m \leq M}$ verifies the inequality (2.4.2) for all $\varepsilon > 0$, with $c(M, \varepsilon) = c_M$ independent of ε. Then, with probability at least $1 - \lambda$ ($\lambda \in (0, 1)$), we have*

   $$-\sqrt{\frac{c_M \log(2/\lambda)}{M}} \leq \frac{1}{M}\sum_{m=1}^{M}(X_m - \mathbb{E}(X_m)) \leq \sqrt{\frac{c_M \log(2/\lambda)}{M}}.$$

   PROOF:
   The level $\varepsilon = \sqrt{c_M \frac{\log(2/\lambda)}{M}}$ verifies $2\exp(-\frac{M\varepsilon^2}{c_M}) = \lambda$, which implies the result. □

   Note that if c_M is independent of M, we get a confidence interval with the normalization $\sqrt{M}$ similar to the central limit theorem.

2. We can also obtain the bounds in L_p-norm for the absolute fluctuation $\left|\frac{1}{M}\sum_{m=1}^{M}(X_m - \mathbb{E}(X_m))\right|$. For this we use the following inequality, valid for any increasing function $g \in \mathcal{C}^1$ and any positive random variable Z:

   $$\mathbb{E}(g(Z)) = g(0) + \int_0^{+\infty} g'(z)\mathbb{P}(Z \geq z)\mathrm{d}z. \qquad (2.4.3)$$

   Taking $g(z) = z^p$ for $p \geq 1$ and supposing again $c(M, \varepsilon) = c_M$, we get

   $$\mathbb{E}\left|\frac{1}{M}\sum_{m=1}^{M}(X_m - \mathbb{E}(X_m))\right|^p \leq \int_0^{+\infty} 2pz^{p-1}\exp\left(-\frac{Mz^2}{c_M}\right)\mathrm{d}z$$

$$= \left(\frac{c_M}{M}\right)^{p/2} \int_0^{+\infty} 2pu^{p-1} \exp(-u^2) du \quad (2.4.4)$$

proving that the error is of order $c_M/\sqrt{M}$ in L_p.

The existence of exponential inequalities as (2.4.1)–(2.4.2) requires, in general, strong integrability conditions on X_m. Indeed, if $c(M,\varepsilon) = c_M$, using (2.4.3) with $g(z) = \exp(z)$ and $M = 1$, we observe, for example, that X_1 has finite exponential moments and even finite quadratic exponential moments. In what follows, we assume one of the following slightly stronger conditions:

– bounded random variables,

– families of bounded random variables,

– random variables that are sub-linear functions of Gaussian random variables.

2.4.2 Concentration inequalities in the case of bounded random variables

Let us start with a rough bound on the variance (see Proposition A.2.2).

Lemma 2.4.2 *Let X be a bounded real-valued random variable, i.e. $\mathbb{P}(X \in [a,b]) = 1$ for two finite real numbers a and b, then $\mathrm{Var}(X) \leq \frac{(b-a)^2}{4}$.*

Now we move to the exponential inequalities; in the case of bounded random variables, there exist several ones, among which the most famous are due to Bernstein (1946), Bennett (1963), and Hoeffding (1963). Certain inequalities can be nicely expressed using the variances of $(X_m)_m$, in general not known precisely in our applications to Monte-Carlo methods. We will state only the Hoeffding inequality, which depends only on the bound of the random variables (see Theorem A.2.5).

Theorem 2.4.3 (Hoeffding inequality) *Let $(X_1, \ldots, X_M)$ be a sequence of independent real-valued random variables such that $\mathbb{P}(X_m \in [a_m, b_m]) = 1$. Then for any $\varepsilon > 0$ we have*

$$\mathbb{P}\left(\left|\frac{1}{M}\sum_{m=1}^M (X_m - \mathbb{E}(X_m))\right| > \varepsilon\right) \leq 2\exp\left(-\frac{2M\varepsilon^2}{\frac{1}{M}\sum_{m=1}^M |b_m - a_m|^2}\right).$$

Consider the case of random variables with the same distribution ($b_m - a_m = b-a$): then by Proposition 2.4.1 with $c_M = (b-a)^2/2$, we deduce that with probability at least $1 - \lambda$ ($\lambda > 0$),

$$-\frac{(b-a)}{2}\sqrt{\frac{2\log(2/\lambda)}{M}} \leq \overline{X}_M - \mathbb{E}X \leq \frac{(b-a)}{2}\sqrt{\frac{2\log(2/\lambda)}{M}}.$$

Let us compare it with the confidence interval of the central limit theorem at the 95% level (i.e., $\lambda = 0.05$): the standard deviation of X_i being upper-bounded by $\frac{(b-a)}{2}$ (Lemma 2.4.2), we compare the usual factor 1.96 with $\sqrt{2\log(2/\lambda)} = \sqrt{2\log(40)} \approx 2.72$. Thus without surprise, the confidence interval is slightly enlarged, but it has the advantage of being valid for all M.

2.4.3 Uniform concentration inequalities

Now we want to reinforce Theorem 2.4.3 considering uniform estimates on a family of random variables of the form $X_m = \varphi(Y_m)$ with φ taken in a class (or dictionary) $\mathcal{G}$ of bounded functions. These much stronger results will be especially useful in Chapter 8 to analyze the numerical solution to the dynamic programming equations using the Monte-Carlo method (see Section 8.3).

More precisely, the generic case we are interested in is where $X_m = \varphi(Y_m)$ with $\varphi : \mathbb{R}^d \mapsto \mathbb{R}$ bounded and $(Y_m)_{m \geq 1}$ are d-dimensional independent random variables with the same distribution μ, and we want to establish deviation probabilities uniformly on $\mathcal{G}$, namely to quantify the probability of the event

$$\left\{ \exists \varphi \in \mathcal{G} : \left| \frac{1}{M} \sum_{m=1}^{M} \varphi(Y_m) - \int_{\mathbb{R}^d} \varphi(y)\mu(dy) \right| > \varepsilon \right\}$$

$$= \bigcup_{\varphi \in \mathcal{G}} \left\{ \left| \frac{1}{M} \sum_{m=1}^{M} \varphi(Y_m) - \int_{\mathbb{R}^d} \varphi(y)\mu(dy) \right| > \varepsilon \right\}.$$

In the following examples, the set of the functions φ under consideration can be assumed countable, which fixes the question of measurability of the above event. Unlike Theorem 2.4.3, we prefer to write the expectation $\mathbb{E}(X) = \mathbb{E}(\varphi(Y))$ in the form $\int \varphi(y)\mu(dy)$ to avoid any ambiguity because the φ that reaches the fluctuation of size ε is a priori random (it depends on $(Y_m)_m$).

Obviously, if the dictionary of functions $\mathcal{G}$ is too rich, it will be impossible to correctly quantify the deviation probability, but if we restrict, for example, to the case of *finite-dimensional vector spaces* (up to a few adjustments), we will be able to provide relevant control. The consequences of these results are to obtain a law of large numbers, occuring simultaneously and uniformly on the function class $\mathcal{G}$. The pioneers in this subject were Vapnik and Chervonenkis with their fundamental works in the 1970s and 1980s, followed by many others, in a domain that is now called *statistical learning*. For references, see [69], [73].

▷ **The first heuristic.** We start with some simple arguments, not optimal, that allow us to better understand how to tackle this problem. The purpose is primarily pedagogical.
— If $\mathcal{G}$ were finite cardinal (denoted by $|\mathcal{G}|$) and its elements were bounded functions with values in $[a, b]$, we would write

$$\mathbb{P}\left(\exists \varphi \in \mathcal{G} : \left|\frac{1}{M}\sum_{m=1}^{M} \varphi(Y_m) - \int_{\mathbb{R}^d} \varphi(y)\mu(dy)\right| > \varepsilon\right)$$

$$\leq \sum_{\varphi \in \mathcal{G}} \mathbb{P}\left(\left|\frac{1}{M}\sum_{m=1}^{M} \varphi(Y_m) - \int_{\mathbb{R}^d} \varphi(y)\mu(dy)\right| > \varepsilon\right)$$

$$\leq 2|\mathcal{G}| \exp\left(-\frac{2M\varepsilon^2}{(b-a)^2}\right), \qquad (2.4.5)$$

applying Theorem 2.4.3. Then, it would not be very difficult to adapt the arguments of Section 2.4.1 to obtain confidence intervals on $\frac{1}{M}\sum_{m=1}^{M} \varphi(Y_m) - \int \varphi(y)\mu(dy)$ valid simultaneously for all $\varphi \in \mathcal{G}$.
— The finiteness condition for $\mathcal{G}$ is too restrictive in practice. A possible improvement for the case $|\mathcal{G}| = +\infty$ is to suppose that for all $\varepsilon > 0$, we can cover $\mathcal{G}$ in uniform norm at the level ε using $n := \mathcal{N}_\infty(\varepsilon, \mathcal{G}) \geq 1$ functions $\{\varphi_1, \ldots, \varphi_n\}$, in the sense where for any $\varphi \in \mathcal{G}$, we can find a function φ_j such that $|\varphi - \varphi_j|_\infty \leq \varepsilon$. The set $\{\varphi_1, \ldots, \varphi_n\}$ is called ε-cover (Figure 2.3). Then a simple triangular inequality using a $\varepsilon/3$-cover gives

$$\left\{\exists \varphi \in \mathcal{G} : \left|\frac{1}{M}\sum_{m=1}^{M} \varphi(Y_m) - \int_{\mathbb{R}^d} \varphi(y)\mu(dy)\right| > \varepsilon\right\}$$

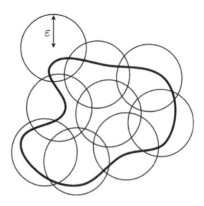

Figure 2.3 An example of ε-cover.

$$\subset \bigcup_{j=1}^{\mathcal{N}_\infty(\varepsilon/3,\mathcal{G})} \left\{ \left| \frac{1}{M} \sum_{m=1}^{M} \varphi_j(Y_m) - \int_{\mathbb{R}^d} \varphi_j(y)\mu(\mathrm{d}y) \right| > \varepsilon/3 \right\}. \quad (2.4.6)$$

We can take φ_j also with values in $[a,b]$ and conclude as for (2.4.5):

$$\mathbb{P}\left(\exists \varphi \in \mathcal{G} : \left| \frac{1}{M} \sum_{m=1}^{M} \varphi(Y_m) - \int_{\mathbb{R}^d} \varphi(y)\mu(\mathrm{d}y) \right| > \varepsilon \right)$$
$$\leq 2\mathcal{N}_\infty(\varepsilon/3,\mathcal{G}) \exp\left(-\frac{2M\varepsilon^2}{9(b-a)^2} \right). \quad (2.4.7)$$

For example, the dictionary $\mathcal{G}$ consisting of the functions with values in $[a,b]$ and piecewise constant on K disjoint subsets $(\mathcal{C}_k)_{1\leq k\leq K}$ of $\mathbb{R}^d$ (i.e. $\varphi \in \mathcal{G} \Rightarrow \varphi(.) = \sum_{k=1}^{K} \alpha_k \mathbf{1}_{\mathcal{C}_k}(.)$ with $\alpha_k \in [a,b]$) fulfills

$$\mathcal{N}_\infty(\varepsilon,\mathcal{G}) = \left(\frac{b-a}{2\varepsilon} \right)^K \vee 1 \quad (2.4.8)$$

(if $\varepsilon \leq (b-a)/2$, take regularly spaced α_k at distance 2ε).

We will not go further in this direction, because in general, the evaluation of the ε-*covering number* $\mathcal{N}_\infty(\varepsilon,\mathcal{G})$ is very delicate or even impossible, and this limits the application of (2.4.7). However, these first computations should appeal to the reader's curiosity, showing that computing the deviation probability uniformly on $\mathcal{G}$ is not absolutely impossible, and may be done under the form of a generalized Hoeffding inequality.

▷ **Covering number L_1.** In the previous analysis leading to (2.4.7), we easily guess that using a covering of $\mathcal{G}$ in uniform norm to quantify the fluctuations of the *average/probability* is likely suboptimal. This leads us to measure the richness (or complexity) of $\mathcal{G}$ evaluating how to cover it by balls of size ε in the sense of L_1-norm (with respect to a certain measure ν on $\mathbb{R}^d$) instead of the L_∞-norm sense. The minimal number of balls for this is called the L_1-*covering number*. The measure ν can also be the empirical measure associated with M points in $\mathbb{R}^d$; this last case is the most commonly used in the following.

Definition 2.4.4 (covering number) *Let $\mathcal{G}$ be a class of functions $\varphi : \mathbb{R}^d \mapsto \mathbb{R}$.*

Consider M points $(y_m)_{1 \leq m \leq M}$ in $\mathbb{R}^d$, and denote by ν^M the associated empirical measure and $|\cdot|_{L_1(\nu^M)}$ the associated empirical L_1 norm: for any function φ,

$$|\varphi|_{L_1(\nu^M)} := \frac{1}{M} \sum_{m=1}^{M} |\varphi(y_m)|.$$

An ε-covering ($\varepsilon > 0$) of $\mathcal{G}$ with respect to the norm $|\cdot|_{L_1(\nu^M)}$ is a finite set of functions $\varphi_1, \ldots, \varphi_n : \mathbb{R}^d \mapsto \mathbb{R}$ such that for any $\varphi \in \mathcal{G}$, we can find an index $j \in \{1, \cdots, n\}$ such that $|\varphi - \varphi_j|_{L_1(\nu^M)} \leq \varepsilon$.

The minimal integer n for which an ε-covering exists is called the ε-covering number and it is denoted by $\mathcal{N}_1(\varepsilon, \mathcal{G}, \nu^M)$.

More generally, we can replace the empirical measure ν^M by a probability measure ν on $\mathbb{R}^d$ and similarly define $\mathcal{N}_1(\varepsilon, \mathcal{G}, \nu)$.

Note that in this definition, the points $(y_m)_{1 \leq m \leq M}$ can be random and de facto, and we will take them equal to a Monte-Carlo sample.

In what follows, we will say that $\mathcal{G}$ can be *covered* for ν if $\mathcal{N}_1(\varepsilon, \mathcal{G}, \nu) < +\infty$ for any $\varepsilon > 0$. We always consider the ε-covers of minimal size.

Here the goal is not to be exhaustive[7] on $\mathcal{G}$, for which we have explicit control of $\mathcal{N}_1(\varepsilon, \mathcal{G}, \nu)$. We rather focus our attention on $\mathcal{G}$ having the form of a properly truncated finite vector space. To proceed to the truncation, we define the *clipping* of a function (see Figure 2.4). This is a Lipschitz transformation, which allows us to reduce the situation to the case of bounded functions.

---

[7] For more details, see [69].

74 ■ Convergences and error estimates

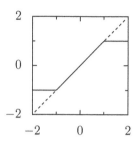

Figure 2.4 Clipping $\mathcal{C}_L$ for $L = 1$.

Definition 2.4.5 (clipped function) *For a real-valued function φ and a level $L \geq 0$, the clipping of φ at the level L is defined by*

$$\mathcal{C}_L \varphi := -L \vee \varphi \wedge L.$$

By extension, the clipping of a class of functions $\mathcal{G}$ is denoted by $\mathcal{C}_L \mathcal{G}$ and is given by

$$\mathcal{C}_L \mathcal{G} = \{\mathcal{C}_L \varphi : \varphi \in \mathcal{G}\}.$$

Quite surprisingly, the clipping of a vector space generated by the functions $(\phi_1, \ldots, \phi_K)$ admits an upper bound on the numbers of ε-covers, independent of the measure ν, of the dimension of the space d and of the form[8] of the $(\phi_j)_j$.

Theorem 2.4.6 (robust upper bound for the L_1 covering number) *Consider*

- $\Phi = \text{Span}.(\phi_1, \ldots, \phi_K) = \{\sum_{k=1}^K \alpha_k \phi_k, \alpha_k \in \mathbb{R}\}$ *with* $\phi_k : \mathbb{R}^d \mapsto \mathbb{R}$;

- $\mathcal{C}_L \Phi$ *the clipping of Φ (for $L > 0$);*

- μ *a probability measure on $\mathbb{R}^d$.*

Then $\mathcal{C}_L \Phi$ can be covered for μ and for all $0 < \varepsilon \leq L/2$, we have

$$\mathcal{N}_1(\varepsilon, \mathcal{C}_L \Phi, \mu) \leq \left(\frac{6L}{\varepsilon}\right)^{2(K+1)}.$$

We recognize that this is a difficult result and we refer the reader to [69,

---

[8]So this is valid for global or local polynomials, Fourier basis or wavelets, etc.

Lemma 9.2, Theorems 9.4 and 9.5].[9] We note that the upper bound for the covering number obviously increases with the dimension of Φ and with the inverse of the covering balls radius. This is similar enough to the estimate (2.4.8) of $\mathcal{N}_\infty(\varepsilon, \mathcal{G})$ when $\mathcal{G}$ consists of piecewise constant and bounded functions.

▷ **Uniform deviations.** With the ε-covers in L_1, we can proceed as for (2.4.6) replacing the union over $\varphi \in \mathcal{G}$ by the finite union on the cover. This is a little more delicate than for L_∞ covers. We will not get into details of the technical computations underlying these ideas, we only give the key asymptotic upper bound (see [69, Theorem 9.1]), similar to the inequality (2.4.7).

Theorem 2.4.7 (uniform deviation probability) *Let $\mathcal{G}$ be a countable set[10] of the functions $\varphi : \mathbb{R}^d \mapsto [0, B]$ with $B > 0$. Let $(Y_m)_{1 \leq m \leq M}$ be a sample of independent random variables with the distribution μ and denote μ^M as the associated empirical measure. Suppose that $\mathcal{G}$ can be covered for μ^M. Then for all $\varepsilon > 0$ we have*

$$\mathbb{P}\Big(\sup_{\varphi \in \mathcal{G}} \Big| \frac{1}{M} \sum_{m=1}^M \varphi(Y_m) - \int_{\mathbb{R}^d} \varphi(y) \mu(dy) \Big| > \varepsilon \Big)$$

$$\leq 8 \mathbb{E}\big(\mathcal{N}_1(\tfrac{\varepsilon}{8}, \mathcal{G}, \mu^M)\big) \exp\big(-\frac{\varepsilon^2 M}{128 B^2}\big).$$

▷ **Applications of uniform control.**
∗ **Exponential estimates of uniform deviations.** Combining the estimates on the covering number and the uniform deviation probability, we get a non-asymptotic estimate of the deviation, uniformly on the clipped Φ.

Theorem 2.4.8 (uniform deviation probability on a clipped vector space) *Assume the hypothesis and notation of Theorem 2.4.6 for Φ and $\mathcal{C}_L\Phi$, with $L > 0$. Let $(Y_m)_{1 \leq m \leq M}$ be a sample of independent random variables with the distribution μ and denote μ^M as the*

---

[9]For the reader who reads carefully the indicated references, we mention that we use an extra upper bound $2e\, x \log(3e\, x) \leq \big(\frac{11x}{5}\big)^2$ for $x \geq 4$, which allows us to simplify the final upper bound without sacrificing accuracy.

[10]Restricting to a countable set ensures that the studied supremum still defines a random variable.

associated empirical measure. Then, for all $\varepsilon > 0$, we have

$$\mathbb{P}\left(\sup_{\varphi \in \mathcal{C}_L \Phi} \left| \frac{1}{M} \sum_{m=1}^M \varphi(Y_m) - \int_{\mathbb{R}^d} \varphi(y)\mu(\mathrm{d}y) \right| > \varepsilon \right)$$

$$\leq 8 \left(\frac{48L}{\varepsilon} \right)^{2(K+1)} \exp\left(-\frac{\varepsilon^2 M}{512 L^2} \right).$$

PROOF:
First, we can take the sup over a countable set by restricting the basis coefficients to be rational, without affecting the definition of the event. Then, we can apply Theorem 2.4.7 taking the class of functions $\mathcal{G} := \mathcal{C}_L \Phi + L = \{\mathcal{C}_L \varphi + L : \varphi \in \Phi\}$, whose elements are non-negative and bounded by $B = 2L$. Of course, we have $\mathcal{N}_1(\varepsilon/8, \mathcal{G}, \mu^M) = \mathcal{N}_1(\varepsilon/8, \mathcal{C}_L \Phi, \mu^M)$ and we conclude by applying Theorem 2.4.6 (we verify that the restriction $\varepsilon/8 \leq L/2$ in this theorem can be relaxed because in the opposite case, the announced upper bound is obvious as the functions φ are bounded by L). □

∗ **Uniform law of large numbers.** The previous upper bounds allow us to show that $\{\mathbb{P}\big(\sup_{\varphi \in \mathcal{C}_L \Phi} \big| \frac{1}{M} \sum_{m=1}^M \varphi(Y_m) - \int_{\mathbb{R}^d} \varphi(y)\mu(\mathrm{d}y)\big| > \varepsilon\big) : M \geq 1\}$ defines a convergent series, for any $\varepsilon > 0$: so by the Borel-Cantelli lemma, with probability 1, for M large enough we have $\sup_{\varphi \in \mathcal{C}_L \Phi} \big| \frac{1}{M} \sum_{m=1}^M \varphi(Y_m) - \int_{\mathbb{R}^d} \varphi(y)\mu(\mathrm{d}y) \big| \leq \varepsilon$. We have proved the law of large numbers, uniform on the class of functions $\mathcal{C}_L \Phi$.

Corollary 2.4.9 (uniform law of large numbers) *With the hypotheses and the notation of Theorem 2.4.8, we have*

$$\sup_{\varphi \in \mathcal{C}_L \Phi} \left| \frac{1}{M} \sum_{m=1}^M \varphi(Y_m) - \int_{\mathbb{R}^d} \varphi(y)\mu(\mathrm{d}y) \right| \xrightarrow[M \to +\infty]{\text{a.s.}} 0.$$

∗ **Uniform L_1 distance between the empirical and exact means.** The following result is very useful for replacing an empirical mean by an exact mean, and vice versa, for measuring the induced error in L_1. The factor $\sqrt{(K+1) \log(4M)}$ is the price that is paid to ensure the uniformity on the function class.

Theorem 2.4.10 (L_1 uniform distance between the empirical and the exact mean) *With the hypotheses and notation of Theorem 2.4.8, we have*

$$\mathbb{E}\left(\sup_{\varphi \in \mathcal{C}_L \Phi} \left| \frac{1}{M} \sum_{m=1}^M \varphi(Y_m) - \int_{\mathbb{R}^d} \varphi(y)\mu(\mathrm{d}y) \right| \right)$$

$$\leq 75 \frac{L}{\sqrt{M}}\sqrt{(K+1)\log(4M)}.$$

PROOF:
Let Z be the random variable for which we compute the expectation. Then, by Theorem 2.4.8 and for $\varepsilon_0 \geq 24\frac{L}{\sqrt{M}}$, we deduce that

$$\mathbb{E}(Z) = \int_0^{+\infty} \mathbb{P}(Z > \varepsilon)d\varepsilon$$

$$\leq \varepsilon_0 + 8\int_{\varepsilon_0}^{+\infty} \left(\frac{48L}{\varepsilon}\right)^{2(K+1)} \exp\left(-\frac{\varepsilon^2 M}{512L^2}\right)d\varepsilon$$

$$\leq \varepsilon_0 + 8(4M)^{(K+1)}\int_{\varepsilon_0}^{+\infty} \frac{\sqrt{M}\varepsilon}{24L} \exp\left(-\frac{\varepsilon^2 M}{512L^2}\right)d\varepsilon$$

$$= \varepsilon_0 + (4M)^{(K+1)}\frac{256L}{3\sqrt{M}} \exp\left(-\frac{\varepsilon_0^2 M}{512L^2}\right).$$

The choice $\varepsilon_0 = \frac{L}{\sqrt{M}}\sqrt{512(K+1)\log(4M)}$ verifies $(4M)^{(K+1)}\exp\left(-\frac{\varepsilon_0^2 M}{512L^2}\right) = 1$ and $\varepsilon_0 \geq 24\frac{L}{\sqrt{M}}$. The estimate

$$\mathbb{E}(Z) \leq \frac{L}{\sqrt{M}}\sqrt{512(K+1)\log(4M)} + \frac{256L}{3\sqrt{M}}$$

follows, which after some elementary computations yields the result. □

2.4.4 Concentration inequalities in the case of Gaussian noise

A powerful approach for getting a concentration inequality of exponential type for very general probability distributions is based on the logarithmic Sobolev inequalities. These tools have experienced some important developments recently; see [103] and [4]. The case of the Gaussian measure goes back to the seminal work of Gross [66].

The application to the exponential concentration inequalities is made for random variables X_m of the form

$$X = f(Y)$$

for a Lipschitz function $f: \mathbb{R}^d \mapsto \mathbb{R}$ and a random vector Y in $\mathbb{R}^d$ (taking a.s. finite values). The induced probability measure is denoted by μ. In this section, it is more convenient to use the notation

$$\mathbb{E}_\mu(f) = \mathbb{E}(f(Y)) = \int_{\mathbb{R}^d} f(y)\mu(dy),$$

which is valid as soon as $f(Y)$ is integrable or f is positive.

For the definition of the logarithmic Sobolev inequality, we must introduce entropy.

Definition 2.4.11 (entropy) *Let μ be a probability measure on $\mathbb{R}^d$. For any measurable function $f : \mathbb{R}^d \mapsto \mathbb{R}^+$, the entropy of f for μ is defined by*

$$\mathrm{Ent}_\mu(f) = \mathbb{E}_\mu(f\log(f)) - \mathbb{E}_\mu(f)\log(\mathbb{E}_\mu(f)). \quad (2.4.9)$$

The measure μ satisfies the logarithmic Sobolev inequality with the constant $C_\mu \in (0, +\infty)$ for the class of functions $\mathcal{A}$, if

$$\mathrm{Ent}_\mu(f^2) \leq C_\mu \mathbb{E}_\mu(|\nabla f|^2), \qquad \forall f \in \mathcal{A}, \quad (2.4.10)$$

where $\nabla f = (\partial_{x_1} f, \ldots, \partial_{x_d} f)$.

In the following, we take for $\mathcal{A}$ the set $\mathcal{C}_b^2(\mathbb{R}^d, \mathbb{R})$ of bounded functions with bounded derivatives. Note the following:

1. $\mathrm{Ent}_\mu(f)$ is finite if and only if $f[\log(f)]_+$ is μ-integrable.

2. $\mathrm{Ent}_\mu(f) \geq 0$: this easily follows from the Jensen inequality applied to the convex function $x \mapsto x\log(x)$. The entropy is zero if f is μ-a.e. constant.

3. $\mathrm{Ent}_\mu(\lambda f) = \lambda \mathrm{Ent}_\mu(f)$ for any $\lambda > 0$.

If μ satisfies a logarithmic Sobolev inequality, then we can deduce[11] an interesting exponential inequality.

Theorem 2.4.12 *Suppose that μ satisfies a logarithmic Sobolev inequality with a constant C_μ. Then for any function f in $\mathcal{A}$ with compact support and such that $|\nabla f| \leq 1$, we have*

$$\mathbb{E}_\mu(e^{\lambda f}) \leq e^{\lambda \mathbb{E}_\mu(f) + \frac{C_\mu \lambda^2}{4}}, \qquad \forall \lambda \in \mathbb{R}.$$

PROOF:
Set $H(\lambda) = \mathbb{E}_\mu(e^{\lambda f})$, which is a finite non-zero quantity, because f has

---

[11] This trick is known as the Herbst argument.

compact support. Let $F = e^{\frac{1}{2}\lambda f}$: as $f \in \mathcal{A}$ and with compact support, F^2 is in $\mathcal{A}$. Applying the logarithmic Sobolev inequality, we get

$$\begin{aligned}\text{Ent}_\mu(F^2) &= \mathbb{E}_\mu(\lambda f e^{\lambda f}) - \mathbb{E}_\mu(e^{\lambda f})\log(\mathbb{E}_\mu(e^{\lambda f})) \\ &= \lambda H'(\lambda) - H(\lambda)\log(H(\lambda)) \\ &\leq C_\mu \mathbb{E}_\mu |\nabla F|^2 = C_\mu \mathbb{E}_\mu |\frac{\lambda}{2}\nabla f e^{\frac{1}{2}\lambda f}|^2 \\ &\leq C_\mu \frac{\lambda^2}{4} H(\lambda).\end{aligned}$$

In other words, the derivative of $K(\lambda) = \frac{1}{\lambda}\log(H(\lambda))$ verifies

$$K'(\lambda) = \frac{1}{\lambda^2}\left(\lambda \frac{H'(\lambda)}{H(\lambda)} - \log(H(\lambda))\right) \leq \frac{C_\mu}{4}.$$

We check easily that $H(0) = 1$, $H'(0) = \mathbb{E}_\mu(f)$ and $K(0) = \mathbb{E}_\mu(f)$; consequently,

$$K(\lambda) = K(0) + \int_0^\lambda K'(u)du \leq \mathbb{E}_\mu(f) + \frac{C_\mu}{4}\lambda,$$

which finishes the proof. $\square$

Before giving a corollary of the previous theorem, we recall that f is Lipschitz if there exists a constant $C \in \mathbb{R}^+$ such that

$$|f(x) - f(y)| \leq C|x-y|, \quad \forall x \in \mathbb{R}^d, y \in \mathbb{R}^d.$$

The Lipschitz constant of f is the minimal C verifying this bound: we denote it by $|f|_{\text{Lip.}}$.

Corollary 2.4.13 (concentration inequality) *Let Y be a random vector in $\mathbb{R}^d$ with the distribution μ. Suppose that μ satisfies a logarithmic Sobolev inequality with constant $C_\mu > 0$. Then, for any Lipschitz function $f: \mathbb{R}^d \mapsto \mathbb{R}$, we have*

$$\mathbb{P}(|f(Y) - \mathbb{E}(f(Y))| > \varepsilon) \leq 2\exp\left(-\frac{\varepsilon^2}{C_\mu |f|^2_{\text{Lip.}}}\right), \quad \forall \varepsilon \geq 0. \quad (2.4.11)$$

PROOF:
▷ Suppose first that f is in $\mathcal{A}$, with compact support and $|\nabla f| \leq 1$. Then the exponential Chebyshev inequality gives (for any $\lambda \geq 0$)

$$\mathbb{P}(f(Y) - \mathbb{E}(f(Y)) > \varepsilon) \leq \mathbb{E}_\mu(e^{\lambda(f - \mathbb{E}_\mu f - \varepsilon)}) \leq e^{\frac{C_\mu \lambda^2}{4} - \lambda\varepsilon}.$$

The upper bound attains its minimum at $\lambda = \frac{2\varepsilon}{C_\mu}$, for which it is equal to $e^{-\frac{\varepsilon^2}{C_\mu}}$. Replacing f by $-f$, we show the same bound for $\mathbb{P}(f(Y) - \mathbb{E}(f(Y)) < -\varepsilon)$. Combining these two bounds, we obtain

$$\mathbb{P}(|f(Y) - \mathbb{E}(f(Y))| > \varepsilon)$$
$$\leq \mathbb{P}(f(Y) - \mathbb{E}(f(Y)) > \varepsilon) + \mathbb{P}(f(Y) - \mathbb{E}(f(Y)) < -\varepsilon)$$
$$\leq 2e^{-\frac{\varepsilon^2}{C_\mu}}. \quad (2.4.12)$$

For f with $|\nabla f|_\infty \neq 0$, replacing f by $f/|\nabla f|_\infty$, we obtain the inequality

$$\mathbb{P}(|f(Y) - \mathbb{E}(f(Y))| > \varepsilon) \leq 2e^{-\frac{\varepsilon^2}{C_\mu |\nabla f|_\infty^2}}, \quad \forall \varepsilon \geq 0. \quad (2.4.13)$$

▷ The rest of the proof is technical and can be skipped at the first reading. Let us extend this inequality to the case where f is no longer $\mathcal{C}^2$, but only Lipschitz, still with compact support (thus bounded). We can suppose that $|f|_{\text{Lip.}} \neq 0$, otherwise the inequality is obvious as f is zero. Let V be a random variable bounded by 1 and with $\mathcal{C}^\infty$ density (supported in $B(0,1)$), set $f_n = f * \xi_n$, the convolution product of f with the density of V/n:

$$f_n(x) = \int_{\mathbb{R}^d} f(u) n^d p(n(x-u)) du$$
$$= \int_{\mathbb{R}^d} f(x-v) n^d p(nv) dv = \mathbb{E}(f(x - V/n)).$$

From these equalities, we easily deduce that f_n is bounded by $\|f\|_\infty$, belongs to $\mathcal{C}^\infty$ and has bounded derivatives (with a bound that can depend on n), thus it is in $\mathcal{A}$. Moreover, $|f_n(x) - f_n(y)| \leq |f|_{\text{Lip.}} |x-y|$ implying that $|\nabla f_n| \leq |f|_{\text{Lip.}}$. Last, since V is bounded, f_n is also bounded with compact support. So, f_n satisfies (2.4.13):

$$\mathbb{P}(|f_n(Y) - \mathbb{E}(f_n(Y))| > \varepsilon) \leq 2e^{-\frac{\varepsilon^2}{C_\mu |f|_{\text{Lip.}}^2}}, \quad \forall \varepsilon \geq 0. \quad (2.4.14)$$

It remains to pass to the limit on the left-hand side. We note that $|f_n(x) - f(x)| \leq |f|_{\text{Lip.}} \frac{1}{n}$. From this we deduce that $f_n(Y) - \mathbb{E}(f_n(Y))$ converges uniformly (hence a.s.) to $f(Y) - \mathbb{E}(f(Y))$ as $n \to +\infty$, in particular, $|f_n(Y) - \mathbb{E}(f_n(Y))|$ converges in distribution to $L := |f(Y) - \mathbb{E}(f(Y))|$, i.e., $\lim_{n \to +\infty} \mathbb{P}(|f_n(Y) - \mathbb{E}(f_n(Y))| > \varepsilon) = \mathbb{P}(|f(Y) - \mathbb{E}(f(Y))| > \varepsilon)$ possibly except for $\varepsilon \in \mathcal{D}$, the set of the discontinuity points for the c.d.f. of L. Summarizing,

$$\mathbb{P}(|f(Y) - \mathbb{E}(f(Y))| > \varepsilon) \leq 2e^{-\frac{\varepsilon^2}{C_\mu |f|_{\text{Lip.}}^2}}, \quad \forall \varepsilon \notin \mathcal{D}. \quad (2.4.15)$$

But $\mathcal{D}$ is at most countable and both above sides are continuous from the right w.r.t. ε: if $\varepsilon \in \mathcal{D}$, it is enough then to take a sequence $(\varepsilon_k)_{k\geq 1}$ outside $\mathcal{D}$ that decreases to ε to get the result (2.4.15) at that $\varepsilon \in \mathcal{D}$.

▷ Let us proceed now to the extension to bounded Lipschitz f, but not necessarily with compact support. Actually, we can exhibit a sequence $(f_n)_{n\geq 1}$ of functions with compact support, bounded uniformly in n, Lipschitz with the same constant as $|f|_{\text{Lip.}}$, and converging pointwise to f. Then the convergence and boundedness ensure that $f_n(Y) - \mathbb{E}(f_n(Y))$ converges a.s. to $f(Y) - \mathbb{E}(f(Y))$, which finally leads to (2.4.11) by arguing as before.

▷ The extension to unbounded Lipschitz f is similar. Set $f_n = -n \vee f \wedge n$ whose Lipschitz constant is bounded by that of f, thus f_n satisfies the inequality (2.4.11) with $|f_n|_{\text{Lip.}} \leq |f|_{\text{Lip.}}$. Following the previous arguments, it is enough to show that $f_n(Y) - \mathbb{E}(f_n(Y))$ converges a.s. to $f(Y) - \mathbb{E}(f(Y))$: obviously, $f_n(Y) \xrightarrow[n\to+\infty]{\text{a.s.}} f(Y)$. Let us show that the convergence also holds in L_1, by showing the uniform integrability of $(f_n(Y))_n$ (see Theorem A.1.2 in the appendix). It is enough to prove that

$$\sup_{n\geq 1} \mathbb{E}(f_n^2(Y)) < +\infty. \tag{2.4.16}$$

As in the proof of (2.4.4) with $p = 2$ and $M = 1$, we deduce from (2.4.11) for the current f_n that

$$\sup_{n\geq 1} \mathbb{E}(f_n(Y) - \mathbb{E}(f_n(Y)))^2 = \sup_{n\geq 1} \left[\mathbb{E}(f_n^2(Y)) - (\mathbb{E}(f_n(Y)))^2\right] < +\infty. \tag{2.4.17}$$

Let us show now that $\sup_{n\geq 1} |\mathbb{E}(f_n(Y))| < +\infty$. Combining the inequality (2.4.11) for f_n with Proposition 2.4.1 (with $M = 1$, $\lambda = \frac{1}{4}$, $c(M,\varepsilon) = C_\mu |f|^2_{\text{Lip.}}$), we can assert that with probability greater than $\frac{3}{4}$, we have

$$|f_n(Y) - \mathbb{E}(f_n(Y))| \leq \sqrt{C_\mu |f|^2_{\text{Lip.}} \log(8)}.$$

For K large enough, $|f_n(Y)| \leq |f(Y)| \leq K$ with probability greater than $\frac{3}{4}$. Then with probability greater than $\frac{1}{2}$,

$$|\mathbb{E}(f_n(Y))| \leq |f_n(Y) - \mathbb{E}(f_n(Y))| + |f_n(Y)| \leq K + \sqrt{C_\mu |f|^2_{\text{Lip.}} \log(8)}.$$

As both sides of the above inequality are deterministic, this gives an upper bound on $|\mathbb{E}(f_n(Y))|$ that is always true (and not with probability $\frac{1}{2}$) and uniform in n, as stated. Combining with (2.4.17), we conclude the proof of (2.4.16). □

An important example of a measure μ satisfying the logarithmic Sobolev inequality is the Gaussian measure $\gamma_d(dx) :=$

$\frac{1}{(2\pi)^{d/2}} \exp(-\frac{1}{2}|x|^2)\mathrm{d}x$. For other measures (Bernoulli, Boltzmann), see [103].

Theorem 2.4.14 (Gaussian distribution in dimension 1) *The Gaussian measure γ_1 in dimension 1 satisfies the logarithmic Sobolev inequality with the constant $C_{\gamma_1} = 2$.*

PROOF:
The tools that we are using now anticipate those presented in Chapter 4, and the proof below can be skipped at the first reading.

If f is zero, the logarithmic Sobolev inequality is obviously true. Now take f^2 in $\mathcal{A}$ and non-zero. We employ a proof based on the stochastic calculus exploiting that W_1 has the distribution γ_1.
▷ To start, suppose that $\mathbb{E}(f^2(W_1)) = 1$. Set $u(t, x) = \mathbb{E}(f^2(W_1)|W_t = x)$: u is positive for all $t < 1$. The Ito formula gives

$$M_t = u(t, W_t) = 1 + \int_0^t u'_x(s, W_s)\mathrm{d}W_s = 1 + \int_0^t 2\mathbb{E}(ff'(W_1)|W_s)\mathrm{d}W_s.$$

Note that $\mathrm{Ent}_{\gamma_1}(f^2) = \mathbb{E}(M_1 \log(M_1))$. Then

$$M_t \log(M_t) = \int_0^t 2(\log(M_s) + 1)\mathbb{E}(ff'(W_1)|W_s)\mathrm{d}W_s$$
$$+ \frac{1}{2}\int_0^t \frac{[2\mathbb{E}(ff'(W_1)|W_s)]^2}{\mathbb{E}(f^2(W_1)|W_s)}\mathrm{d}s.$$

By carefully localizing the processes and passing to the expectation in the previous decomposition, we get

$$\mathrm{Ent}_{\gamma_1}(f^2) \leq 2\int_0^1 \mathbb{E}\Big[\frac{(\mathbb{E}(ff'(W_1)|W_s))^2}{\mathbb{E}(f^2(W_1)|W_s)}\Big]\mathrm{d}s.$$
$$\leq 2\int_0^1 \mathbb{E}\Big[\frac{\mathbb{E}(f^2(W_1)|W_s)\mathbb{E}((f')^2(W_1)|W_s)}{\mathbb{E}(f^2(W_1)|W_s)}\Big]\mathrm{d}s \quad \text{(Cauchy-Schwarz)}$$
$$= 2\mathbb{E}((f')^2(W_1)) = \mathbb{E}_{\gamma_1}(|f'|^2).$$

▷ If $\mathbb{E}(f^2(W_1)) \neq 0$, then taking $g = f/\sqrt{\mathbb{E}(f^2(W_1))}$ brings us back to the previous case: indeed, because $\mathbb{E}(g^2(W_1)) = 1$ and $\mathrm{Ent}_{\gamma_1}(f^2) = \mathbb{E}(f^2(W_1))\mathrm{Ent}_{\gamma_1}(g^2) \leq \mathbb{E}(f^2(W_1))C_{\gamma_1}\mathbb{E}(|\nabla g|^2(W_1)) = C_{\gamma_1}\mathbb{E}(|\nabla f|^2(W_1))$. □

The strength of a logarithmic Sobolev inequality lies in its natural capacity to pass without difficulty to higher dimensions, using tensorization.

Lemma 2.4.15 (tensorization) *Let $n \geq 2$. If each μ_i (for $1 \leq i \leq n$) satisfies the logarithmic Sobolev inequality with the constant C_{μ_i}, then the product measure $\mu^{\otimes n}$ — associated with the distribution of $(Y_1, \ldots, Y_n)$, i.e. a vector with independent components with i-th component having the distribution μ_i, — satisfies again the logarithmic Sobolev inequality with constant $C_{\mu^{\otimes n}} = \max_{1 \leq i \leq n} C_{\mu_i}$.*

PROOF:
  Beforehand, let us present an equivalent characterization of the entropy,
$$\text{Ent}_\mu(f) = \sup\{\mathbb{E}_\mu(fg) : \mathbb{E}_\mu(e^g) = 1\}. \tag{2.4.18}$$
This equality is clear if $\mathbb{E}_\mu(f) = 0$ ($f = 0$ μ-a.e.).

In the case $\mathbb{E}_\mu(f) > 0$, as both sides of (2.4.18) are linear in f, it is enough to prove the equality when $\mathbb{E}_\mu(f) = 1$, i.e. the case for which $\text{Ent}_\mu(f) = \mathbb{E}_\mu(f \log(f))$. Let $f \geq 0$, verifying $\mathbb{E}_\mu(f) = 1$.

- On the one hand, the function $g = \log(f)\mathbf{1}_{f>0} - c\mathbf{1}_{f=0}$ verifies $\mathbb{E}_\mu(e^g) = 1$ for a certain explicit c. For this choice, we have $f \log(f) = fg$, which shows that the left-hand side in (2.4.18) is less than the right-hand side.
- On the other hand, let us start from the inequality[12] $uv \leq u\log(u) - u + e^v$ for $u \geq 0$ and $v \in \mathbb{R}$; it implies $\mathbb{E}_\mu(fg) \leq \mathbb{E}_\mu(f\log(f)) - \mathbb{E}_\mu(f) + \mathbb{E}_\mu(e^g) = \text{Ent}_\mu(f)$ for any g such that $\mathbb{E}_\mu(e^g) = 1$. This finishes the proof of (2.4.18).

Now take g such that $\mathbb{E}_{\mu^{\otimes n}}(e^g) = 1$, and write
$$g = \sum_{i=1}^n g_i := g - \log\left(\int e^g d\mu_1(x_1)\right) + \sum_{i=2}^n \log \frac{\int e^g d\mu_1(x_1) \ldots d\mu_{i-1}(x_{i-1})}{\int e^g d\mu_1(x_1) \ldots d\mu_i(x_i)}.$$

Each g_i, defined as above, verifies $\int e^{g_i} d\mu_i(x_i) = 1$. Denoting by ∇_i the gradient with respect to the i-th variable, we deduce

$$\mathbb{E}_{\mu^{\otimes n}}(f^2 g) = \sum_{i=1}^n \mathbb{E}_{\mu^{\otimes n}}(f^2 g_i) = \sum_{i=1}^n \mathbb{E}_{\mu^{\otimes n}}\left(\mathbb{E}_{\mu_i}(f^2 g_i)\right)$$

$$\leq \sum_{i=1}^n \mathbb{E}_{\mu^{\otimes n}}\left(\text{Ent}_{\mu_i}(f^2)\right) \leq \sum_{i=1}^n C_{\mu_i} \mathbb{E}_{\mu^{\otimes n}}\left(\mathbb{E}_{\mu_i}(|\nabla_i f|^2)\right)$$

$$\leq \max_{1 \leq i \leq n} C_{\mu_i} \sum_{i=1}^n \mathbb{E}_{\mu^{\otimes n}}(|\nabla_i f|^2) = \max_{1 \leq i \leq n} C_{\mu_i} \mathbb{E}_{\mu^{\otimes n}}(|\nabla f|^2).$$

In view of (2.4.18), the result is proved. □

---

[12] This can be justified by observing that firstly, $u \mapsto g_v(u) = u\log(u) - u + e^v - uv$ (v is fixed) is convex, secondly $g'_v(e^v) = 0$, and finally $g_v(u) \geq g_v(e^v) \geq 0$.

Taking $\mu_i = \gamma_1$, we immediately deduce that γ_n satisfies the logarithmic Sobolev inequality with constant 2.

Corollary 2.4.16 (Gaussian distribution in dimension n) *The Gaussian measure γ_n in dimension n satisfies the logarithmic Sobolev inequality with the constant $C_{\gamma_n} = 2$.*

We are now in a position to state an important result, giving an exponential concentration inequality for empirical means while calculating $\mathbb{E}X$ when $X = f(Y)$, with a Lipschitz function f of a standard Gaussian random vector Y.

Theorem 2.4.17 (concentration inequality in the Gaussian case) *Let $(Y_1, \ldots, Y_M)$ be a sequence of independent random vectors, having the standard Gaussian d-dimensional distribution. Then for any Lipschitz function $f : \mathbb{R}^d \mapsto \mathbb{R}$, we have*

$$\mathbb{P}\left(\left|\frac{1}{M}\sum_{m=1}^{M} f(Y_m) - \mathbb{E}(f(Y))\right| > \varepsilon\right) \leq 2\exp\left(-\frac{M\varepsilon^2}{2|f|_{\text{Lip.}}^2}\right).$$

PROOF:
Apply Corollary 2.4.13 to the distribution of $(Y_1, \ldots, Y_m)$ verifying the logarithmic Sobolev inequality with constant 2 and to the function $g(y_1, \ldots, y_M) = \frac{1}{M}\sum_{m=1}^{M} f(y_m)$. Its Lipschitz constant is bounded by $|f|_{\text{Lip.}}/\sqrt{M}$ since

$$|g(y_1, \ldots, y_M) - g(y_1', \ldots, y_M')| \leq \frac{1}{M}\sum_{m=1}^{M} |f|_{\text{Lip.}} |y_m - y_m'|$$

$$\leq \frac{|f|_{\text{Lip.}}}{\sqrt{M}} \sqrt{\sum_{m=1}^{M} |y_m - y_m'|^2}.$$

This finishes the proof. □

With such an inequality in hand we can deduce non-asymptotic confidence intervals as in Proposition 2.4.1. This will be useful when simulating stochastic differential equations in Chapter 6.

2.5 EXERCISES

Exercise 2.1 (central limit theorem with various rates) Let $(X_m)_{m\geq 1}$ be a sequence of i.i.d. random variables having the symmetric Pareto distribution with parameter $\alpha > 0$, whose density is

$$\frac{\alpha}{2|z|^{\alpha+1}} 1_{|z|\geq 1}.$$

Set $\overline{X}_M := \frac{1}{M}\sum_{m=1}^M X_m$.

i) For $\alpha > 1$, justify the a.s. convergence of $\overline{X}_M$ to 0 as $M \to +\infty$.

ii) For $\alpha > 2$, prove a central limit theorem for $\overline{X}_M$ at rate $\sqrt{M}$.

iii) For $\alpha \in (1,2]$, determine the rate $u_M \to +\infty$ such that $u_M \overline{X}_M$ converges in distribution to a limit.

Hint: to distinguish the cases $\alpha = 2$ and $\alpha \in (1,2)$, use the Levy criterion (Theorem A.1.3) with the representation of the characteristic function

$$\mathbb{E}(e^{iuX}) = 1 + \alpha|u|^\alpha \int_{|u|}^{+\infty} \frac{\cos(t)-1}{t^{\alpha+1}} dt.$$

Exercise 2.2 (substitution method) We aim at estimating the Laplace transform

$$\phi(u) := \mathbb{E}(e^{uX}) = e^{\sigma^2 u^2/2}$$

of a Gaussian random variable $X \stackrel{d}{=} \mathcal{N}(0,\sigma^2)$, using a sample of size M of i.i.d. copies of X. The parameter σ^2 is unknown. Which procedure is the most accurate?

i) Computing the empirical mean $\phi_{1,M}(u) := \frac{1}{M}\sum_{m=1}^M e^{uX_m}$;

or

ii) Estimating σ^2 by the empirical variance σ_M^2, then estimating $\phi(u)$ by $\phi_{2,M}(u) := e^{\frac{u^2}{2}\sigma_M^2}$.

We will compare estimators using related confidence intervals.

Exercise 2.3 (central limit theorem, substitution method) Consider the setting of Proposition 2.2.6, with the estimation of $\max(\mathbb{E}X, a)$.

i) Assume $a > \mathbb{E}X$. Prove that both the upper estimator $\overline{f}_M$ and the lower estimator $\underline{f}_M$ converge to $\max(\mathbb{E}X, a)$ in L_1 at the rate M.

ii) Assume $a < \mathbb{E}X$. Establish a central limit theorem at rate $\sqrt{M}$ for $\overline{f}_M - \max(\mathbb{E}X, a)$, for $\underline{f}_M - \max(\mathbb{E}X, a)$, and for the pair $(\overline{f}_M - \max(\mathbb{E}X, a), \underline{f}_M - \max(\mathbb{E}X, a))$.

iii) Investigate the case $a = \mathbb{E}X$.

Exercise 2.4 (sensitivity formulas, exponential distribution)
Let $Y \stackrel{d}{=} \mathcal{E}xp(\lambda)$ with $\lambda > 0$ and set $F(\lambda) = \mathbb{E}(f(Y))$ for a given bounded function f.

i) Use the likelihood ratio method to represent F' as an expectation.

ii) Use the pathwise differentiation method to get another representation for smooth f (use the formula from Proposition 1.2.2).

iii) By integrating by parts, show that both formulas coincide.

Exercise 2.5 (sensitivity formulas, multidimensional Gaussian distribution) Extend Example 2.2.11 to the multidimensional case for the sensitivity with respect to the mean $\mathbb{E}(Y)$ and to the covariance matrix $\mathbb{E}(YY^\mathsf{T})$ (assumed to be invertible).

Hint: for the sensitivity w.r.t. elements of $\mathbb{E}(YY^\mathsf{T})$, one has to perturb the matrix $\mathbb{E}(YY^\mathsf{T})$ in a symmetric way to keep it symmetric.

Exercise 2.6 (sensitivity formulas, resimulation method) We aim at illustrating the benefit of using Common Random Numbers (CRN) in (2.2.6) and the impact of the smoothness of f on the estimator variance. We consider the Gaussian model $Y \stackrel{d}{=} \mathcal{N}(\theta, 1)$.

i) Denote by $(G_1, \ldots, G_M, G'_1, \ldots, G'_M)$ i.i.d. copies of $\mathcal{N}(0,1)$ and consider a smooth function f, bounded with bounded derivatives. Compute the variance of the estimator with different random numbers

$$\frac{1}{M} \sum_{m=1}^{M} \frac{f(\theta + \varepsilon + G_m) - f(\theta - \varepsilon + G'_m)}{2\varepsilon}$$

as $\varepsilon \to 0$ (M fixed).

ii) Compare it with that of the CRN estimator

$$\frac{1}{M} \sum_{m=1}^{M} \frac{f(\theta + \varepsilon + G_m) - f(\theta - \varepsilon + G_m)}{2\varepsilon}.$$

iii) Analyze the variance of the CRN estimator when $f(x) = \mathbf{1}_{x \geq 0}$. What is its dependency w.r.t. $\varepsilon \to 0$?

Write a simulation program illustrating these features.

Exercise 2.7 (concentration inequality, maximum of Gaussian variables)

i) Use the Gaussian concentration inequality of Corollary 2.4.13 to establish the Borell inequality (1975): for any centered d-dimensional Gaussian vector $Y = (Y_1, \ldots, Y_d)$, we have

$$\mathbb{P}\left(|\max_{1 \leq i \leq d} Y_i - \mathbb{E}(\max_{1 \leq i \leq d} Y_i)| > \varepsilon\right) \leq 2 \exp\left(-\frac{\varepsilon^2}{2\sigma^2}\right), \qquad \forall \varepsilon \geq 0$$

where[13] $\sigma^2 = \max_{1 \leq i \leq d} \mathbb{E}(Y_i^2)$.

Hint: first assume that $(Y_i)_i$ are i.i.d. standard Gaussian random variables. To prove the general case, use the representation of Proposition 1.4.1.

ii) We consider the case $d \to +\infty$ and assume that $\sigma^2 = \max_{1 \leq i \leq d} \mathbb{E}(Y_i^2)$ is bounded as $d \to +\infty$. Assuming that $\mathbb{E}(\max_{1 \leq i \leq d} Y_i) \to +\infty$, and deduce that

$$\frac{\max_{1 \leq i \leq d} Y_i}{\mathbb{E}(\max_{1 \leq i \leq d} Y_i)} \xrightarrow[d \to +\infty]{\text{Prob.}} 1.$$

Application: in the standard i.i.d. case, since $\mathbb{E}(\max_{1 \leq i \leq d} Y_i) \sim \sqrt{2 \log(d)}$ as $d \to +\infty$ (see [49]), we obtain a nice deterministic equivalent (in probability) of $\max_{1 \leq i \leq d} Y_i$.

---

[13] Observe that the constants do not depend much on the dimension, thus passing to infinite dimension is possible.

CHAPTER 3

Variance reduction

In the previous chapter we saw that the confidence intervals for controlling the error in Monte-Carlo computations have a length proportional to the standard deviation of the sampled random variable X. Reducing the variance (either by modifying the simulation procedure, or transforming the problem) may help to accelerate the convergence of the Monte-Carlo method. Decreasing the variance by a factor 10 is equivalent - asymptotically - to decreasing the number of simulations (and hence the computational time) by a factor of 10 to achieve a given accuracy. In this chapter we study some techniques of variance reduction, with a particular focus on importance sampling.

3.1 ANTITHETIC SAMPLING

The method is based on the Chebyshev covariance inequality.

Lemma 3.1.1 *Let Y be a real-valued random variable. Consider a non-increasing function $f : \mathbb{R} \mapsto \mathbb{R}$ and a non-decreasing function $g : \mathbb{R} \mapsto \mathbb{R}$. Suppose that $f(Y)$ and $g(Y)$ are square integrable. Then the covariance between $f(Y)$ and $g(Y)$ is non-positive:*

$$\mathbb{E}[f(Y)g(Y)] \leq \mathbb{E}[f(Y)]\mathbb{E}[g(Y)].$$

PROOF:
Let Y' be an independent copy of Y. Set $C := \operatorname{Cov}(f(Y) - f(Y'), g(Y) - g(Y'))$.

On the one hand, $\mathbb{E}(f(Y) - f(Y')) = \mathbb{E}(g(Y) - g(Y')) = 0$, which implies $C = \mathbb{E}([f(Y) - f(Y')][g(Y) - g(Y')])$. From the monotonicity of f and g, we deduce that $C \leq 0$.

On the other hand, $C = \operatorname{Cov}(f(Y), g(Y)) + \operatorname{Cov}(f(Y'), g(Y')) =$

$2\mathbb{C}\mathrm{ov}(f(Y),g(Y))$. It readily follows that $C = 2\mathbb{C}\mathrm{ov}(f(Y),g(Y)) \le 0$.
□

In particular, if f is monotone, then
$$\mathbb{C}\mathrm{ov}[f(Y), f(\varphi(Y))] \le 0,$$
for any given non-increasing function φ. Concerning its use in Monte-Carlo methods, this inequality is interesting if $\varphi(Y)$ and Y have the same distribution, giving thus the possibility to save one generation of Y by simply taking $\varphi(Y)$ (*antithetic variable*).

Example 3.1.2

1. If Y is uniformly distributed on $[0,1]$, then $\varphi(Y) := 1 - Y \stackrel{\mathrm{d}}{=} Y$.

2. If Y is a centered Gaussian random variable, then $\varphi(Y) := -Y \stackrel{\mathrm{d}}{=} Y$. More generally, for any random variable with a symmetric distribution $(-Y \stackrel{\mathrm{d}}{=} Y)$, the choice $\varphi(x) = -x$ is suitable.

3. If Y has the Cauchy distribution with the parameter σ, then $\varphi(Y) := \frac{\sigma^2}{Y} \stackrel{\mathrm{d}}{=} Y$. However, $\varphi(y) = \sigma^2/y$ is non-increasing only if y is restricted to positive or negative values. Thus, the method below will work with that φ if f is constant on $\mathbb{R}^+$, or $\mathbb{R}^-$, or if f is even.

Generally, we can design an *antithetic* function φ with the help of the c.d.f. of Y and its inverse.

Lemma 3.1.3 *Suppose that the c.d.f. $y \mapsto F(y) = \mathbb{P}(Y \le y)$ is continuous. Denote by F^{-1} its generalized inverse, defined in Proposition 1.2.1. Then $\varphi(y) = F^{-1}(1 - F(y))$ is a non-increasing function such that*
$$\varphi(Y) \stackrel{\mathrm{d}}{=} Y;$$
hence $\varphi(Y)$ is an antithetic variable associated with Y.

Proof:
Clearly, φ is non-increasing. By Proposition 1.2.1, we have
$$\varphi(Y) = F^{-1}(1 - F(Y)) = F^{-1}(1 - U) \stackrel{\mathrm{d}}{=} F^{-1}(U) \stackrel{\mathrm{d}}{=} Y$$
where U is a random variable, uniform on $[0,1]$.
□

This construction of antithetic variables allows us to recover the first two cases of Example 3.1.2, but not the one with the Cauchy distribution.

Applications. Suppose we want to calculate $\mathbb{E}(f(Y))$ using a Monte-Carlo method. Which estimator should we use:

$$I_{1,M} := \frac{1}{M} \sum_{m=1}^{M} f(Y_m) \quad \text{or} \quad I_{2,M} := \frac{1}{M} \sum_{m=1}^{M} \frac{f(Y_m) + f(\varphi(Y_m))}{2}?$$

Both estimators converge a.s. to $\mathbb{E}(f(Y))$; they require us to simulate the same number of points Y, thus they have the same computational cost if we assume that the cost of the numerical evaluation of $\varphi(Y), f(Y), f(\varphi(Y))$ is small compared to the cost of the simulation of Y. Their accuracy can be analyzed via their respective variance:

$$M \operatorname{Var}(I_{1,M}) = \operatorname{Var}(f(Y)),$$

$$M \operatorname{Var}(I_{2,M}) = \frac{1}{4}[\operatorname{Var}(f(Y) + f(\varphi(Y)))]$$

$$= \frac{1}{2}[\operatorname{Var}(f(Y)) + \operatorname{Cov}(f(Y), f(\varphi(Y)))].$$

Because $\operatorname{Cov}(f(Y), f(\varphi(Y))) \leq \sqrt{\operatorname{Var}(f(Y))}\sqrt{\operatorname{Var}(f(\varphi(Y)))} = \operatorname{Var}(f(Y))$, we always have $\operatorname{Var}(I_{1,M}) \geq \operatorname{Var}(I_{2,M})$: the antithetic sampling is at least as good as the usual sampling. The performance ratio is defined as

$$\frac{\operatorname{Var}(I_{1,M})}{\operatorname{Var}(I_{2,M})} = \frac{2}{1 + \operatorname{Cor}(f(Y), f(\varphi(Y)))} := R \geq 1.$$

If f is *monotone*, from Lemma 3.1.1 we have $R \geq 2$: the performance of the antithetic sampling is at least twice as good as the usual sampling.

Since it always gives better results, this technique is rather popular, even though its gain is not always significant. Finally, let us mention that the confidence interval must be computed keeping in mind the dependence of $f(Y_m)$ and $f(\varphi(Y_m))$.

Example 3.1.4 Suppose that Y is uniformly distributed on $[0,1]$ and $f(y) = y^2$. A simple calculation gives $\operatorname{Var}(f(Y)) = \frac{1}{5} - (\frac{1}{3})^2 = \frac{4}{45}$ and $\operatorname{Var}(f(Y)) + \operatorname{Cov}(f(Y), f(\varphi(Y))) = [\frac{1}{5} - (\frac{1}{3})^2] + [\frac{1}{3} + \frac{1}{5} - \frac{2}{4} - (\frac{1}{3})^2] = \frac{1}{90}$; the performance ratio is $R = 16$.

3.2 CONDITIONING AND STRATIFICATION

We continue considering the problem of the evaluation of $\mathbb{E}X$ by Monte-Carlo method.

3.2.1 Conditioning technique

Suppose moreover that for some random variable Z, we know $g(z) = \mathbb{E}(X|Z = z)$ explicitly. Then by generating independent copies of $(Z_1, \ldots, Z_m, \ldots)$, we obtain a new Monte-Carlo estimator

$$\frac{1}{M} \sum_{m=1}^{M} g(Z_m).$$

By the law of large numbers, it converges a.s. to $\mathbb{E}(g(Z)) = \mathbb{E}(\mathbb{E}(X|Z)) = \mathbb{E}X$. The accuracy of this estimator can be evaluated using the variance that verifies $\mathrm{Var}(g(Z)) \leq \mathrm{Var}(X)$.[1] This *conditioning technique* provides a systematic variance reduction. Implementing this approach is not always possible because the efficient evaluation of the conditional expectation is in general a tough problem (see later in Chapter 8). Here is an example that works.

Example 3.2.1 Suppose that $X = 1_{U_1^2 + U_2^2 \leq 1}$ for two independent random variables uniformly distributed on $[0, 1]$, which gives $\mathbb{E}X = \frac{\pi}{4}$. Then $\mathbb{E}(X|U_1) = \sqrt{1 - U_1^2}$. The variance gain is equal to

$$\frac{\mathrm{Var}(X)}{\mathrm{Var}(\mathbb{E}(X|U_1))} = \frac{\frac{\pi}{4}(1 - \frac{\pi}{4})}{1 - \frac{1}{3} - (\frac{\pi}{4})^2} \approx 3,38.$$

3.2.2 Stratification technique

As mentioned before, having an explicit formula for $g(z) = \mathbb{E}(X|Z = z)$ is usually restrictive in practice. An alternative approach based on the conditioning is the stratification technique,[2] which assumes that the set of the values of Z is decomposable into k strata $S_1, \ldots, S_k$ such that

  - for each stratum S_j, $\mathbb{P}(Z \in S_j) = p_j$ is known and $\sum_{j=1}^{k} p_j = 1$;

---

[1] We use the classic decomposition $\mathrm{Var}(X) = \mathbb{E}(\mathrm{Var}(X|Z)) + \mathrm{Var}(\mathbb{E}(X|Z))$.
[2] In opinion polls, the method is also known as the *quota method*.

– the conditional distribution of X given $Z \in \mathcal{S}_j$ can be simulated (efficiently enough for the algorithm to be interesting). Denote by X_j a random variable having this distribution.

For each stratum we generate M_j independent simulations of X_j, independent of the other strata. The stratification estimator is

$$I^{\text{strat.}}_{M_1,\ldots,M_k} = \sum_{j=1}^{k} p_j \frac{1}{M_j} \sum_{m=1}^{M_j} X_{j,m}. \tag{3.2.1}$$

If all M_j tend to infinity, it converges a.s. to $\mathbb{E}X$:

$$I^{\text{strat.}}_{M_1,\ldots,M_k} \xrightarrow{\text{a.s.}} \sum_{j=1}^{k} p_j \mathbb{E}(X_j) = \sum_{j=1}^{k} \mathbb{E}(X|Z \in \mathcal{S}_j)\mathbb{P}(Z \in \mathcal{S}_j) = \mathbb{E}X.$$

Here again, it is possible to prove a central limit theorem. Thus the error analysis of this technique can be in principle reduced to the analysis of its variance: denoting $\sigma_j^2 = \text{Var}(X_j) = \text{Var}(X|Z \in \mathcal{S}_j)$, we have

$$\text{Var}(I^{\text{strat.}}_{M_1,\ldots,M_k}) = \sum_{j=1}^{k} p_j^2 \frac{\sigma_j^2}{M_j}. \tag{3.2.2}$$

There exist several strategies to allocate the computational effort within the strata, under the constraint on the global computational budget

$$M = \sum_{j=1}^{k} M_j. \tag{3.2.3}$$

▷ **Allocation proportional to the probability of the strata.** This strategy corresponds to $\frac{M_j}{M} = p_j$ (neglecting the rounding effects), for which

$$\text{Var}(I^{\text{strat.,prop.alloc.}}_{M_1,\ldots,M_k}) = \frac{1}{M} \sum_{j=1}^{k} p_j \sigma_j^2$$

$$= \frac{1}{M} \sum_{j=1}^{k} \text{Var}(X|Z \in \mathcal{S}_j)\mathbb{P}(Z \in \mathcal{S}_j)$$

$$= \frac{\mathbb{E}(\text{Var}(X|I))}{M},$$

where I is a discrete random variable defined by $\{I = j\} = \{Z \in \mathcal{S}_j\}$. As $\mathbb{E}(\text{Var}(X|I)) \leq \text{Var}(X)$, this strategy always gives variance reduction with respect to the method without stratification.

▷ **Optimal allocation.** After writing equations for minimizing the variance $\sum_{j=1}^{k} p_j^2 \frac{\sigma_j^2}{M_j}$ under the budget constraint (3.2.3), we find that the optimal number of simulations per stratum is $M_j^* = M \frac{p_j \sigma_j}{\sum_{i=1}^{k} p_i \sigma_i}$. Then the minimal variance equals

$$\mathrm{Var}(I_{M_1,\ldots,M_k}^{\mathrm{strat.,opt.alloc.}}) = \frac{1}{M}(\sum_{j=1}^{k} p_j \sigma_j)^2.$$

Owing to the Jensen inequality, we observe that this variance is less than the one obtained using proportional allocation. However, this strategy cannot be realized directly because it requires us to know $\sigma_j's$, which have to be evaluated beforehand. Besides the optimization of the number of simulations per stratum, we could also optimize the strata, which is in general a much more difficult problem.

Example 3.2.2 (stratification of Gaussian vectors) *If $Y = (G_1, \ldots, G_d)$ is a Gaussian vector, we can take $Z = \beta \cdot Y$ as a stratification variable: indeed, the distribution of Z is Gaussian with explicit characteristics and an exact calculation of $(p_j = \mathbb{P}(Z \in \mathcal{S}_j))_j$ follows. The conditional simulation of Y given $\{Z \in \mathcal{S}_j\}$ can be realized as follows:*

1. *Simulate Z given $\{Z \in \mathcal{S}_j\}$, using the inversion method (see Proposition 1.2.9).*

2. *Simulate Y given Z, which by the properties of Gaussian vectors can be reduced to the simulation of Gaussian variables with a mean and a variance depending on Z.*

See Exercise 3.4. For applications, see [55].

3.3 CONTROL VARIATES

3.3.1 Concept

Definition 3.3.1 *We define a control variate, for a problem of computation of $\mathbb{E}X$, as a centered square integrable random variable Z, possibly d-dimensional, which can be simulated jointly with X. For $\beta \in \mathbb{R}^d$, we denote*

$$Z(\beta) := \beta \cdot Z = \sum_{j=1}^{d} \beta_i Z_i;$$

$Z(\beta)$ is again a centered square integrable random variable.

The interest in control variates comes from the fact that adding $Z(\beta)$ to the simulation of X does not change the expectation. On the other hand, by optimizing the weight β, we hope to reduce the variance. The Monte-Carlo estimator of $\mathbb{E}X$ with a control variate is thus given by

$$I_{\beta,M}^{\text{Cont.Var.}} = \frac{1}{M} \sum_{m=1}^{M} (X_m - Z_m(\beta)). \qquad (3.3.1)$$

It converges a.s. to $\mathbb{E}(X - Z(\beta)) = \mathbb{E}(X)$. The asymptotic variance of the renormalized error is equal to

$$M\text{Var}(I_{\beta,M}^{\text{Cont.Var.}}) = \text{Var}(X - Z(\beta)).$$

If this variance is smaller than $\text{Var}(X) = \text{Var}(X - Z(\beta))|_{\beta=0}$, the estimator $I_{\beta,M}^{\text{Cont.Var.}}$ will asymptotically perform better than the one without the control variate.

3.3.2 Optimal choice

In this approach, the two main questions are:

- For a given problem, how can we produce a control variate Z?

- When Z is selected, how can we determine the best parameter β?

There does not exist a universal answer to the first question; we rather consider it case by case according to the problem at hand. The analysis of the second question helps us to better understand what is a good choice. Minimization of $\text{Var}(X - Z(\beta))$ reduces to minimization of $\mathbb{E}(X - Z(\beta))^2$ because the expectation does not depend on β: from this point of view, it boils down to a classic statistical linear regression problem in data processing. To choose the variable Z well, we shall find the best *explanatory variables* to explain the variability of X.

Example 3.3.2 If $X = f(U)$ with U uniformly distributed, $Z = U - \frac{1}{2}$ (here $d = 1$) is a control variate. If f is locally linear, this is a priori a good choice of a control variate.

The optimal parameter β comes from the minimization of

$$\mathbb{E}(X - Z(\beta))^2 = \mathbb{E}(X^2) - 2\beta \cdot \mathbb{E}(XZ) + \beta \cdot \mathbb{E}(ZZ^\mathsf{T})\beta,$$

which is quadratic in β. Supposing that the components of Z are not redundant ($\mathbb{E}(ZZ^T)$ invertible), the solution is unique and given by

$$\beta^* = [\mathbb{E}(ZZ^T)]^{-1}\mathbb{E}(XZ).$$

This shows that if Z is uncorrelated with X (i.e. $\mathbb{E}(XZ) = 0$), the method will not work better in comparison to the usual one. On the contrary, the method is much more efficient when the correlation of X with the components of Z is significant.

The exact calculation of β^* is not possible (a priori more difficult than that for $\mathbb{E}X$) and we can evaluate it numerically using the same simulations:

$$\beta_M^* = \left[\frac{1}{M}\sum_{m=1}^M Z_m Z_m^T\right]^{-1} \frac{1}{M}\sum_{m=1}^M X_m Z_m. \qquad (3.3.2)$$

In practice, the evaluation using a smaller sample may be sufficient to obtain a variance reduction as expected. Finally, for the evaluation of $\mathbb{E}X$, we calculate $I_{\beta_M^*,M}^{\text{Cont.Var.}}$: the renormalized error of this estimator again satisfies the central limit theorem[3] with the limit variance $\text{Var}(X - Z(\beta^*))$. We leave the proof to the reader.

Example 3.3.3 Take $X = \exp(U)$, with U having uniform distribution, and $Z = U - \frac{1}{2}$. The optimal parameter β^* (calculated numerically on 1000 simulations) is approximately 1.69. For this value, the factor of improvement on variance is

$$\frac{\text{Var}(X)}{\text{Var}(X - Z(\beta^*))} \approx \frac{0.24}{0.0040} = 60.$$

We finish by giving some generic examples of control variates.

Example 3.3.4 Consider the case where $X = f(G)$ and G is a standard Gaussian random variable.

- The Hermite polynomials $H_k(x) = (-1)^k e^{x^2/2} \partial_x^k (e^{-x^2/2})$ are such that $\mathbb{E}(H_k(G)) = 0$. Hence $Z = (H_1(G), \ldots, H_d(G))$ is a control variate.

- Inspired by the explicit Laplace transform, we can also take $Z = e^{\lambda G} - e^{\frac{1}{2}\lambda^2}$, for different λ's.

---

[3]This is a little less obvious to show than before, because $(X_m - Z_m(\beta_M^*))_m$ are not independent anymore, due to β_M^* random but converging to β^*.

Obviously, this can be generalized to other distributions for G, as soon as the new random variable G has explicit moments or an explicit Laplace transform.

3.4 IMPORTANCE SAMPLING

This technique of variance reduction is based on a modification of the sampling distributions, in order to sample outputs that we consider more relevant for the problem at hand. This distribution modification is simply a change of probability, which is commonly used in statistics and probability. In statistics for example, the theory of the estimation by likelihood maximization is based on the study of the density of a parametric model with respect to a reference probability measure. This density function, called *likelihood*, encodes the dependence of the model with respect to the parameters. In probability, one can find spectacular applications of the change of probability in the theory of Gaussian spaces and in the study of martingales. In financial mathematics, this is a basic tool in the valuation of financial contracts. In the theory of Monte-Carlo methods, this underlies all the algorithms used to efficiently simulate rare events.

In the sequel, after some general information, we study the changes of probability based on exponential transformations (known as *Esscher transforms*). Finally, we provide adaptive algorithms to determine automatically the optimal parameters of a change of probability.

For discrete random variables, the changes of probability measure are often elementary; thus we pay more attention to the case of continuous random variables in our presentation.

3.4.1 Changes of probability measure: basic notions and applications to Monte-Carlo methods

▷ Definition and general information. Let $(\Omega, \mathcal{F}, \mathbb{P})$ be a reference probability space, which for instance stands for the probabilistic model to simulate.

Definition 3.4.1 *A probability measure $\mathbb{Q}$ on the space $(\Omega, \mathcal{F})$ defines a change of probability measure (with respect to $\mathbb{P}$), if there exists a positive random variable L such that*

$$\mathbb{Q}(A) = \mathbb{E}[L\mathbf{1}_A] \quad \text{if } A \in \mathcal{F}. \tag{3.4.1}$$

The random variable L is called the density or the likelihood of $\mathbb{Q}$ with respect to $\mathbb{P}$. *In general we denote*

$$\mathbb{Q} = L \cdot \mathbb{P}, \quad \text{or} \quad d\mathbb{Q} = Ld\mathbb{P}, \quad \text{or} \quad \frac{d\mathbb{Q}}{d\mathbb{P}} = L.$$

We say that the probability measures $\mathbb{P}$ and $\mathbb{Q}$ are equivalent, if the random variable L is strictly positive $\mathbb{P}$-a.s., which implies that $\mathbb{P}$ is a change of probability measure with respect to $\mathbb{Q}$ with density L^{-1}.

For $\mathbb{Q}$ to be a probability measure (to have mass 1), it is necessary that $\mathbb{E}_\mathbb{P}(L) = 1$. Conversely, with the positivity assumption, this condition is sufficient to properly define a probability measure $\mathbb{Q}$ (the axioms of mass 1 and of σ-additivity are thus verified).

When the probability measures are equivalent — which will always be the case in the following — the density L is often represented in the form e^l, where l is called the *log-likelihood*. In the case of equivalent probability measures, an event which is negligible for $\mathbb{P}$ is also negligible for $\mathbb{Q}$, and vice versa: $\mathbb{P}(A) = 0 \Rightarrow \mathbb{Q}(A) = \mathbb{E}_\mathbb{P}(L\mathbf{1}_A) = 0$ because $\mathbf{1}_A = 0$ $\mathbb{P}$-a.s. and conversely, $\mathbb{Q}(A) = 0 \Rightarrow \mathbb{P}(A) = \mathbb{E}_\mathbb{Q}(L^{-1}\mathbf{1}_A) = 0$. Hence, an a.s. equality or an a.s. limit takes place simultaneously under both probabilities, and it is not necessary to add $\mathbb{P}$-a.s. or $\mathbb{Q}$-a.s. .

In practice, to calculate an expectation under the new probability measure, we use the following formulas which allow us to pass from calculations under $\mathbb{P}$ to calculations under $\mathbb{Q}$, and vice versa; this follows from the definition of $\mathbb{Q}$ with respect to $\mathbb{P}$. For any measurable bounded function f (or appropriately integrable)

$$\mathbb{E}_\mathbb{Q}(f(Y)) = \mathbb{E}_\mathbb{P}(Lf(Y)), \qquad \mathbb{E}_\mathbb{P}(f(Y)) = \mathbb{E}_\mathbb{Q}(L^{-1}f(Y)). \quad (3.4.2)$$

Example 3.4.2 Suppose that Y is a random variable with density $p(.)$ on $\mathbb{R}^d$ — with respect to the Lebesgue measure — and let $q(.)$ be another probability density on $\mathbb{R}^d$: we suppose in addition that $\{x \in \mathbb{R}^d : q(x) = 0\} = \{x \in \mathbb{R}^d : p(x) = 0\}$.

Then $L = \frac{q}{p}(Y)$ well defines the likelihood of a new probability measure $\mathbb{Q}$ equivalent to $\mathbb{P}$: indeed, L is positive and $\mathbb{E}_\mathbb{P}(L) = \int_{\mathbb{R}^d} \frac{q}{p}(y)p(y)dy = 1$.

The distribution of Y under $\mathbb{Q}$ has q as density: indeed, for any bounded measurable function f, we have

$$\mathbb{E}_\mathbb{Q}(f(Y)) = \mathbb{E}_\mathbb{P}(Lf(Y)) = \int_{\mathbb{R}^d} f(y)\frac{q}{p}(y)p(y)dy = \int_{\mathbb{R}^d} f(y)q(y)dy.$$

This example leads to several simple and important remarks.

1. If the components of Y are independent under $\mathbb{P}$ (i.e., the density p is written in the form of a product $p(y) = p_1(y_1)\ldots p_d(y_d)$) and $q(y) = q_1(y_1)\ldots q_d(y_d)$ so that

$$L = \prod_{i=1}^{d} \frac{q_i}{p_i}(Y_i),$$

then the random variables $(Y_i)_i$ are independent under $\mathbb{Q}$ and the distribution of Y_i under $\mathbb{Q}$ has the density $q_i(.)$. This is a common situation.

2. A change of probability measure may not preserve independence: indeed, if the components of Y are independent under $\mathbb{P}$ ($p(y) = p_1(y_1)\ldots p_d(y_d)$), they can be no longer independent under $\mathbb{Q}$ (q is not necessarily written as a product over coordinates) and vice versa. However, if the likelihood depends on additional random variables independent of the others, the independence is preserved.

Proposition 3.4.3 (independence) *Let Y and Z be two random variables, independent under $\mathbb{P}$ and let $L = \Phi(Y)$ be a likelihood defining a new equivalent probability measure $\mathbb{Q}$. Then, Y and Z are again independent under $\mathbb{Q}$.*

PROOF:
For any continuous bounded functions g_1 and g_2, we directly check

$$\begin{aligned}\mathbb{E}_{\mathbb{Q}}(g_1(Y)g_2(Z)) &= \mathbb{E}_{\mathbb{P}}(\Phi(Y)g_1(Y)g_2(Z)) \\ &= \mathbb{E}_{\mathbb{P}}(\Phi(Y)g_1(Y))\mathbb{E}_{\mathbb{P}}(g_2(Z)) \\ &= \mathbb{E}_{\mathbb{P}}(\Phi(Y)g_1(Y))\mathbb{E}_{\mathbb{P}}(\Phi(Y)g_2(Z)) \\ &= \mathbb{E}_{\mathbb{Q}}(g_1(Y))\mathbb{E}_{\mathbb{Q}}(g_2(Z)).\end{aligned}$$

$\square$

3. As the choice of q is very flexible, the mean and the covariance of Y may a priori change from $\mathbb{P}$ to $\mathbb{Q}$. In the Gaussian case, the changes of parameters will be much easier to identify.

4. If the random variable $f(Y)$ is square integrable under $\mathbb{P}$, it may not be square integrable under $\mathbb{Q}$. It is easy to find such examples.

▷ **Applications to Monte-Carlo methods.** Before going further in the analysis of the effects of change of probability measure, we give an application to the numerical computation of $\mathbb{E}X = \mathbb{E}_{\mathbb{P}}(X)$, in which we indicate explicitly the reference to $\mathbb{P}$ to avoid any ambiguity. Starting from (3.4.2), we derive another Monte-Carlo estimator for $\mathbb{E}_{\mathbb{P}}(X)$.

Proposition 3.4.4 *Let $\mathbb{Q}$ be a probability measure equivalent to $\mathbb{P}$, associated with a likelihood L. Suppose that (L, X) can be simulated jointly under $\mathbb{Q}$ and that $(L_m, X_m)_{m \geq 1}$ is a sequence of independent random variables with the same distribution as (L, X) under $\mathbb{Q}$. Then for any random variable X, integrable under $\mathbb{P}$, we have*

$$I_{\mathbb{Q},M}^{\text{Imp.Samp.}} = \frac{1}{M} \sum_{m=1}^{M} L_m^{-1} X_m \xrightarrow[M \to +\infty]{a.s.} \mathbb{E}_{\mathbb{P}}(X) = \mathbb{E}X. \quad (3.4.3)$$

It is important to note in (3.4.3) that the simulation outputs are no longer equally weighted with the weights $1/M$ as before: this is to account for the change of measure that modifies the likelihood of samples.

We highlight a minor drawback: the evaluation of the expectation of a constant random variable — say 1 — is no more exact (as a difference with the usual Monte-Carlo method): indeed, it is likely that $\frac{1}{M} \sum_{m=1}^{M} L_m^{-1} \neq 1$. Similarly, the empirical average of $X \in [0, 1]$ may give an estimator larger than 1, although it converges as $M \to +\infty$ to $\mathbb{E}X \in [0, 1]$.

PROOF:
Let us check that $L^{-1}X$ is integrable under $\mathbb{Q}$. Taking the truncation $|X| \wedge n \underset{n \to +\infty}{\uparrow} |X|$ and applying (3.4.2), we have

$$\mathbb{E}_{\mathbb{P}}(|X|) = \lim_{n \to +\infty} \mathbb{E}_{\mathbb{P}}(|X| \wedge n) = \lim_{n \to +\infty} \mathbb{E}_{\mathbb{Q}}(L^{-1}(|X| \wedge n))$$
$$\geq \mathbb{E}_{\mathbb{Q}}(\liminf_{n \to +\infty} L^{-1}(|X| \wedge n)) = \mathbb{E}_{\mathbb{Q}}(L^{-1}|X|)$$

by the monotone convergence theorem and the Fatou lemma. Thus the law of large numbers can be applied to $\mathbb{Q}$: the limit is $\mathbb{E}_{\mathbb{Q}}(L^{-1}X) = \mathbb{E}_{\mathbb{P}}(X)$. □

In order to derive confidence intervals, the next stage consists of an application of the central limit theorem, with a limit variance equal to

$$\text{Var}_{\mathbb{Q}}(L^{-1}|X|) = \mathbb{E}_{\mathbb{Q}}(L^{-2}|X|^2) - (\mathbb{E}X)^2 = \mathbb{E}_{\mathbb{P}}(L^{-1}|X|^2) - (\mathbb{E}X)^2. \quad (3.4.4)$$

▷ Is this variance smaller than $\mathrm{Var}_\mathbb{P}(X) = \mathbb{E}_\mathbb{P}(X^2) - (\mathbb{E}X)^2$? If yes, the importance sampling gives an asymptotically more accurate Monte-Carlo estimator of $\mathbb{E}X$. But actually any situation regarding the variance may take place: the new variance can be less, or greater, even infinite. Indeed, in view of (3.4.4) we observe that supposing X to be square integrable under $\mathbb{P}$ is not sufficient for $L^{-1}X$ to be square integrable under $\mathbb{Q}$; see the example below.

Example 3.4.5 (of the estimator $I_{\mathbb{Q},M}^{\mathrm{Imp.Samp.}}$ with infinite variance) Let $X = \exp(Y/3)$ with Y distributed as $\mathcal{E}\mathrm{xp}(1)$, satisfying

$$\mathbb{E}(X^2) = \int_0^{+\infty} \exp(2y/3) \exp(-y) dy < +\infty.$$

Let $q(.)$ be the density of the distribution $\mathcal{E}\mathrm{xp}(2)$: then, $L = \frac{q}{p}(Y)$ and

$$\begin{aligned}\mathbb{E}_\mathbb{Q}(|L^{-1}X|^2) &= \mathbb{E}_\mathbb{P}(L^{-1}|X|^2) \\ &= \int_0^\infty \frac{\exp(-y)}{2\exp(-2y)} \exp(2y/3) \exp(-y) dy = +\infty.\end{aligned}$$

▷ Guide to a good change of measure. However, there exists an optimal change of probability measure.

Proposition 3.4.6 *Suppose that X is a positive real-valued random variable. Then $L = \frac{X}{\mathbb{E}X}$ defines a new probability measure $\mathbb{Q}$ equivalent to $\mathbb{P}$ such that the variance of $I_{\mathbb{Q},M}^{\mathrm{Imp.Samp.}}$ is zero: only one sample is sufficient to estimate $\mathbb{E}X$.*

PROOF:
L is a likelihood with respect to $\mathbb{P}$ because it is positive and with expectation 1 under $\mathbb{P}$. Moreover, the sampled random variable is $L^{-1}X = \mathbb{E}X$, thus constant and $I_{\mathbb{Q},M}^{\mathrm{Imp.Samp.}} = \mathbb{E}X$ for all $M \geq 1$. □

This result is of course a utopia, because in practice an effective sampling under $\mathbb{Q}$ is out of reach: indeed, without the knowledge of $\mathbb{E}X$, it seems to be very difficult to simulate L and thus a fortiori (L, X).

Proposition 3.4.6 is very important to understand how to choose L. For example, if $X = \mathbf{1}_A$ for a certain event A with non-zero probability — the goal is thus to calculate $\mathbb{P}(A)$ — then we set

$$L_\varepsilon = \frac{\mathbf{1}_A + \varepsilon}{\mathbb{P}(A) + \varepsilon} \qquad \text{with } \varepsilon > 0, \qquad (3.4.5)$$

to properly define a new probability measure $\mathbb{Q}_\varepsilon$. If $\varepsilon = 0$, we recover the optimal likelihood given by Proposition 3.4.6. We have $\mathbb{Q}_\varepsilon(A) = \mathbb{E}_\mathbb{P}(L_\varepsilon 1_A) = \frac{\mathbb{P}(A)(1+\varepsilon)}{\mathbb{P}(A)+\varepsilon} \to 1$ as $\varepsilon \to 0$. So under $\mathbb{Q}_\varepsilon$ with $\varepsilon \ll \mathbb{P}(A)$, A occurs with probability very close to 1: the "optimal" change of probability measure has the effect of making the event very likely while it is not under $\mathbb{P}$. In other terms, *the changes of probability measure, which will significantly reduce the variance in the Monte-Carlo evaluation of $\mathbb{E}X$, are those which sample the most important outputs for X.*

▷ The dilemma: explicit likelihood or explicit sampling. For implementing an efficient change of probability measure, two key elements are necessary:

- an explicit form of the likelihood L, useful to weight the outputs in (3.4.3), with the likelihood close to optimal;

- an explicit distribution of the simulations under $\mathbb{Q}$, to be able to effectively generate the sequence $(X_m)_{m \geq 1}$.

To fulfill both conditions is delicate in practice. To illustrate this discussion, consider the case $X = f(Y)$. On the one hand, a good choice of the likelihood L must be approximately proportional to $f(Y)$: by a numerical or theoretical study of the function f, we may hope to propose a good candidate $\widetilde{L}(Y)$ for the likelihood. We must note that, unfortunately, this choice does not lead systematically to a simple and tractable procedure to generate Y — thus X — under $\mathbb{Q}$. By default, an acceptance-rejection method may be used.

Conversely, specifying first the distribution of Y under $\mathbb{Q}$ is certainly rather convenient from the point of view of the simulation of X; but even if L is explicit, nothing ensures that it will be close to the optimal $\frac{X}{\mathbb{E}X}$. While not perfect, this is the approach that will be used in general.

That is why a good adjustment of L requires a good deal of intuition and modeling for the problem. In Paragraph 3.4.4 we describe a robust algorithm of the parameter adjustment for change of probability measure.

Example 3.4.7 (rare event) To illustrate the type of the heuristic we may consider, suppose $X = 1_{Y \geq y}$ with Y a centered random variable. X is a Bernoulli random variable with parameter $p = \mathbb{P}(Y \geq y)$

and variance $p(1-p)$. The *relative* error of a simple Monte-Carlo method, with probability 95%, equals
$$\frac{1.96\sqrt{p(1-p)/M}}{p} = \frac{1.96\sqrt{1-p}}{\sqrt{p\,M}}.$$

Suppose now that $y \gg 1$, so that $p \ll 1$ (rare event): the relative error is of order $1/\sqrt{p\,M}$, and the accuracy deteriorates rapidly as the event becomes rarer (a very small proportion of simulations of Y gives a value greater than y). To guarantee a given relative accuracy, the number of simulations and the sought probability must evolve inversely proportionally.

Concerning the "optimal" change of probability $\mathbb{Q}_\varepsilon$ (defined in (3.4.5)), it shall make $\{Y \geq y\}$ appear with large probability. An efficient choice of $\mathbb{Q}$ — though not optimal — consists certainly of making this event more likely, for example $\mathbb{Q}(Y \geq y) \approx \frac{1}{2}$. Modifying the mean and/or the variance of Y may achieve this goal.

3.4.2 Changes of probability measure by affine transformations

We continue by describing classic families of changes of probability measures, specifying the form of likelihood and giving the distributions of new random variables.

The following result shows how to easily modify mean and variance. This is a simple change of probability measure, which offers the advantage to keep an elementary simulation under the new probability (affine transformation of the initial random variable) with an explicit likelihood.

Proposition 3.4.8 *Suppose that Y is a random variable in $\mathbb{R}^d$, with the density $p(.)$, positive on $\mathbb{R}^d$.*

- **(Change of mean)** *For any $\theta_\mu \in \mathbb{R}^d$, the likelihood*
$$L = \frac{p(Y - \theta_\mu)}{p(Y)}$$
defines an equivalent probability measure $\mathbb{Q}$ under which Y has the same distribution as $Y + \theta_\mu$ under $\mathbb{P}$.

- **(Change of mean and variance)** *More generally, for any $\theta_\mu \in \mathbb{R}^d$ and any square matrix θ_σ of size d, the likelihood*
$$L = \frac{1}{|\det.(\theta_\sigma)|} \frac{p(\theta_\sigma^{-1}(Y - \theta_\mu))}{p(Y)}$$

defines a probability measure equivalent to $\mathbb{Q}$ under which Y has the same distribution as $\theta_\sigma Y + \theta_\mu$ under $\mathbb{P}$.

PROOF:
In the general case, we remark that by the example (3.4.2), the distribution of Y under $\mathbb{Q}$ has the density $q(y) = \frac{1}{|\det.(\theta_\sigma)|} p(\theta_\sigma^{-1}(y - \theta_\mu))$, which leads to the conclusion. $\square$

In the case where Y is a standard Gaussian random variable in dimension 1 ($p(y) = \frac{1}{\sqrt{2\pi}} e^{-\frac{y^2}{2}}$), the result takes the following form.

Corollary 3.4.9 (Gaussian one-dimensional variable) Let $Y \stackrel{\text{law}(\mathbb{P})}{=} \mathcal{N}(0,1)$.

- **(Change of mean)** For all $\theta_\mu \in \mathbb{R}$, the likelihood

$$L = \exp(\theta_\mu Y - \frac{1}{2}\theta_\mu^2)$$

leads to $Y \stackrel{\text{law}(\mathbb{Q})}{=} \mathcal{N}(\theta_\mu, 1)$.

- **(Change of mean and variance)** For all $\theta_\mu \in \mathbb{R}$ and $\theta_\sigma > 0$, the likelihood

$$L = \frac{1}{\theta_\sigma} \exp\left(\frac{1}{2}(1 - \frac{1}{\theta_\sigma^2})Y^2 + \frac{Y\theta_\mu}{\theta_\sigma^2} - \frac{\theta_\mu^2}{2\theta_\sigma^2}\right)$$

leads to $Y \stackrel{\text{law}(\mathbb{Q})}{=} \mathcal{N}(\theta_\mu, \theta_\sigma^2)$.

The changes of probability measure with a log-likelihood that is a polynomial of order 2 preserve the Gaussian distribution. This is again true for multidimensional Gaussian vectors. We establish this for the change of mean only.

Proposition 3.4.10 (multidimensional Gaussian variable)
Let $(Y_1, \ldots, Y_d, Z)$ be a Gaussian vector of dimension $d+1$. Under the probability measure $\mathbb{Q}$, having the density

$$L = \exp\left(Z - \mathbb{E}_\mathbb{P}(Z) - \frac{1}{2}\mathrm{Var}_\mathbb{P}(Z)\right)$$

with respect to $\mathbb{P}$, the vector $(Y_1, \ldots, Y_d)$ is Gaussian, with the same covariance matrix under $\mathbb{P}$ and under $\mathbb{Q}$, and with the expectation vector

$$\mathbb{E}_\mathbb{Q}[Y_i] = \mathrm{Cov}_\mathbb{P}(Y_i, Z) + \mathbb{E}_\mathbb{P}[Y_i].$$

PROOF:
Instead of working with densities as in the previous proofs, we efficiently use the Laplace transform of the linear combinations of the random variables. Setting $Y = \sum_{i=1}^{d} a_i Y_i$, which with Z forms a Gaussian vector, we have

$$\mathbb{E}_\mathbb{Q}[\exp(\sum_{i=1}^{d} a_i Y_i)]$$
$$= \mathbb{E}_\mathbb{P}[\exp(Z - \mathbb{E}_\mathbb{P}(Z) - \frac{1}{2}\mathbb{V}\mathrm{ar}_\mathbb{P}(Z))\exp(Y)]$$
$$= \exp[\mathbb{E}_\mathbb{P}(Y+Z) - \mathbb{E}_\mathbb{P}(Z) + \frac{1}{2}(\mathbb{V}\mathrm{ar}_\mathbb{P}(Y+Z) - \mathbb{V}\mathrm{ar}_\mathbb{P}(Z))]$$
$$= \exp[\mathbb{E}_\mathbb{P}(Y) + \mathbb{C}\mathrm{ov}_\mathbb{P}(Y,Z) + \frac{1}{2}\mathbb{V}\mathrm{ar}_\mathbb{P}(Y)].$$

The Laplace transform of the vector $(Y_1, \ldots, Y_d)$ under the probability measure $\mathbb{Q}$ is that of the Gaussian vector with the same covariance matrix as under $\mathbb{P}$, and with expectation given by $\mathbb{E}_\mathbb{Q}[\sum_{i=1}^{d} a_i Y_i] = \sum_{i=1}^{d} a_i (\mathbb{E}_\mathbb{P}[Y_i] + \mathbb{C}\mathrm{ov}_\mathbb{P}(Y_i, Z))$. □

y	Exact value of $\mathbb{E}X$ $= \mathcal{N}(-y)$	Empirical mean (simple MC)	Half-width Confid. Interv. (at 95%)	Empirical mean (Import. Samp.)	Half-width Confid. Interv. (at 95%)	Ratio of variances
1	1.59E-1	1.48E-1	6.96E-3	1.58E-1	3.74E-3	3.46
2	2.28E-2	2.12E-2	2.82E-3	2.24E-2	6.77E-4	17.4
3	1.35E-3	1.70E-3	8.07E-4	1.34E-3	4.84E-5	279
4	3.17E-5	0.00E+0	0.00E+0	3.24E-5	1.34E-6	∞
5	2.87E-7	0.00E+0	0.00E+0	2.90E-7	1.35E-8	∞

Table 3.1 Comparison of half-width of confidence intervals of the simple Monte-Carlo method and of that with importance sampling; for the second, the mean of the Gaussian distribution is adjusted to y.

Example 3.4.11 Consider the computation of $\mathbb{E}X$ for $X = 1_{Y \geq y}$, with $Y \stackrel{d}{=} \mathcal{N}(0,1)$. We compare the estimation by the simple Monte-Carlo method and the importance sampling Monte-Carlo method. The results are obtained with 10,000 simulations. Under the new probability measure, Y is distributed as $\mathcal{N}(y,1)$ so that $\mathbb{Q}(Y \geq y) = \frac{1}{2}$. The numerical values are presented in Table 3.1. Beyond $y = 4$ the simple

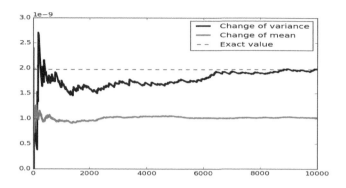

Figure 3.1 Monte-Carlo evaluation of $\mathbb{P}(|Y| \geq 6)$ by a naive method and by importance sampling changing the mean $\theta_\mu = 6$ or the variance $\theta_\sigma = 9$: only the empirical means are represented.

Monte-Carlo estimator often returns 0 as none of the 10,000 simulations exceeds y: obviously the estimation of $\mathbb{E}X$ is completely wrong. In the last column, the ratio of the variances is given, providing in this way the factor of improvement in the computational time — for a given expected accuracy — by using the importance sampling scheme rather than the simple method. The greater is y, the rarer is the event and the more significant is the gain; this is a general phenomenon commonly observed when the importance sampling method is compared to the simple method.

In the previous example, increasing the mean of Y gives a good result, because, due to the form of $X = 1_{Y \geq y}$, we need to sample very large positive values of Y. Now let us consider $X = 1_{|Y| \geq y}$, with a special focus on the very positive or very negative values of Y. Changing the mean does not give good results at all. Even worse, taking $y = 6$ gives the impression of convergence but it evaluates only half of the distribution (because $\mathbb{Q}(Y \leq -6) = \mathbb{P}(Y \leq -12)$ is extremely small). In Figure 3.1, we plot the empirical means. In this case, increasing the variance of Y (with $\theta_\sigma = 9$) works very well. This illustrates that the choice of a relevant change of probability measure requires that we first carefully analyze the considered problem.

3.4.3 Change of probability measure by Esscher transform

A random variable Y with exponential moments allows us to easily define the change of probability, known as the Esscher transform.

Definition and Proposition 3.4.12 *Let Y be a real-valued random variable such that $\mathbb{E}(e^{r|Y|}) < +\infty$ for a certain $r > 0$. The moment-generating function of Y is defined by*

$$M : \theta \in \mathbb{R} \mapsto \mathbb{E}(e^{\theta Y}) > 0.$$

It is finite on an open interval I containing 0:

- *M is C^∞ on I, with $M^{(k)}(\theta) = \mathbb{E}(Y^k e^{\theta Y})$;*
- *its logarithm $\Gamma(\theta) = \log(M(\theta))$ is convex.*

These properties are classic, and we leave their proofs to the reader. From this we deduce the following.

Corollary 3.4.13 *Let Y be as in Definition 3.4.12. For $\theta \in I$, set $L = e^{\theta Y - \Gamma(\theta)}$: L defines an equivalent probability measure $\mathbb{Q}_\theta$ under which Y is the random variable having the moment-generating function*

$$\mathbb{E}_{\mathbb{Q}_\theta}(e^{zY}) = e^{\Gamma(\theta+z) - \Gamma(\theta)}.$$

Its expectation and its variance under $\mathbb{Q}_\theta$ are, respectively,

$$\mathbb{E}_{\mathbb{Q}_\theta}(Y) = \Gamma'(\theta), \qquad \text{Var}_{\mathbb{Q}_\theta}(Y) = \Gamma''(\theta).$$

PROOF:
By the definition of Γ, L is centered at 1 under $\mathbb{P}$; in addition, $\mathbb{E}_{\mathbb{Q}_\theta}(e^{zY}) = \mathbb{E}_{\mathbb{P}}(e^{\theta Y - \Gamma(\theta)} e^{zY}) = e^{\Gamma(\theta+z) - \Gamma(\theta)}$. Second, the moment-generating function uniquely characterizes the distribution of the random variable. Finally, $\mathbb{E}_{\mathbb{Q}_\theta}(Y) = \mathbb{E}_{\mathbb{P}}(Y e^{\theta Y - \Gamma(\theta)}) = \frac{M'(\theta)}{M(\theta)} = \Gamma'(\theta)$ and $\text{Var}_{\mathbb{Q}_\theta}(Y) = \mathbb{E}_{\mathbb{P}}(Y^2 e^{\theta Y - \Gamma(\theta)}) - [\frac{M'(\theta)}{M(\theta)}]^2 = \frac{M''(\theta) M(\theta) - [M'(\theta)]^2}{M^2(\theta)} = \Gamma''(\theta)$.
□

In the case where Y is Gaussian, we recover the likelihood changing the mean (Corollary 3.4.9). Here are some other examples that are easy to verify:

- **(Exponential distribution $\mathcal{E}xp(\lambda)$):** we have $\Gamma(\theta) = \ln(\frac{\lambda}{\lambda - \theta}) \mathbf{1}_{\theta < \lambda} + \infty \mathbf{1}_{\theta \geq \lambda}$ and under $\mathbb{Q}_\theta$ (for $\theta < \lambda$), $Y \stackrel{d}{=} \mathcal{E}xp(\lambda - \theta)$.

108 ■ Variance reduction

– **(Poisson distribution $\mathcal{P}(\lambda)$):** we have $\Gamma(\theta) = \lambda(e^\theta - 1)$ and under $\mathbb{Q}_\theta$, Y has again the Poisson distribution with the parameter λe^θ.

In addition, if we want Y to have a given mean μ (respectively, a given variance σ^2) under the new probability measure, we simply have to choose θ so that $\Gamma'(\theta) = \mu$ (respectively, $\Gamma''(\theta) = \sigma^2$).

The next example is devoted to a random sum of independent random variables, which serves as a basis for the processes with jumps (compound Poisson process).

Definition and Proposition 3.4.14 (random sum of random variables) *Let N be a Poisson random variable with the parameter λ and let $(Z_n)_{n \geq 1}$ be a sequence of independent random variables with the same distribution as Z, and having exponential moments. N and $(Z_n)_{n \geq 1}$ are supposed independent. Then the moment-generating function of $Y = \sum_{n=1}^{N} Z_n$ is given by*

$$\mathbb{E}(e^{\theta Y}) = e^{\lambda \mathbb{E}(e^{\theta Z} - 1)}, \qquad \forall \theta \in \mathbb{R}. \tag{3.4.6}$$

PROOF:
Using the independence of N and $(Z_n)_{n \geq 1}$, $\mathbb{E}(e^{\theta Y})$ is equal to

$$\sum_{k \geq 0} e^{-\lambda} \frac{\lambda^k}{k!} \mathbb{E}(e^{\theta \sum_{n=1}^{N} Z_n} | N = k) = \sum_{k \geq 0} e^{-\lambda} \frac{\lambda^k}{k!} \mathbb{E}(e^{\theta \sum_{n=1}^{k} Z_n})$$

$$= \sum_{k \geq 0} e^{-\lambda} \frac{[\lambda \mathbb{E}(e^{\theta Z})]^k}{k!}.$$

□

As in Corollary 3.4.13 we can easily deduce explicit changes of probability measure, by playing with the parameter of N and the distribution of Z. This offers great flexibility. This shows that, as in the Gaussian models, the models based on the Poisson distribution can be stable under the change of probability measure.

Proposition 3.4.15 *Let $Y = \sum_{n=1}^{N} Z_n$ with N and $(Z_n)_{n \geq 1}$ as in Definition 3.4.14. Let $\varphi : \mathbb{R} \mapsto \mathbb{R}$ be a measurable function with a linear growth at most.*[4] *Set*

$$L = \exp\left(\sum_{n=1}^{N} \varphi(Z_n) - \lambda \mathbb{E}_\mathbb{P}(e^{\varphi(Z)} - 1)\right):$$

---

[4] $\sup_{x \in \mathbb{R}} \frac{|\varphi(x)|}{1+|x|} < +\infty.$

L defines a probability measure $\mathbb{Q}$ equivalent to $\mathbb{P}$ under which Y has the same distribution as

$$Y' = \sum_{n=1}^{N'} Z'_n$$

with

- N' and $(Z'_n)_{n\geq 1}$ are independent,
- N' is a random variable having the Poisson distribution with parameter $\lambda' = \lambda \mathbb{E}_\mathbb{P}(e^{\varphi(Z)})$,
- the random variables $(Z'_n)_{n\geq 1}$ are independent and with the same distribution as Z' with $\mathbb{Q}(Z' \in \mathrm{d}z) = \frac{e^{\varphi(z)}}{\mathbb{E}_\mathbb{P}(e^{\varphi(Z)})}\mathbb{P}(Z \in \mathrm{d}z)$.

PROOF:

L is a well-defined likelihood as it is positive with mean 1 under $\mathbb{P}$, using the equality (3.4.6) with $\theta = 1$ and $\varphi(Z_n)$ instead of Z_n (also having exponential moments by the hypothesis on φ).

Then let us calculate the moment-generating function of Y under $\mathbb{Q}$: applying again the equality (3.4.6), we have

$$\mathbb{E}_\mathbb{Q}(e^{\theta Y}) = \mathbb{E}_\mathbb{P}(L e^{\theta Y}) = \mathbb{E}_\mathbb{P}\left(\exp\left(\sum_{n=1}^{N}(\varphi(Z_n) + \theta Z_n) - \lambda \mathbb{E}_\mathbb{P}(e^{\varphi(Z)} - 1)\right)\right)$$

$$= \exp\left(\lambda \mathbb{E}_\mathbb{P}(e^{\theta Z + \varphi(Z)} - 1) - \lambda \mathbb{E}_\mathbb{P}(e^{\varphi(Z)} - 1)\right)$$

$$= \exp\left(\lambda' \mathbb{E}_\mathbb{P}\left(\frac{e^{\theta Z + \varphi(Z)}}{\mathbb{E}_\mathbb{P}(e^{\varphi(Z)})} - 1\right)\right) = \exp\left(\lambda' \int (e^{\theta z} - 1)\mathbb{Q}(Z' \in \mathrm{d}z)\right).$$

We conclude the proof by identifying the law of Y under $\mathbb{Q}$, owing to the form (3.4.6). □

Example 3.4.16 Let $X = 1_{Y \geq 6}$ with Y following the Poisson distribution with parameter $\lambda = 1$. An explicit calculation shows that $\mathbb{E}X = 0.000594$. A simple Monte-Carlo procedure gives, with 1000 simulations, a confidence interval - at 95 % - equal to $[-0.00096, 0.00296]$. Using an importance sampling method based on Proposition 3.4.15 with $\lambda' = 6$, the confidence interval becomes $[0.000507, 0.000647]$. The ratio of variances is approximately 800, representing an equivalent gain in the computational time.

3.4.4 Adaptive methods

Once the type of change of probability measure is chosen according to the user's expertise, nonetheless it remains to adjust several parameters to get a numerically efficient algorithm. We can obviously tune them by hand but an automatic tuning is beneficial for robustness issues. This is a question which motivates current research and we present an approach based on [82], valid in the Gaussian case with a change of the mean.

We want to evaluate $\mathbb{E}X$ with $X = f(Y_1, \ldots, Y_d)$, i.e., a function of a Gaussian $Y = (Y_1, \ldots, Y_d)$, with components possibly independent. The change of probability measure is given by Proposition 3.4.10: this requires us to specify a variable Z that forms with $Y_1, \ldots, Y_d$ a Gaussian vector. For example:

- If X depends mainly on the sum $Y_1 + \cdots + Y_d$, it can be relevant to take $Z = \theta(Y_1 + \cdots + Y_d)$ with a parameter $\theta \in \mathbb{R}$ to optimize.

- If X depends mainly on the two first components Y_1 and Y_2, the choice $Z = \theta_1 Y_1 + \theta_2 Y_2$ is reasonable, with a two-dimensional parameter $(\theta_1, \theta_2) \in \mathbb{R}^2$ to optimize.

- We can also take $Z = \sum_{i=1}^{d} \theta_i Y_i = \theta \cdot Y$ if we prefer not to account for specific relations.

We encompass all these different cases in the notation $Z = \theta \cdot W$, with $\theta \in \mathbb{R}^{d'}$ and W a random vector of dimension d' such that (Y, W) is a Gaussian vector.

Without loss of generality, we suppose that W is centered, otherwise we replace W by $W - \mathbb{E}_\mathbb{P}(W)$. If we denote by K the covariance of W, the likelihood of $\mathbb{Q}_\theta$ with respect to $\mathbb{P}$ is written as

$$L_\theta = \exp\left(\theta \cdot W - \frac{1}{2}\theta \cdot K\theta\right).$$

From Proposition 3.4.10, under $\mathbb{Q}_\theta$ the vector (Y, W) is Gaussian with an unchanged covariance and a modified mean

$$\begin{cases} \mathbb{E}_{\mathbb{Q}_\theta}(Y) = (\mathbb{E}_{\mathbb{Q}_\theta}(Y_i))_i = (\mathbb{E}_\mathbb{P}(Y_i) + \underbrace{\mathbb{E}(Y_i W^\mathsf{T})}_{:=\mu_i}\theta)_i, \\ \mathbb{E}_{\mathbb{Q}_\theta}(W) = \mathbb{E}_\mathbb{P}(W) + K\theta, \end{cases}$$

writing $\mu = \mathbb{E}(YW^\mathsf{T})$. We suppose that this matrix of size $d \times d'$ is known. Thus to simulate (Y, W) under $\mathbb{Q}_\theta$, it is enough to simulate (Y, W) under $\mathbb{P}$ and then to add $(\mu\theta, K\theta)$: we use this remark in the following.

The standard Monte-Carlo estimator is $\dfrac{1}{M}\sum_{m=1}^{M} f(Y_m)$ whereas the importance sampling Monte-Carlo estimator is (see (3.4.3))

$$I^{\text{Imp.Samp.}}_{\mathbb{Q}, M} = \frac{1}{M}\sum_{m=1}^{M} \exp\Big(-\theta \cdot W_m - \frac{1}{2}\theta \cdot K\theta\Big) f(Y_m + \theta\mu)$$

where $(Y_m, W_m)_{m \geq 1}$ have the same distribution as (Y, W) under $\mathbb{P}$. The main problem is still the choice of the parameter θ minimizing the variance of $L^{-1}X$ under $\mathbb{Q}_\theta$. From (3.4.4), θ minimizes

$$v(\theta) = \mathbb{E}_\mathbb{P}(L_\theta^{-1}|X|^2) = \mathbb{E}_\mathbb{P}\Big(\exp\Big(-\theta \cdot W + \frac{1}{2}\theta \cdot K\theta\Big) f^2(Y)\Big),$$

where to avoid the problems with integrability, we suppose that $\mathbb{E}(f^2(Y)e^{\theta \cdot W}) < +\infty$ for any θ.

Performing a little differential calculus shows that v is then $\mathcal{C}^2$ and its Hessian is equal to

$$\nabla^2 v(\theta) = \mathbb{E}\Big((K + (K\theta - W)(K\theta - W)^\mathsf{T})$$
$$\times \exp\Big(-\theta \cdot W + \frac{1}{2}\theta \cdot K\theta\Big) f^2(Y)\Big) \geq 0.$$

The function v is thus convex, in fact strictly convex as soon as $\mathbb{P}(f(Y) \neq 0) > 0$, which we now suppose. In fact, we can even show in certain cases — namely $W = Y$ — that v is strongly convex. We note that this reasoning does not require any hypothesis on the regularity of f. The strong convexity implies the existence of a unique θ^* which minimizes the variance of the importance sampling Monte-Carlo estimator.

To determine it numerically, it is natural to minimize the empirical version of the function $v(\theta)$:

$$v_M(\theta) := \frac{1}{M}\sum_{m=1}^{M} \exp\Big(-\theta \cdot W_m + \frac{1}{2}\theta \cdot K\theta\Big) f^2(Y_m).$$

By the same arguments as before, we verify that v_M is convex in θ and an optimization procedure of Newton type enables us to find θ_M^*. Additional work is necessary to show that the empirical quantities converge a.s. to the expected limits ($\theta_M^* \xrightarrow[M \to +\infty]{\text{a.s.}} \theta^*$ and $v_M(\theta_M^*) \xrightarrow[M \to +\infty]{\text{a.s.}} v(\theta^*)$), with some additional controls of fluctuations at rate $\sqrt{M}$ (central limit theorem); for the details of this analysis, we refer the reader to [82].

3.5 EXERCISES

Exercise 3.1 (antithetic sampling) *Describe carefully how to define the confidence intervals of the antithetic estimator $I_{2,M}$, where the standard deviation is computed on the sample.*

Exercise 3.2 (antithetic sampling, Cauchy distribution) *Let Y be a standard Cauchy variable. Write a simulation program to compare the antithetic transformation $Y \to -Y$ and the semi-antithetic transformation $Y \to 1/Y$ for the computations of $\mathbb{E}(f(Y))$ in the three cases*

i) $f(y) = \sin(y)$,

ii) $f(y) = \cos(y)$,

iii) $f(y) = (y)_+^{1/4}$.

In each case, discuss and explain the possible variance improvement in comparison with the standard procedure.

Exercise 3.3 (stratification, optimal allocation)

i) Prove that the optimal allocation is indeed given by $M_j^* = M \frac{p_j \sigma_j}{\sum_{i=1}^k p_i \sigma_i}$.

ii) For such a choice, derive a central limit theorem for the estimator $I_{M_1,\ldots,M_k}^{\text{strat.,opt.alloc.}}$ as $M \to +\infty$.

Exercise 3.4 (stratification of Gaussian vectors) *We aim at describing in detail the generation of random variables in steps 1 and 2 of Example 3.2.2. We assume that Y is a standard d-dimensional Gaussian vector and that β is normalized to 1, i.e. $|\beta| = 1$.*

i) What is the distribution of $Z = \beta \cdot Y$ and its parameters?

ii) *Assume that* $\mathcal{S}_j = [-x_{j-1}, x_j)$ *with* $-\infty := x_0 < \cdots < x_j < \cdots < x_k := +\infty$. *Compute* $p_j = \mathbb{P}(Z \in \mathcal{S}_j)$ *as a function of the* $(x_i)_i$. *Derive the cumulative distribution function of* Z *given* $\{Z \in \mathcal{S}_j\}$. *Deduce an algorithm to generate such a distribution (assuming that* $\mathcal{N}^{-1}(\cdot)$ *is known).*

iii) *Compute explicitly the distribution of* $Y - \beta Z$. *Show that it is independent on* Z.

iv) *Deduce a generation scheme of* Y *given* $\{Z \in \mathcal{S}_j\}$.

Exercise 3.5 (control variates) *Establish a central limit theorem for the control variate estimator* $I^{\text{Cont.Var.}}_{\beta^*_M, M}$ *that uses the optimal empirical weights* β^*_M *defined in* (3.3.2).

Check that the limit variance is $\text{Var}(X - Z(\beta^*))$.

Exercise 3.6 (importance sampling, Gaussian vectors) *Extend the formula of Corollary 3.4.9 to the multidimensional case, by changing both the mean and the covariance.*

Exercise 3.7 (Esscher transform, Gaussian and exponential distributions) *In Proposition 3.4.15, explicitly characterize the distributions of* Z *under* $\mathbb{Q}$, *when we take* $\varphi(z) = z$ *and when under* $\mathbb{P}$, Z *has either a Gaussian distribution or an exponential one.*

Exercise 3.8 (importance sampling, Poisson distribution) *Write a simulation program for computing* $\mathbb{P}(Y \geq x)$, *by importance sampling, when* Y *has a Poisson distribution with parameter 1 and* x *is large (see Example 3.4.16).*

PART B: SIMULATION OF LINEAR PROCESSES

CHAPTER 4

Stochastic differential equations and Feynman-Kac formulas

In this chapter, we make explicit the relations between stochastic processes and partial differential equations (PDE): namely, the solution to a PDE is written as an expectation of a functional of the process. We refer to *probabilistic representations* for partial differential equations, or *Feynman-Kac formulas*.

Historically, the first studied example is the heat equation, which is connected to Brownian motion. This random process, whose trajectories are very erratic, was described first by Robert Brown in 1827, then studied in other contexts by Louis Bachelier in 1900 and Albert Einstein in 1905. Einstein used it to describe the random movement of a physical particle that diffuses, and found that its probability density of being at position x at time t is the Gaussian density, which is called fundamental solution of the heat equation. Later, Mark Kac (1914–1984) extended this relation to functionals of Brownian trajectories, instead of functions of the position at a given time, see [84]: in fact, this new point of view echoed the PhD work by Richard Feynman[1] in 1942, about the Schrödinger equation in quantum mechanics, where he introduced path integrals. For an introductory presentation, see [86].

With the spectacular development of stochastic tools during the second half of the 20th century, this connection has been extended to

---

[1] (1918–1988), Nobel Prize in physics in 1965.

more stochastic processes and more partial differential equations, under the generic name of the *Feynman-Kac formulas*. This representation of the solution to partial differential equations as an expectation of a functional of a stochastic process can take various forms: see for instance the monographs by Friedmann [45, 46], Durrett [33], Freidlin [43] or the books [27], [88], [1]. This is the starting point for fructuous developments. This double point of view — analysis and probability — is very useful

- to solve theoretical questions by giving two sets of tools: the usual observation shows that these two approaches are very complementary;

- to tackle numerical issues by providing two ways of resolution that are quite different (Monte-Carlo simulation or discretization of partial differential equation): each one has advantages, specific to the situations at hand.

In this text — and this is obviously incomplete — we emphasize the approach with *probabilistic algorithms*, enabling us to provide a numerical solution to partial differential equation.

The plan of this chapter is the following. We start with Brownian motion, linked with the heat equation. Brownian motion is used as a building block for more complex stochastic models. In the presentation, we develop the properties that are the most useful for simulating or approximating Brownian motion. Then we introduce the stochastic differential equations to treat a more general class of linear partial differential equations. This extension requires the use of *stochastic calculus*, where we point out the main results without proof: for more details, the reader may consult the standard works on the subject [88] or [127].

Regarding applications, the stochastic differential equations serve as building block for numerous models in biology [101], chemistry [95], dynamics and genetics of populations [39], finance and economy [119], random mechanics [96], physics [27] ... other references will given throughout the chapter. This part **B** is dedicated to linear problems, while nonlinear processes are studied in part **C**.

4.1 Brownian motion

4.1.1 A brief history

Brownian motion was first described in 1827 by the English botanist Robert Brown as he observed the erratic movement of fine organic particles in suspension in a gas or a fluid. In 1900, Louis Bachelier introduced Brownian motion to model the random changes in stock prices at the stock exchange. In 1905, Albert Einstein found the Brownian motion by building a probabilistic model to describe the movement of a diffusive particle. The rigorous construction of Brownian motion is due to Norbert Wiener in 1923. We give his proof here.

4.1.2 Definition

The Brownian motion can be seen as the limit of symmetric random walks defined by $S_n = \sum_{i=1}^{n} X_i$ with independent random variables $(X_i)_i$ with Rademacher distribution $\mathbb{P}(X_i = \pm 1) = \frac{1}{2}$. As the random variable X_i is centered and with unit variance, an application of the central limit theorem proves that $\frac{1}{\sqrt{n}} S_n \underset{n \to +\infty}{\Longrightarrow} \mathcal{N}(0,1)$. We can look at the convergence at several large times n and by doing so, we obtain a stochastic process: for this, we have to renormalize in time and space $(S_n)_{n \geq 1}$ by setting

$$W_t^n = \frac{1}{\sqrt{n}} \sum_{i=1}^{\lfloor nt \rfloor} X_i. \quad (4.1.1)$$

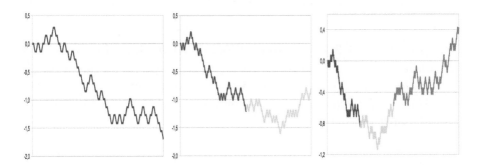

Figure 4.1 The random walk after renormalization in time and space. From left to right: the processes W^n for $n = 50, 100, 200$. The parts with the same color are generated with the same X_i.

120 ■ Stochastic differential equations and Feynman-Kac formulas

Defined as above, W^n is piece-wise constant: to make it continuous, we interpolate it linearly between times of the form i/n. The central limit theorem proves once again that $W_t^n \underset{n\to+\infty}{\Longrightarrow} \mathcal{N}(0,t)$. Moreover, the independence property and the common distribution of $(X_i)_i$ enable to show that the time increments of W^n are asymptotically independent and with limit distribution $W_t^n - W_s^n \underset{n\to+\infty}{\Longrightarrow} \mathcal{N}(0, t-s)$ (for $t > s$). These arguments constitute the core of the proof of the *Donkser theorem* (see Breiman [19]), which states the convergence in distribution of the process W^n toward a limit process called *Brownian motion*. We do not go into details of this convergence in distribution at the level of the stochastic process and we simply take the limiting characteristics as a definition for Brownian motion.

Definition 4.1.1 (of Brownian motion in dimension 1) *A standard Brownian motion in dimension 1 is a continuous-time stochastic process $\{W_t; t \geq 0\}$ with continuous paths, such that*

- $W_0 = 0$;

- *the time increment $W_t - W_s$ ($0 \leq s < t$) has the Gaussian distribution with zero mean and variance $(t-s)$;*

- *for any $0 = t_0 < t_1 < t_2..... < t_n$, the increments $\{W_{t_{i+1}} - W_{t_i}; 0 \leq i \leq n-1\}$ are independent.*

The property of independent increments induces that W is a Markov process. One can also define Brownian motion as a Gaussian process,[2] and the proof of the equivalence between both definitions is left to the reader.

Proposition 4.1.2 (Brownian motion as a Gaussian process) *A continuous-time stochastic process $(W_t)_{t\geq 0}$ with continuous paths is a Brownian motion if and only if it is a Gaussian process centered ($\mathbb{E}(W_t) = 0$ for any $t \geq 0$) with covariance function $\mathrm{Cov}(W_t, W_s) = \min(s,t)$ for any $s,t \geq 0$.*

Because W_t has the Gaussian distribution with zero mean and standard deviation equal to $\sqrt{t}$, we obtain $|W_t| \leq 1.96\sqrt{t}$ with probability

---

[2]$(X_t)_t$ is a Gaussian process if any linear combination $\sum_i a_i X_{t_i}$ has a Gaussian distribution; $(X_t)_t$ is characterized by its mean function $t \mapsto \mathbb{E}(X_t)$ and covariance function $(t,s) \mapsto \mathrm{Cov}(X_t, X_s)$, [127, Chapter 1].

95% for a given time t: this shows the diffusive behavior of Brownian motion, a property that is also satisfied in general by subsequent stochastic differential equations.

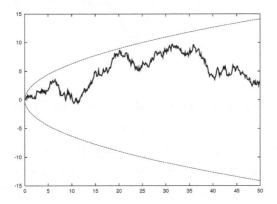

Figure 4.2 A Brownian path and the two curves $f(t) = \pm 1.96\sqrt{t}$.

Theorem 4.1.3 *Brownian motion exists!*

PROOF:
There are several proofs of existence of Brownian motion. Here, we follow the constructive and very explicit approach by Wiener, an approach that is sometimes used in numerical schemes. Set

$$W_t = \frac{t}{\sqrt{\pi}} G_0 + \sqrt{\frac{2}{\pi}} \sum_{m \geq 1} \frac{\sin(mt)}{m} G_m$$

where $(G_m)_{m \geq 0}$ is a sequence of independent standard Gaussian random variables $\mathcal{N}(0, 1)$: we show below that it is a standard Brownian motion on the interval $[0, \pi]$, by using the characterization of Brownian motion as a Gaussian process. To build a Brownian motion on $\mathbb{R}^+$, it is then enough to put such processes end to end: on $[i\pi, (i+1)\pi]$, we set $W_t = W_{i\pi} + \frac{(t-i\pi)}{\sqrt{\pi}} G_0^{(i)} + \sqrt{\frac{2}{\pi}} \sum_{m \geq 1} \frac{\sin(mt)}{m} G_m^{(i)}$ where $(G_m^{(i)})_{i,m \geq 0}$ are other independent standard Gaussian random variables.

On several occasions in the following, we use the following property (see Lemma A.1.4).

Lemma 4.1.4 *Suppose that the real-valued random variable $X_n \stackrel{d}{=} \mathcal{N}(\mu_n, \sigma_n^2)$ converges in distribution. Then the sequence of parameters $(\mu_n, \sigma_n^2)_n$ converges and the limit distribution is Gaussian with parameters $(\lim_n \mu_n, \lim_n \sigma_n^2)$.*

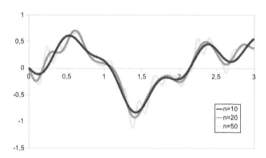

Figure 4.3 Wiener approximation W^n for $n = 10, 20, 50$.

Set $W_t^{(n)} = \frac{t}{\sqrt{\pi}} G_0 + \sqrt{\frac{2}{\pi}} \sum_{1 \leq m \leq n} \frac{\sin(mt)}{m} G_m$, which is written as a sum of independent random variables that has a uniformly (in n) bounded variance since $\sum_{m \geq 1} \text{Var}(\frac{\sin(mt)}{m} G_m) < +\infty$: similar to (2.1.2), we show that $W_t^{(n)}$ converges a.s. (and thus in distribution) as $n \to +\infty$. Moreover, $W_t^{(n)}$ has a centered Gaussian distribution. By Lemma 4.1.4, W_t has a centered Gaussian distribution. Finally, $W^{(n)}$ is a Gaussian process since any linear combination $\sum_i a_i W_{t_i}^{(n)}$ has the Gaussian distribution; therefore at the limit $\sum_i a_i W_{t_i}$ also has the Gaussian distribution, i.e., W is a Gaussian process. The covariance function $\text{Cov}(W_t, W_s)$ can be computed as the limit of covariances $\text{Cov}(W_t^{(n)}, W_t^{(s)})$, i.e.,

$$\text{Cov}(W_t, W_s) = \frac{ts}{\pi} + \frac{2}{\pi} \sum_{m \geq 1} \frac{\sin(mt)}{m} \frac{\sin(ms)}{m}.$$

The above series is equal to $\min(s,t)$ for (s,t) in $[0, \pi]$: this can be shown by computing the Fourier coefficients of the function $t \in [-\pi, \pi] \mapsto \min(s,t)$ (s fixed).

To conclude, it remains to establish that $(W_t)_{0 \leq t \leq \pi}$ built as above is continuous: this is justified by proving the uniform convergence of $\sum_{0 \leq m \leq n} \frac{\sin(mt)}{m} G_m$ for a *certain sequence* $n \to +\infty$. Consider directly the complex-valued series and set $\mathcal{S}_{n,p} := \sup_{t \leq \pi} |\sum_{n \leq m \leq n+p} \frac{e^{imt}}{m} G_m|$. We have

$$\mathcal{S}_{n,p}^2 = \sup_{t \leq \pi} \left| \sum_{n \leq m, m' \leq n+p} \frac{e^{-imt}}{m} \frac{e^{im't}}{m'} G_m G_{m'} \right|$$

$$\leq \sum_{n \leq m \leq n+p} \frac{G_m^2}{m^2} + 2 \sup_{t \leq \pi} \left| \sum_{1 \leq l \leq p} \sum_{n \leq m \leq n+l-1} e^{ilt} \frac{G_m G_{m+l}}{m(m+l)} \right|$$

$$\leq \sum_{n \leq m \leq n+p} \frac{G_m^2}{m^2} + 2 \sum_{1 \leq l \leq p} \left| \sum_{n \leq m \leq n+l-1} \frac{G_m G_{m+l}}{m(m+l)} \right|,$$

$$\mathbb{E}(S_{n,p}^2) \leq \sum_{n\leq m\leq n+p} \frac{1}{m^2} + 2\sum_{1\leq l\leq p} \sqrt{\mathbb{E}\Big(\sum_{n\leq m\leq n+l-1} \frac{G_m G_{m+l}}{m(m+l)}\Big)^2}$$

$$= \sum_{n\leq m\leq n+p} \frac{1}{m^2} + \sum_{1\leq l\leq p} \sqrt{\sum_{n\leq m\leq n+l-1} \frac{1}{m^2(m+l)^2}}.$$

It readily follows that $\mathbb{E}(S_{n,p}^2) \leq C(n^{-1} + pn^{-3/2})$ for some constant C. Thus, we obtain that $\mathbb{E}(\sum_{k\geq 1} S_{2^k, 2^k - 1}) \leq \sum_{k\geq 1} \sqrt{C(2^{-k} + 2^{-k/2})} < +\infty$, which shows that, a.s., the series of functions with general term $(\sum_{2^k \leq m \leq 2^{k+1}-1} \frac{\sin(mt)}{m} G_m)_{k\geq 1}$ is uniformly convergent on $[0,\pi]$: as a consequence, we get the continuity of the limit, and thus that of W. □

What is the application regarding the simulation of Brownian motion? The previous proof has the advantage of providing a convergent, explicit, and simple approximation of Brownian motion on $[0, \pi]$:

$$W_t^{(n)} = \frac{t}{\sqrt{\pi}} G_0 + \sqrt{\frac{2}{\pi}} \sum_{1\leq m\leq n} \frac{\sin(mt)}{m} G_m.$$

It has the advantage of being quickly computable at any points, up to the generation of n independent standard Gaussian random variables. On the other hand, the convergence in L_2 is of order $n^{-1/2}$ (the square root of the remainder of the series $\sum_{m\geq n} m^{-2}$): thus, the truncation of the series must be made with n large to expect an accurate approximation. Moreover, for all n the approximation $W^{(n)}$ is $\mathcal{C}^\infty$ (see Figure 4.3), whereas the Brownian motion is nowhere differentiable, monotonic on any interval (Proposition 4.1.5). Thus, the irregularity of the Brownian trajectory is badly translated by the Wiener approximation, which is a major drawback in some problems — for example for the simulation of extrema of the Brownian motion.

Proposition 4.1.5 (absence of monotonicity) *We have*

$$\mathbb{P}(t \mapsto W_t \text{ is monotonic on an interval}) = 0.$$

PROOF:
Define $M_{s,t}^\uparrow = \{\omega : u \mapsto W_u(\omega) \text{ as non-decreasing in the interval }]s,t[\}$ and $M_{s,t}^\downarrow$ similarly. We notice that

$$M := \{\omega : t \mapsto W_t(\omega) \text{ is monotonic on an interval}\}$$
$$= \bigcup_{s,t\in\mathbb{Q}, 0\leq s<t} (M_{s,t}^\uparrow \cup M_{s,t}^\downarrow),$$

and since the union is countable, it is enough to show $\mathbb{P}(M_{s,t}^\uparrow) = \mathbb{P}(M_{s,t}^\downarrow) = 0$ in order to conclude $\mathbb{P}(M) \leq \sum_{s,t \in \mathbb{Q}, 0 \leq s < t}[\mathbb{P}(M_{s,t}^\uparrow) + \mathbb{P}(M_{s,t}^\downarrow)] = 0$. For a given n, put $t_i = s + i(t-s)/n$, then $\mathbb{P}(M_{s,t}^\uparrow) \leq \mathbb{P}(W_{t_{i+1}} - W_{t_i} \geq 0, 0 \leq i < n)$. The increments being independent and centered, we have $\mathbb{P}(M_{s,t}^\uparrow) \leq \prod_{i=0}^{n-1} \mathbb{P}(W_{t_{i+1}} - W_{t_i} \geq 0) = \frac{1}{2^n}$. The integer n being arbitrary, we obtain $\mathbb{P}(M_{s,t}^\uparrow) = 0$. The same arguments yield $\mathbb{P}(M_{s,t}^\downarrow) = 0$. □

Now we state some properties of the Brownian path: for the proofs, the reader may refer to [88] or [127]. These properties are very useful for the simulation.

Proposition 4.1.6 *Let W be a Brownian motion.*

i) SYMMETRY PROPERTY. $-W$ *is a Brownian motion.*

ii) SCALING PROPERTY. *For any $c > 0$ $\{W_t^c = c^{-1} W_{c^2 t}; t \in \mathbb{R}^+\}$ is a Brownian motion.*

iii) TIME REVERSAL. *The process with time reversed from time T, i.e. $\widehat{W}_t^T = W_T - W_{T-t}$, is a Brownian motion on $[0,T]$.*

iv) DISTRIBUTION OF THE MAXIMUM. *For any $y \geq 0$ and any $x \leq y$, we have:*

$$\mathbb{P}\left(\sup_{t \leq T} W_t \geq y, W_T \leq x\right) = \mathbb{P}(W_T \geq 2y - x), \quad (4.1.2)$$

$$\mathbb{P}\left(\sup_{t \leq T} W_t \geq y\right) = \mathbb{P}(|W_T| \geq y). \quad (4.1.3)$$

4.1.3 Simulation

Forward procedure. If the aim is to generate the skeleton $(W_{t_i})_{0 \leq i \leq n}$ for a set of predefined increasing times $0 = t_0 < \cdots < t_i < \cdots < t_n$, it is enough to compute iteratively $W_{t_{i+1}} = W_{t_i} + \sqrt{t_{i+1} - t_i}\, G_i$ for $i = 0, \ldots, n-1$ with $(G_i)_i$ independent standard Gaussian random variables.

In some algorithms, it may be necessary to add one value W at a different date t:

- If $t > t_n$, we continue the forward procedure by generating $W_t \stackrel{\mathrm{d}}{=} W_{t_n} + \sqrt{t - t_n}\, \mathcal{N}(0,1)$.

– If $t \in (t_i, t_{i+1})$, one has to insert a new value W_t accounting for the other previously generated values (refinement of path). This is the Brownian bridge technique, i.e., it consists of simulating the Brownian path when one knows some intermediate values.

The procedure relies on the following result, which proves that one has to take into account only the values W_{t_i} and $W_{t_{i+1}}$ to generate W_t for $t \in (t_i, t_{i+1})$. It is interesting for the sequel to state a slightly more general version for Brownian motion with drift μ.

Lemma 4.1.7 (Brownian bridge) *Let $\mu \in \mathbb{R}$ and set $X_t = W_t + \mu t$. Let $0 \leq u \leq v$, the distribution of $(X_t)_{u \leq t \leq v}$ conditionally on $(X_s : s \leq u; X_s : s \geq v)$ coincides with the distribution of $(X_t)_{u \leq t \leq v}$ conditionally on X_u and X_v. This conditional distribution is that of*

$$\left(X_u + (X_v - X_u)\frac{t-u}{v-u} + B_t^{u,v}\right)_{u \leq t \leq v}$$

where $(B_t^{u,v})_{u \leq t \leq v}$ is a Gaussian process, centered, with covariance function

$$\mathrm{Cov}(B_t^{u,v}, B_s^{u,v}) = \frac{(s-u)(v-t)}{(v-u)}, \quad v \geq t \geq s \geq u,$$

and independent of $(X_s : s \leq u; X_s : s \geq v)$.

Since $(B_t^{u,v})_{u \leq t \leq v}$ depends *neither on X_u, nor on X_v, nor on the drift μ*, it has the same distribution as a Brownian motion $(W_t)_{u \leq t \leq v}$ conditioned to start from $W_u = 0$ and to arrive at $W_v = 0$, i.e. a so-called *standard Brownian bridge*, between times u and v. Therefore, owing to the previous result, *a general Brownian bridge is the independent superposition of a standard Brownian bridge and of the affine function connecting (u, X_u) to (v, X_v).*

PROOF:
Let us start with the first result on conditional distributions. Let us consider the three bounded continuous functionals $\varphi_u, \varphi_{uv}, \varphi_v$ and set $\Phi_u = \varphi_u(X_s : s \leq u)$, $\Phi_{uv} = \varphi_{uv}(X_s : u \leq s \leq v)$, $\Phi_v = \varphi_v(X_s : s \geq v)$. A repeated application of the tower property for conditional expectations and of the Markov property of Brownian motion yields

$$\begin{aligned}\mathbb{E}(\Phi_u \Phi_{uv} \Phi_v) &= \mathbb{E}(\Phi_u \mathbb{E}(\Phi_{uv} \mathbb{E}(\Phi_v | X_s : s \leq v) | X_s : s \leq u)) \\ &= \mathbb{E}(\Phi_u \mathbb{E}(\Phi_{uv} \mathbb{E}(\Phi_v | X_v) | X_u)) \\ &= \mathbb{E}(\,\mathbb{E}(\Phi_u | X_u)\, \mathbb{E}(\mathbb{E}(\Phi_{uv} | X_u, X_v)\, \mathbb{E}(\Phi_v | X_v)\, | X_u)\,)\end{aligned}$$

$$= \mathbb{E}(\Phi_u \mathbb{E}(\Phi_{uv}|X_u, X_v)\Phi_v), \quad \text{(perform the computations back)}$$

which shows indeed that the distribution of $(X_t)_{u \leq t \leq v}$ conditionally to $(X_s : s \leq u, s \geq v)$ coincides with that conditionally to X_u and X_v.

The definition of $B^{u,v}$ is equivalent to

$$B_t^{u,v} := X_t - X_u - (X_v - X_u)\frac{t-u}{v-u}; \quad (4.1.4)$$

this process is clearly Gaussian, centered, since $\mathbb{E}(B_t^{u,v}) = \mu t - \mu u - (\mu v - \mu u)\frac{t-u}{v-u} = 0$, and a computation of covariance gives

$$\mathbb{C}\text{ov}(B_t^{u,v}, X_r) = t \wedge r - u \wedge r - (v \wedge r - u \wedge r)\frac{t-u}{v-u}.$$

We obtain a zero covariance for $r \leq u$ or $r \geq v$; owing to properties of Gaussian distributions, this shows the announced independence.

Last, the covariance function of the process $B^{u,v}$ easily follows from (4.1.4); the computation is left to the reader. □

Recursive procedure. Consequently, the distribution of W_t conditionally to W_{t_i} and $W_{t_{i+1}}$ is Gaussian, centered on the line connecting W_{t_i} and $W_{t_{i+1}}$ and with explicit variance. For the midpoint $t = \frac{1}{2}(t_{i+1}+t_i)$, its characteristics are pretty simple: conditionally to W_{t_i} and $W_{t_{i+1}}$, we have

$$W_{\frac{t_{i+1}+t_i}{2}} \stackrel{d}{=} \mathcal{N}\left(\frac{W_{t_i}+W_{t_{i+1}}}{2}, \frac{t_{i+1}-t_i}{4}\right).$$

By simulating in the order $W_1, W_{\frac{1}{2}}, (W_{\frac{1}{4}}, W_{\frac{3}{4}}), (W_{\frac{1}{8}}, W_{\frac{3}{8}}, W_{\frac{5}{8}}, W_{\frac{7}{8}}), \ldots$ and by performing linear interpolation between points, we obtain Figure 4.4. This construction is due to Paul Lévy (1886–1971), who used it to build the Brownian motion.

At which frequency should one sample the Brownian trajectory? It depends much on the problem to solve.

- If the objective is to simulate W_1, which is useless obviously to discretize time and only one Gaussian random variable is enough to generate W_1.

- If the interest is in the integral $\int_0^1 W_s ds$, one can use a method of rectangles and approximate it by $\frac{1}{n}\sum_{i=1}^n W_{\frac{i}{n}}$: one can show

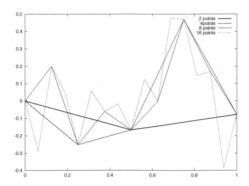

Figure 4.4 Iterative construction of Brownian motion by the Brownian bridge technique.

that the error is of order $1/n$ in L_2.

But there is a simpler way: $\int_0^1 W_s ds$ has a Gaussian distribution (because this is the limit of the Riemann sums which have Gaussian distribution — see Lemma A.1.4) of zero mean and variance $\frac{1}{3}$.

- If we now consider $M_1 = \max_{t \in [0,1]} W_t$, approximating it by $\max_{i \leq n} W_{\frac{i}{n}}$ is natural but quite rough. It is noticed that the sought random variable is thus underestimated: as the time increments of W are about the square root of the time step, one may guess that the error decreases like $1/\sqrt{n}$. This intuition is exact but difficult to prove: see [6] for a complete justification. Anyhow, this under-estimation yields important skews, which decrease slowly with n. Actually, the distribution of M_1 is equal to that $|W_1|$ (see equality (4.1.3)) and this trick is much simpler for the simulation of M_1.

If we are now interested in $(W_1, M_1 = \max_{T \in [0,1]} W_t)$, one can also generate an exact simulation via (4.1.2). One starts by simulating W_1, then M_1 conditionally on W_1. For this second stage, one is based on the explicit expression of the conditional c.d.f.:

$$\mathbb{P}(M_1 \leq y | W_1 = x) = \frac{\partial_x \mathbb{P}(M_1 \leq y, W_1 \leq x)}{\partial_x \mathbb{P}(W_1 \leq x)}$$

$$= 1 - \frac{\partial_x \mathbb{P}(W_1 \geq 2y - x)}{\partial_x \mathbb{P}(W_1 \leq x)}$$

128 ■ Stochastic differential equations and Feynman-Kac formulas

$$= 1 - \exp\left(-\frac{1}{2}(2y-x)^2 + \frac{1}{2}x^2\right)$$
$$= 1 - \exp\left(-2y(y-x)\right). \qquad (4.1.5)$$

By the inversion method (Proposition 1.2.1) and using a uniformly distributed random variable U, we obtain

$$\frac{1}{2}\left(x + \sqrt{x^2 - 2\log(1-U)}\right) \stackrel{d}{=} M_1 | W_1 = x.$$

4.1.4 Heat equation

▷ *One-dimensional case.* The distribution of $x + W_t$ is Gaussian with mean x and variance t: for $t > 0$, its density at point y is

$$g(t, x, y) := \frac{1}{\sqrt{2\pi t}} \exp(-(y-x)^2/2t),$$

called the *heat kernel* in that context. A direct computation shows that $g(t, x, y)$ satisfies the partial differential equation

$$g'_t(t, x, y) = \frac{1}{2} g''_{xx}(t, x, y), \quad t > 0.$$

Then, multiply this by $f(y)$ — for a measurable function f which we assume to be bounded to simplify — and integrate in y: it gives

$$\int_\mathbb{R} g'_t(t, x, y) f(y) \, dy = \frac{1}{2} \int_\mathbb{R} g''_{xx}(t, x, y) f(y) \, dy.$$

If we set $u(t, x) = \mathbb{E}(f(x + W_t)) = \int_\mathbb{R} g_t(t, x, y) f(y) \, dy$ and we apply the Lebesgue differentiation theorem, it is not difficult to justify that (for $t > 0$) the integral on the left-hand side is $u'_t(t, x)$ while that on the right-hand-side is $\frac{1}{2} u''_{xx}(t, x)$. We have established a first Feynman-Kac formula, in dimension 1 of space.

Theorem 4.1.8 (heat equation, $d = 1$) *Let $f : \mathbb{R} \mapsto \mathbb{R}$ be a bounded measurable function. The function $u : (t, x) \in \mathbb{R}^+ \times \mathbb{R} \mapsto u(t, x) = \mathbb{E}[f(x + W_t)]$ is the solution to the heat equation*

$$u'_t(t, x) = \frac{1}{2} u''_{xx}(t, x), \quad u(0, x) = f(x). \qquad (4.1.6)$$

As there is a unique solution to such equations, the solution is thus given by $\mathbb{E}[f(x + W_t)]$.

▷ *Multidimensional case.* The extension to the dimension $d > 1$ is similar. For this we introduce a d-dimensional Brownian motion $W = \begin{pmatrix} W_1 \\ \vdots \\ W_d \end{pmatrix}$, each component being a one-dimensional Brownian motion and the components being independent. The distribution of $x + W_t$

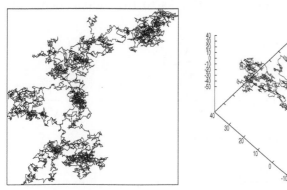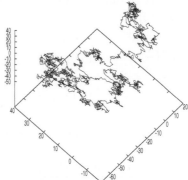

Figure 4.5 Brownian motion in dimension 2 and 3.

is Gaussian and has density $g(t, x, y) = \frac{1}{(2\pi t)^{d/2}} \exp(-|y-x|^2/2t)$, that solves the heat equation

$$g'_t(t, x, y) = \frac{1}{2} \sum_{k=1}^{d} g''_{x_k, x_k}(t, x, y) = \frac{1}{2} \Delta g(t, x, y),$$

where we use the notation Δ for the d-dimensional Laplacian. As before, we obtain the following:

Theorem 4.1.9 (heat equation, $d \geq 1$) *Let $f : \mathbb{R}^d \mapsto \mathbb{R}$ be a bounded measurable function. The function $u : (t, x) \in \mathbb{R}^+ \times \mathbb{R}^d \mapsto u(t, x) = \mathbb{E}[f(x + W_t)]$ is the solution to the heat equation*

$$u'_t(t, x) = \frac{1}{2} \Delta u(t, x), \quad u(0, x) = f(x). \tag{4.1.7}$$

Link between random walk and finite difference scheme for partial differential equation. Consider the random walk in dimension 1, defined in (4.1.1),

$$W_t^n = \frac{1}{\sqrt{n}} \sum_{i=1}^{\lfloor nt \rfloor} X_i;$$

the discussion in the multidimensional case is similar. As $n \to +\infty$, W^n converges in distribution to a Brownian motion and this implies the convergence of

$$u^n(t, x) = \mathbb{E}(f(x + W_t^n))$$

to $u(t,x)$ solution to the heat equation (4.1.7), provided that f is continuous and bounded. For $t = \frac{i}{n}$, thanks to the independence of the $(X_i)_i$ we write

$$u^n\left(\frac{i}{n}, x\right) = \mathbb{E}\left(f\left(x + W^n_{\frac{i-1}{n}} + \frac{X_i}{\sqrt{n}}\right)\right)$$
$$= \frac{1}{2} u^n\left(\frac{i-1}{n}, x + \frac{1}{\sqrt{n}}\right) + \frac{1}{2} u^n\left(\frac{i-1}{n}, x - \frac{1}{\sqrt{n}}\right).$$

By rearranging the terms, we obtain

$$\frac{u^n\left(\frac{i}{n}, x\right) - u^n\left(\frac{i-1}{n}, x\right)}{\frac{1}{n}}$$
$$= \frac{1}{2} \frac{u^n\left(\frac{i-1}{n}, x + \frac{1}{\sqrt{n}}\right) - 2 u^n\left(\frac{i-1}{n}, x\right) + u^n\left(\frac{i-1}{n}, x - \frac{1}{\sqrt{n}}\right)}{\left(\frac{1}{\sqrt{n}}\right)^2}$$

and we retrieve a finite difference scheme for the heat equation, see [2, Chapter 2].

Application in Monte-Carlo method. To evaluate the solution of the heat equation (4.1.7), one can apply the Monte-Carlo method for computing the expectation by generating independent simulations of $x + W_t$. The convergence and the error control are described by the results of Part **A**. It should be noted that these simulations make it possible to compute the solution pointwise, at the point (t, x) say, whereas a PDE scheme gives the solution on a full grid in time/space. To have the solution at other space points or other times by Monte-Carlo method, one shall keep the same Brownian drawings $(W_t^m)_m$, then shift them in space or cut/extend their path in time; for other processes, this trick is not possible in general.

Summary. The main comparative features between a numerical approach based on a partial differential equation or on the Monte-Carlo method are the following.

1. **Numerical approach by partial differential equation.**
   (a) Advantages: computation of a global solution (or in a large domain); good convergence rates (related to the choice of the time and space steps); possible refinement of mesh.
   (b) Drawbacks: large linear system to invert (whose size grows exponentially with the dimension d); convergence under conditions about the model coefficients (stability condition of CFL type for explicit scheme, ellipticity condition in the case of nonconstant coefficients).

2. **Numerical approach by Monte-Carlo method.**
   (a) Advantages: unconditional convergence (universality of the law of large numbers, no condition related to the model non-degeneracy); computational code that is simple (to write and to update); complexity nonsensitive to the dimension; a posteriori error estimates (confidence intervals).
   (b) Drawbacks: computation of a pointwise solution; slow convergence (given by the central limit theorem); random error (nonreproducible).

It is commonly admitted that in general, a PDE method is preferable in dimension lower than 3, and a Monte-Carlo approach becomes better in a larger dimension. In Chapter 8, the Monte-Carlo methods developed for non-linear equations provide a numerical solution everywhere and they become sensitive to the dimension.

4.1.5 Quadratic variation

Brownian motion has the astonishing property of finite and non-zero quadratic variation. Here, Brownian motion is one-dimensional.

Definition and Proposition 4.1.10 (quadratic variation) *For a given subdivision $\pi = \{t_0 = 0 < \cdots < t_i < \cdots\}$ of the time interval, the quadratic variation of W along π is defined by*

$$V_t^\pi = \sum_{t_i \in \pi \cap [0,t]} (W_{t_{i+1} \wedge t} - W_{t_i})^2.$$

Then, for any sequence of subdivisions π with size $|\pi| = \sup_i |t_{i+1} - t_i| \to 0$, V_t^π converges to t in L_2:

$$\lim_{|\pi| \to 0} V_t^\pi \stackrel{L_2}{=} t.$$

Let us recall that a continuously differentiable function has a quadratic variation equal to 0 at the limit: it is a crucial difference when it is a question of writing a specific differential calculus.

PROOF:
Denote by $n(t)$ the integer such that $t_{n(t)} \leq t < t_{n(t)+1}$ and write $V_t^\pi - t = \sum_{i=0}^{n(t)-1} Z_i$ with $Z_i = (W_{t_{i+1} \wedge t} - W_{t_i \wedge t})^2 - (t_{i+1} \wedge t - t_i \wedge t)$. The random variables $(Z_i)_i$ are independent, centered and square integrable (since the Gaussian distribution has finite moments of order 4): moreover, the scaling property of Proposition 4.1.6 ensures that $\mathbb{E}(Z_i^2) = C_2 (t_{i+1} \wedge t - t_i \wedge t)^2$ for a positive constant C_2. Then $\mathbb{E}(V_t^\pi - t)^2 = \sum_{i=0}^{n(t)-1} \mathbb{E}(Z_i^2) = C_2 \sum_{i=0}^{n(t)-1} (t_{i+1} \wedge t - t_i \wedge t)^2 \leq C_2 \, t \, |\pi| \to 0$. □

Had W been continuously differentiable, with the usual differential calculus rules, we would have written $W_t^2 = 2 \int_0^t W_s dW_s = \lim_{|\pi| \to 0} \sum_{t_i \in \pi \cap [0,t]} 2 W_{t_i} (W_{t_{i+1} \wedge t} - W_{t_i})$. The previous property of non-zero quadratic variation modifies this rule and we obtain a formula for the square that is different because of an extra "t" term.

Proposition 4.1.11 (A first Itô formula) *We have*

$$W_t^2 \stackrel{L_2}{=} t + \lim_{|\pi| \to 0} \sum_{t_i \in \pi \cap [0,t]} 2 W_{t_i} (W_{t_{i+1} \wedge t} - W_{t_i}).$$

PROOF:
It is enough to write $W_t^2 = \sum_{t_i \leq t} (W_{t_{i+1}}^2 - W_{t_i}^2) = \sum_{t_i \leq t} (W_{t_{i+1}} - W_{t_i})^2 + 2 \sum_{t_i \leq t} W_{t_i} (W_{t_{i+1}} - W_{t_i})$ and then to apply Proposition 4.1.10. □

The purpose of the next section is to provide general properties of the above limit, which can still be written as $\int_0^t 2 W_s dW_s$ by notational abuse, and which is called the *stochastic integral*. We observe that the term $\int_0^t 2 W_s dW_s$ has zero expectation since $\mathbb{E}(W_t^2) = t$.

4.2 STOCHASTIC INTEGRAL AND ITÔ FORMULA

We give in this section the minimum mathematical results related to stochastic calculus to allow us to continue the study of the simulation of stochastic differential equations. Their proofs are not given, only some heuristics, and the reader can refer to the usual textbooks on the subject, for instance [88] or [127].

4.2.1 Filtration and stopping times

On the probability space $(\Omega, \mathcal{F}, \mathbb{P})$ on which we suppose to have defined a Brownian motion W (possibly in dimension d), we associate to $t \geq 0$ the sigma-field $\mathcal{F}_t^0 = \sigma(W_s : 0 \leq s \leq t)$, i.e., the smallest sigma-field making measurable the random variables W_s for $0 \leq s \leq t$. In the following, we consider the natural filtration of Brownian motion completed[3] with negligible sets, i.e., the non-decreasing sequence of sigma-fields $(\mathcal{F}_t)_{t \geq 0}$ where $\mathcal{F}_t$ is the sigma-field generated by the sets of $\mathcal{F}_t^0$ and those of zero probability of the sigma-field $\mathcal{F}$.

Definition 4.2.1 *An $\mathbb{R}$-valued stochastic process $(X_t)_{t \geq 0}$ is measurable if $X : (t, \omega) \in (\mathbb{R}^+ \times \Omega) \mapsto X_t(\omega) \in \mathbb{R}$ is $\mathcal{B}(\mathbb{R}^+) \otimes \mathcal{F}$-measurable.*

It is adapted to $(\mathcal{F}_t)_{t \geq 0}$ (or simply adapted) if for any $t \geq 0$, X_t is $\mathcal{F}_t$-measurable.

Thus we verify that $(W_t)_{t \geq 0}$ and $(W_t^2 - t)_{t \geq 0}$ are martingales with respect to $(\mathcal{F}_t)_t$: indeed, for any $t > s$,

i) $\mathbb{E}(W_t | \mathcal{F}_s) = W_s + \mathbb{E}(W_t - W_s | \mathcal{F}_s) = W_s$;

ii) $\mathbb{E}(W_t^2 - t | \mathcal{F}_s) = W_s^2 - s + \mathbb{E}((W_t - W_s)(2W_s + (W_t - W_s)) | \mathcal{F}_s) - (t - s) | \mathcal{F}_s) = W_s^2 - s$.

These martingale properties will be transmitted to the stochastic integrals in the sequel.

Definition 4.2.2 *A non-negative random variable τ is a stopping time (implicitly for the filtration $(\mathcal{F}_t)_{t \geq 0}$) if for any $t \geq 0$, the event $\{\tau \leq t\}$ is in $\mathcal{F}_t$.*

Therefore,

- the first hitting time of $y > 0$ by the scalar Brownian motion is a stopping time;

- the last passage time at 0 before time 1 is not.

---

[3]This completion is *technical* and ensures that negligible sets are also $\mathcal{F}_t$-measurable.

4.2.2 Stochastic integral and its properties

For a given time $T > 0$ (possibly $T = +\infty$), we define several sets of processes.

- $\mathbb{H}_T^2 = \{\phi$ adapted process such that $\|\phi\|_{\mathbb{H}_T^2}^2 := \mathbb{E}\int_0^T |\phi_s|^2 \mathrm{d}s < +\infty\}$. The process can be scalar or vector valued (as a row).
- $\mathbb{H}_T^{2,\mathrm{loc.}} = \{\phi$ adapted process such that $\int_0^T |\phi_s|^2 \mathrm{d}s < +\infty$ a.s. $\}$.
- $\mathbb{H}^{\mathrm{elem.}}$ stands for the set of elementary adapted processes (step processes), i.e., $\phi_t = \phi_{t_i}$ for $t_i < t \leq t_{i+1}$ and for some time grid $\pi = \{0 = t_0 < \cdots < t_i < \ldots\}$.

▷ *The case of elementary process.* If ϕ is elementary, with values in $\mathbb{R}^d$ and square integrable (i.e., $\phi \in \mathbb{H}_T^2 \cap \mathbb{H}^{\mathrm{elem.}}$), then the stochastic integral with respect to the d-dimensional Brownian motion is defined by

$$\int_0^T \phi_s \mathrm{d}W_s = \sum_{t_i \leq T} \phi_{t_i}(W_{T \wedge t_{i+1}} - W_{t_i})$$

$$= \sum_{t_i \leq T} \sum_{k=1}^d \phi_{k,t_i}(W_{k,T \wedge t_{i+1}} - W_{k,t_i}). \quad (4.2.1)$$

Its two first moments can be computed explicitly, by taking advantage of the fact that ϕ is adapted, that the Brownian increments are independent, centered, and with explicit variance, and that the Brownian motions $(W_k)_k$ are independent.

- The stochastic integral is centered:

$$\mathbb{E}(\int_0^T \phi_s \mathrm{d}W_s) = \sum_{t_i \leq T}\sum_{k=1}^d \mathbb{E}(\phi_{k,t_i}\mathbb{E}(W_{k,T \wedge t_{i+1}} - W_{k,t_i}|\mathcal{F}_{t_i})) = 0.$$

- The stochastic integral is square integrable with an explicit L_2-norm:

$$\mathbb{E}(\int_0^T \phi_s \mathrm{d}W_s)^2$$

$$= 2\sum_{t_i < t_j \leq T}\mathbb{E}(\phi_{t_i}(W_{T\wedge t_{i+1}} - W_{t_i})(W_{T\wedge t_{j+1}} - W_{t_j})^\mathsf{T}\phi_{t_j}^\mathsf{T})$$

$$(4.2.2)$$

$$+ \sum_{t_i \leq T} \mathbb{E}(\phi_{t_i}(W_{T \wedge t_{i+1}} - W_{t_i})(W_{T \wedge t_{i+1}} - W_{t_i})^\mathsf{T} \phi_{t_i}^\mathsf{T})$$

$$= \sum_{t_i \leq T} \mathbb{E}(\phi_{t_i} \phi_{t_i}^\mathsf{T})(T \wedge t_{i+1} - t_i) = \mathbb{E}(\int_0^T |\phi_s|^2 \mathrm{d}s). \quad (4.2.3)$$

By using a polarization identity, we obtain for another $\varphi \in \mathbb{H}^{\text{elem}}$.

$$\mathbb{E}\Big[(\int_0^T \phi_s \mathrm{d}W_s)(\int_0^T \varphi_s \mathrm{d}W_s)\Big] = \mathbb{E}\Big[\int_0^T \phi_s \cdot \varphi_s \mathrm{d}s\Big].$$

We easily observe that the stochastic integral $(I_T^\phi = \int_0^T \phi_s \mathrm{d}W_s)_{T \geq 0}$ as a process indexed by time is continuous, adapted, and with finite quadratic variation given by

$$\lim_{|\pi| \to 0} \sum_{t_i \leq T} (I_{t_{i+1} \wedge T}^\phi - I_{t_i}^\phi)^2 = \int_0^T |\phi_s|^2 \mathrm{d}s.$$

Finally, a simple calculus similar to that for the two first moments gives the following martingale properties (by analogy with the Brownian motion):

$$\mathbb{E}\big(I_T^\phi | \mathcal{F}_t\big) = I_t^\phi, \quad \mathbb{E}\big((I_T^\phi)^2 - \int_0^T |\phi_s|^2 \mathrm{d}s | \mathcal{F}_t\big) = (I_t^\phi)^2 - \int_0^t |\phi_s|^2 \mathrm{d}s.$$

▷ *The general case.* The extension of the stochastic integral for $\phi \in \mathbb{H}_T^2$ relies on the equality (4.2.3), which defines an isometry between $L_2(d\mathbb{P})$ and $L_2(\mathrm{d}t \otimes \mathrm{d}\mathbb{P})$, and on the existence of elementary processes in $\mathbb{H}^{\text{elem}}$ approximating — in the norm $\|\cdot\|_{\mathbb{H}_T^2}$ — the process in $\mathbb{H}_T^2$. By doing so, the previous properties remain true in the general case and the construction can be done to preserve the continuity properties of the stochastic integral process. We synthesize all these properties in the form of a theorem, ready for use.

Theorem 4.2.3 *Let W be a Brownian motion in dimension d and let $\phi \in \mathbb{H}_T^2$ be valued in $\mathbb{R}^d$ (written as a row). The stochastic integral $(I_t^\phi = \int_0^t \phi_s \mathrm{d}W_s)_{0 \leq t \leq T}$ is a continuous adapted process that fulfills the following properties:*

- **(Martingale)** *$(I_t^\phi)_{0 \leq t \leq T}$ and $((I_t^\phi)^2 - \int_0^t |\phi_s|^2 \mathrm{d}s)_{0 \leq t \leq T}$ are martingales.*

- **(Covariance)** *for any $\varphi \in \mathbb{H}_T^2$ with values in $\mathbb{R}^d$*

$$\mathbb{E}\Big[(\int_0^T \phi_s dW_s)(\int_0^T \varphi_s dW_s)\Big] = \mathbb{E}\Big[\int_0^T \phi_s \cdot \varphi_s ds\Big].$$

- **(Doob maximal inequality)** *There exists a universal positive constant[4] c_2 such that*

$$\mathbb{E}\Big(\sup_{0 \leq t \leq T} |I_t^\phi|^2\Big) \leq c_2 \sup_{0 \leq t \leq T} \mathbb{E}\Big(|I_t^\phi|^2\Big) = c_2 \, \mathbb{E}\Big(\int_0^T |\phi_s|^2 ds\Big).$$

- **(Localization)** *for any stopping time τ, $I_{t \wedge \tau}^\phi = I_t^{\phi 1_{\cdot < \tau}}$ ($= \int_0^t 1_{s < \tau} \phi_s dW_s$).*

- **(Quadratic variation)** $\lim_{|\pi| \to 0} \sum_{t_i \leq T} (I_{t_{i+1} \wedge T}^\phi - I_{t_i}^\phi)^2 = \int_0^T |\phi_s|^2 ds.$

▷ *Some ramifications.*

- In general, the distribution of a stochastic integral is not explicit: the remarkable exception corresponds to the case where ϕ is deterministic. In that case, I_t^ϕ has a Gaussian distribution, centered, and with variance $\int_0^t |\phi_s|^2 ds$: I_t^ϕ is called the *Wiener integral*.

- One can also define a stochastic integral for $\phi \in \mathbb{H}_T^{2,\text{loc.}}$, but the martingale properties and the equalities on the two first moments are not necessarily satisfied.

▷ *Simulation of a stochastic integral.* To simulate $\int_0^T \phi_s dW_s$ for $\phi \in \mathbb{H}_T^2$, in general there does not exist an exact method because its distribution is not explicit. One can approximate it by a Riemann sum $\sum_{i=0}^{n-1} \phi_{t_i}(W_{t_{i+1}} - W_{t_i})$ along an equidistant time grid $t_i = i\frac{T}{n}$. One can show (but this is quite technical) that the error converges to 0 in probability: exhibiting convergence rates in L_2 is quite difficult in full generality and the rates depend strongly on the choice of discretization times and on the regularity properties of ϕ. See Exercise 4.3.

---

[4] $c_2 = 4$.

4.2.3 Itô process and Itô formula

Definition 4.2.4 (Itô process) *Let $(b_t)_{t\geq 0}$ and $(\sigma_t = [\sigma_{1,t}, \ldots, \sigma_{d,t}])_t$ be two adapted processes satisfying $\int_0^t (|b_s| + |\sigma_s|^2) \mathrm{d}s < +\infty$ a.s. for any $t \geq 0$, the first one taking values in $\mathbb{R}^q$ and the second one in $\mathbb{R}^q \otimes \mathbb{R}^d$. The Itô process $(X_t)_{t\geq 0}$ with drift coefficient b and diffusion coefficient σ starting from x_0 is the q-dimensional process defined by*

$$X_t = x_0 + \int_0^t b_s \mathrm{d}s + \int_0^t \sigma_s \mathrm{d}W_s$$

$$= x_0 + \int_0^t b_s \mathrm{d}s + \sum_{k=1}^d \int_0^t \sigma_{k,s} \mathrm{d}W_{k,s}, \quad t \geq 0.$$

The integral related to b is $\mathcal{C}^1$ in time, thus its quadratic variation is zero: therefore the quadratic variation of X is that of $\int_0^{\cdot} \sigma_s \mathrm{d}W_s$, i.e., $\int_0^t |\sigma_s|^2 \mathrm{d}s$ at time t. This point is important because the infinitesimal decomposition of $f(X_t)$ via a Taylor formula reveals a term with a second derivative implying the quadratic variation of X. It is to some extent the trademark of stochastic calculus, with the famous Itô formula.

Theorem 4.2.5 (Itô formula) *Let $f : \mathbb{R}^+ \times \mathbb{R}^q \mapsto \mathbb{R}$ be a function of class $\mathcal{C}^{1,2}$ and let X be a q-dimensional Itô process with coefficients b and σ.*

Then $Y_t = f(t, X_t)$ defines again an Itô process, in dimension 1, whose coefficients are given by

$$f(t, X_t) = f(0, x_0) + \int_0^t \partial_t f(s, X_s) \mathrm{d}s + \int_0^t \nabla_x f(s, X_s) \, b_s \mathrm{d}s$$

$$+ \int_0^t \nabla_x f(s, X_s) \, \sigma_s \mathrm{d}W_s + \frac{1}{2} \sum_{k,l=1}^q \int_0^t f''_{x_k,x_l}(s, X_s) \, [\sigma_s \sigma_s^\mathsf{T}]_{k,l} \mathrm{d}s.$$

In other words, a smooth transformation of the Itô process remains an Itô process, with an explicit decomposition. We will repeatedly use this result in what follows.

4.3 STOCHASTIC DIFFERENTIAL EQUATIONS

4.3.1 Definition, existence, uniqueness

A *stochastic differential equation* is an ordinary differential equation $\dot{x}_t = b(t, x_t)$ to which we add a Brownian perturbation.

Theorem 4.3.1 *Let W be a d-dimensional Brownian motion. Let $b : \mathbb{R}^+ \times \mathbb{R}^d \mapsto \mathbb{R}^d$ and $\sigma : \mathbb{R}^+ \times \mathbb{R}^d \mapsto \mathbb{R}^d \otimes \mathbb{R}^d$ be two continuous functions satisfying the following regularity and boundedness conditions[5]: for a finite positive constant $C_{b,\sigma}$ we have*

i) $|b(t, x) - b(t, y)| + |\sigma(t, x) - \sigma(t, y)| \leq C_{b,\sigma} |x - y|$ *for all* $(t, x, y) \in [0, T] \times \mathbb{R}^d \times \mathbb{R}^d$;

ii) $\sup_{0 \leq t \leq T} (|b(t, 0)| + |\sigma(t, 0)|) \leq C_{b,\sigma}$.

For a given $x_0 \in \mathbb{R}^d$, consider the following stochastic differential equation with drift coefficient b *and* diffusion coefficient σ:

$$X_t = x_0 + \int_0^t b(s, X_s) \mathrm{d}s + \int_0^t \sigma(s, X_s) \mathrm{d}W_s. \quad (4.3.1)$$

Then, there exists a unique adapted solution X in $\mathbb{H}_T^2$; the solution is continuous and verifies $\mathbb{E}(\sup_{0 \leq t \leq T} |X_t|^2) < C(1 + |x_0|^2)$ for a constant C depending on T and $C_{b,\sigma}$.

By analogy with the Brownian motion and the heat equation, this stochastic process is also called a *diffusion process* or simply *diffusion*. We refer to the bibliography for the detailed proof. Let us discuss only the principle in the purpose of simulation. The idea is, as for an ordinary differential equation, to show that some Picard iterations are converging (i.e. a certain map is contracting in a certain Hilbert space with an appropriate norm). Namely, we initialize with $X^{(0)} = x_0$, then we set

$$X_t^{(n+1)} = x_0 + \int_0^t b(s, X_s^{(n)}) \mathrm{d}s + \int_0^t \sigma(s, X_s^{(n)}) \mathrm{d}W_s, \quad (4.3.2)$$

thus $(X^{(n)})_n$ converges to X. In fact, this scheme is not efficient at all

---

[5] The reader familiar with the techniques of stochastic calculus will note that we take sufficient assumptions, but not minimal, for the sake of simplicity.

regarding the simulation since it requires us to iteratively simulate the stochastic integral $\int_0^t \sigma(s, X_s^{(n)})\mathrm{d}W_s$ for any n; indeed, this is possible only in an approximate manner (see our discussion at the end of Section 4.2.2) and the time-discretization errors cumulate along iterations. In Chapter 5 we will study a simpler scheme (Euler scheme).

4.3.2 Flow property and Markov property

Instead of solving the stochastic differential equation from time 0, we can similarly solve it from time $t_0 \geq 0$

$$X_t = x_0 + \int_{t_0}^t b(s, X_s)\mathrm{d}s + \int_{t_0}^t \sigma(s, X_s)\mathrm{d}W_s, \quad t \geq t_0, \qquad (4.3.3)$$

and we then obtain a solution adapted to the filtration of the shifted Brownian motion $(W_s^{t_0} := W_s - W_{t_0})_{s \geq t_0}$: we denote this solution by X^{t_0, x_0} to emphasize its initial condition. We now state two quite intuitive results (but not so straitgthforward to establish).

Proposition 4.3.2 *Assume that the assumptions of Theorem 4.3.1 are in force. Then*

- *(flow property) with probability 1, for any $t_0 \leq t \leq s \leq T$ and any $x_0 \in \mathbb{R}$, we have*

$$X_s^{t_0, x_0} = X_s^{t, X_t^{t_0, x_0}};$$

- *(Markov property) for any continuous functional Φ, we have*

$$\mathbb{E}(\Phi(X_s : t \leq s \leq T)|\mathcal{F}_t) = \mathbb{E}(\Phi(X_s : t \leq s \leq T)|X_t).$$

4.3.3 Examples

▷ *Arithmetic Brownian motion.* It is defined by the *constant coefficients* $b(t, x) = b$ and $\sigma(t, x) = \sigma$, so that

$$X_t = x_0 + bt + \sigma W_t.$$

In the landscape of stochastic differential equations, this is the first variation around the Brownian motion. It can be simulated as easily as Brownian motion itself.

▷ *Geometric Brownian motion.* In dimension 1, it is the exponential of an arithmetic Brownian motion:
$$X_t = x_0 \exp\left((b - \frac{1}{2}\sigma^2)t + \sigma W_t\right).$$
This serves as a basic model in finance, for modeling stock prices by some positive processes, see [131]. The Itô formula (Theorem 4.2.5) leads to *linear coefficients*
$$X_t = x_0 + \int_0^t bX_s ds + \int_0^t \sigma X_s dW_s,$$
i.e., $b(t,x) = bx$ and $\sigma(t,x) = \sigma x$. The distribution of X_t is log-normal and its simulation is handled via the simulation of its logarithm, and then applying an exponential transformation.

▷ *Ornstein-Uhlenbeck process.* This is a Gaussian process that exhibits a mean-reverting behavior for some parameter values: it is used in physics (also called *physical Brownian motion*), random mechanics (see [96]), and economy and finance (modeling the inflation rate or the short-term interest rate [141]). In dimension 1, the Ornstein-Uhlenbeck process is the solution of
$$X_t = x_0 - a\int_0^t (X_s - \theta)ds + \sigma W_t, \qquad (4.3.4)$$
i.e., $b(t,x) = -a(x - \theta)$ and $\sigma(t,x) = \sigma$. It can be solved explicitly, thanks to the Itô formula applied to $e^{at}X_t$ which gives
$$X_t = \theta + (x_0 - \theta)e^{-at} + \sigma \int_0^t e^{-a(t-s)} dW_s.$$
The above stochastic integral is a Wiener integral, thus it defines a Gaussian process. Its mean equals $\theta + (x_0 - \theta)e^{-at}$ and its covariance function is $\mathrm{Cov}(X_t, X_s) = e^{-a(t-s)}\frac{\sigma^2}{2a}(1 - e^{-2as})$ for $t > s$. Therefore, the simulation of X at many dates readily follows from generating Gaussian random variables.

We observe that for $a > 0$, the Gaussian distribution of X_t converges to $\mathcal{N}(\theta, \frac{\sigma^2}{2a})$ as $t \to +\infty$, with a lapse of memory of the initial condition in large time. That illustrates the mean-reverting effect.

Lastly, let us note that the Ornstein-Uhlenbeck process can be defined in a multidimensional version without additional difficulty, except those due to the vector and matrix notations,

$$\mathcal{X}_t = x_0 + \int_0^t (\Theta - A\mathcal{X}_s)\mathrm{d}s + \Sigma W_t \tag{4.3.5}$$

where A, σ are matrices and $\mathcal{X}, \Theta, W$ are vectors. The following example is a simple case, where the noise is degenerate on a component. In random mechanics [96], X as in (4.3.4) is the equation for the speed of a system — for example, a spring — subjected to random forces, modeled via Newton's second law of motion. The position of the system is the speed antiderivative: $P_t = p_0 + \int_0^t X_s \mathrm{d}s$. In this case, we can still write $\mathcal{X} = \begin{pmatrix} X \\ P \end{pmatrix}$ as a solution of an Ornstein-Uhlenbeck equation in dimension 2, of the form (4.3.5) but without noise in the second coordinate, i.e.

$$A = \begin{pmatrix} a & 0 \\ -1 & 0 \end{pmatrix}, \Theta = \begin{pmatrix} \theta/a \\ 0 \end{pmatrix}, \Sigma = \begin{pmatrix} \sigma & 0 \\ 0 & 0 \end{pmatrix}.$$

▷ *Square root process in dimension 1.* This process exhibits a mean-reverting effect but with the specificity to stay non-negative: it is defined by

$$X_t = x_0 - a\int_0^t X_s \mathrm{d}s + \int_0^t \sigma\sqrt{X_s}\mathrm{d}W_s,$$

i.e., $b(t,x) = -ax$ and $\sigma(t,x) = \sigma\sqrt{x}$. Although σ is not Lipschitz continuous, it is possible to show the existence and uniqueness of the solution, but it is more delicate. This model is often used in finance to model stochastic volatility (Heston model [77]) and interest rates (CIR model [30]). In population dynamics [39], the equation of X is called the *Feller equation* and functions like the limit model of the birth-death process. The distribution of X_t is known in the form of special functions (in particular Bessel functions) and its simulation is not simple; see [22].

In genetic model, the Fisher-Wright process describing the asymptotic proportion of individuals carrying a given allele is an example of the square root process in the form

$$X_t = x_0 + \int_0^t (rX_s(1-X_s) - \beta_1 X_s + (1-X_s)\beta_2)\mathrm{d}s + \int_0^t \sqrt{X_s(1-X_s)}\mathrm{d}W_s,$$

where the drift coefficient is aimed at accounting for the rare phenomenon of selection and mutation.

▷ *Equilibrium model.* Denote by π a probability density on $\mathbb{R}^d$, positive, and set

$$X_t = x_0 + \frac{1}{2}\int_0^t \partial_x[\log(\pi)](X_s)\mathrm{d}s + W_t.$$

The Lipschitz conditions on the drift coefficient can be verified case by case, according to the expression of π. This model has the advantage of having (under some conditions on π) an ergodic behavior: the distribution of X_t converges to the distribution with density π as $t \to +\infty$. In molecular chemistry, this process represents the energy of a molecule as its configuration changes; see [95]. The case $\pi(x) = \text{Cste} \times \exp(-ax^2)$ in dimension 1 with $a > 0$ boils down to the previous Ornstein-Uhlenbeck equation (4.3.4) with $\theta = 0$ and $\sigma = 1$.

▷ *General case.* Apart from these isolated cases, generally the distribution of X_t is not explicit (neither at a given t, nor in large time). It is necessary to resort to an approximation scheme for the simulation; see Chapter 5.

4.4 PROBABILISTIC REPRESENTATIONS OF PARTIAL DIFFERENTIAL EQUATIONS: FEYNMAN-KAC FORMULAS

To relate the Brownian motion to the heat equation, we have taken advantage of the knowledge of the density of $x + W_t$ and we have worked out a relation between its partial derivatives in time and space, a relation that has been transferred to those of $\mathbb{E}(f(x+W_t))$ by integrating the relation against f. In the case of a stochastic differential equation, the distribution is not explicit and it is necessary to find another approach: the Itô formula provides the right tool.

4.4.1 Infinitesimal generator

We start by letting the differential operator of the subsequent partial differential equation appear.

Definition 4.4.1 (infinitesimal generator) *Let X be the solution* to the stochastic differential equation with coefficients (b,σ) under the assumptions of Theorem 4.3.1. The infinitesimal generator associated with X and denoted by $\mathcal{L}^X_{b,\sigma\sigma^\mathsf{T}}$ is given by

$$\mathcal{L}^X_{b,\sigma\sigma^\mathsf{T}} = \frac{1}{2}\sum_{i,j=1}^d [\sigma\sigma^\mathsf{T}]_{i,j}(t,x)\partial^2_{x_ix_j} + \sum_{i=1}^d b_i(t,x)\partial_{x_i}.$$

When there is no ambiguity, we simply write $\mathcal{L}$ instead of $\mathcal{L}^X_{b,\sigma\sigma^\mathsf{T}}$.

Proposition 4.4.2 *With the assumptions and notations of Theorem 4.3.1, for any function $f \in \mathcal{C}^2$ with compact support and any initial condition (t,x), we have*

$$\frac{\mathbb{E}(f(X^{t,x}_{t+h})) - f(x)}{h} \xrightarrow[h\to 0^+]{} \mathcal{L}^X_{b,\sigma\sigma^\mathsf{T}} f(t,x).$$

PROOF:
Write $\mathcal{L} = \mathcal{L}^X_{b,\sigma\sigma^\mathsf{T}}$ to simplify. Let (t,x) be fixed; write $X_s = X^{t,x}_s$ for $s \geq t$. Apply the Itô formula to $f(X_s)$ between $s = t+h$ and $s = t$ in order to get

$$\begin{aligned}
f(X_{t+h}) - f(x) &= \int_t^{t+h} \nabla_x f(X_s)\, b(s,X_s)\mathrm{d}s + \int_t^{t+h} \nabla_x f(X_s)\, \sigma(s,X_s)\mathrm{d}W_s \\
&\quad + \frac{1}{2}\sum_{k,l=1}^d \int_t^{t+h} f''_{x_k,x_l}(X_s)\,[\sigma\sigma^\mathsf{T}]_{k,l}(s,X_s)\mathrm{d}s \\
&= \int_t^{t+h} \mathcal{L}f(s,X_s)\mathrm{d}s + \int_t^{t+h} \nabla_x f(X_s)\,\sigma(s,X_s)\mathrm{d}W_s. \quad (4.4.1)
\end{aligned}$$

The stochastic integral of $\phi_s = \nabla_x f(X_s)\,\sigma(s,X_s)$ is in $\mathbb{H}^2_T$, owing to assumptions on σ and f with compact support (say in $B(0,R)$): indeed

$$|\phi_s| \leq \sup_{|x|\leq R} |\nabla_x f(x)\,\sigma(s,x)|$$

$$\leq |\nabla_x f|_\infty (|\sigma(s,0)| + C_{b,\sigma}R) \leq |\nabla_x f|_\infty C_{b,\sigma}(R+1).$$

Therefore, by passing to the expectation, the term with the stochastic integral vanishes and it remains

$$\frac{\mathbb{E}(f(X_{t+h})) - f(x)}{h} = \mathbb{E}\Big(\frac{1}{h}\int_t^{t+h} \mathcal{L}f(s,X_s)\mathrm{d}s\Big).$$

144 ■ Stochastic differential equations and Feynman-Kac formulas

We easily verify that $\frac{1}{h}\int_t^{t+h}\mathcal{L}f(s,X_s)ds$ converges a.s. to $\mathcal{L}f(t,x)$, owing to the continuity of the paths of X and to the continuity of $(s,y) \mapsto \mathcal{L}(s,y)$, while being bounded (here we use that f has compact support). The dominated convergence theorem allows us to conclude the proof. □

4.4.2 Linear parabolic partial differential equation with Cauchy condition

We are now in a position to establish the link between expectation of functionals of diffusion and partial differential equations. We start with the case where only one condition in T is predetermined (known as the Cauchy problem). Imposing a condition on the PDE values at the boundary of a domain corresponds to the Dirichlet problem, which we study later. One could also impose values at the boundary related to the first and second derivatives of the PDE: these are the so-called Neumann and Robin problems, respectively, associated with processes with reflection or diffusion at the boundary; these cases go largely beyond the framework of this monograph and the interested reader may refer to [43].

These equations are *linear*[6] because if we add two PDE conditions (f_1, g_1) and (f_2, g_2) all things being equal, the resulting solution is the sum of individual solutions $u_1 + u_2$.

Theorem 4.4.3 *Let $T > 0$ be fixed. Assume that*

i) *the diffusion process X with coefficients b and σ fulfills the notations and assumptions of Theorem 4.3.1;*

ii) *$f : \mathbb{R}^d \mapsto \mathbb{R}$ and $g, k : [0,T] \times \mathbb{R}^d \mapsto \mathbb{R}$ are three continuous functions verifying*

$$\sup_{x \in \mathbb{R}^d} \frac{|f(x)|}{1+|x|^2} + \sup_{0 \leq t \leq T,\ x \in \mathbb{R}^d} \left(\frac{|g(t,x)|}{1+|x|^2} + |k(t,x)| \right) < +\infty;$$

iii) *there exists a function $u : [0,T] \times \mathbb{R}^d \mapsto \mathbb{R}$, of class $\mathcal{C}^{1,2}$ on any open set of $[0,T[\times\mathbb{R}^d$, which fulfills*

$$\begin{cases} \partial_t u(t,x) + \mathcal{L}u(t,x) - k(t,x)u(t,x) + g(t,x) = 0, & t < T, x \in \mathbb{R}^d, \\ u(T,x) = f(x), & x \in \mathbb{R}^d; \end{cases}$$

(4.4.2)

---

[6]Some non-linear cases are studied in part **C**.

iv) *u is continuous on* $[0,T] \times \mathbb{R}^d$ *and verifies* $\sup_{0 \leq t \leq T,\ x \in \mathbb{R}^d} \dfrac{|u(t,x)|}{1+|x|^2} < +\infty.$

Then u is given by the probabilistic representation

$$u(t,x) = \mathbb{E}\bigg[f(X_T^{t,x})e^{-\int_t^T k(r,X_r^{t,x})\mathrm{d}r} + \int_t^T g(s, X_s^{t,x})e^{-\int_t^s k(r,X_r^{t,x})\mathrm{d}r}\mathrm{d}s\bigg]. \quad (4.4.3)$$

The attentive reader has noticed that the sign in front of ∂_t has changed compared to the heat equation (Theorem 4.1.9): actually, this is just a time reversal $t \mapsto T-t$ and a change of condition at $t=0$[7] for the heat equation into a condition at $t=T$[8] for this PDE.

Regarding the terminology, the function f is called the *terminal condition* (at T), g the *source term*, and k is the *discount factor (or attenuation factor)*. The convention on the sign $-k$ comes from the problem in infinite time horizon $T = +\infty$ which requires us to have k positive (or suitably bounded from below); see Theorem 4.4.4.

Regarding the applications to *numerical schemes*, it immediately makes it possible to solve the partial differential equation by Monte-Carlo method, by generating M independent simulations of the diffusion X and by taking the average of functional samples

$$\bigg(f(X_T^{t,x,m})e^{-\int_t^T k(r,X_r^{t,x,m})\mathrm{d}r}$$

$$+ \int_t^T g(s, X_s^{t,x,m})e^{-\int_t^s k(r,X_r^{t,x,m})\mathrm{d}r}\mathrm{d}s\bigg)_{1 \leq m \leq M}.$$

We will see later in Chapter 5 how to discretize the diffusion process X and the above functional.

PROOF:

Let (t,x) be fixed and to simplify, put $X_s = X_s^{t,x}$.

▷ **A custom-tailored Itô formula.** The Itô formula applied to the smooth function v, to the process X between times s_0 and s, gives

$$v(s, X_s) = v(s_0, X_{s_0}) + \int_{s_0}^s [\partial_t + \mathcal{L}]v(r, X_r)\mathrm{d}r$$

$$+ \int_{s_0}^s \nabla_x v(r, X_r)\sigma(r, X_r)\mathrm{d}W_r \quad (4.4.4)$$

---

[7] PDE convention
[8] Probabilistic convention

in an analogous way to the equality (4.4.1). The Itô formula for the product function $(x, y) \mapsto xy$ applied to a generic two-dimensional Itô process

$$(Y_s, Z_s) = \left(y_0 + \int_0^s b_r^Y \, dr + \int_0^s \sigma_r^Y \, dW_r, e^{\int_0^s c_r dr}\right)$$

for some adapted coefficients (b^Y, σ^Y, c) gives

$$Y_s e^{\int_0^s c_r dr} = Y_0 + \int_0^s e^{\int_0^r c_{s_1} ds_1}(c_r Y_r + b_r^Y) dr + \int_0^s e^{\int_0^r c_{s_1} ds_1} \sigma_r^Y dW_r. \tag{4.4.5}$$

Combining (4.4.4) and (4.4.5) with $Y_s = v(s, X_s)$ and $c_s = -k(s, X_s)\mathbf{1}_{s \geq t}$ yields

$$v(s, X_s)e^{-\int_t^s k(s_1, X_{s_1})ds_1} = v(s_0, X_{s_0})e^{-\int_t^{s_0} k(s_1, X_{s_1})ds_1}$$
$$+ \int_{s_0}^s e^{-\int_t^r k(s_1, X_{s_1})ds_1}\left(-k(r, X_r)v(r, X_r) + [\partial_t + \mathcal{L}]v(r, X_r)\right)dr$$
$$+ \int_{s_0}^s e^{-\int_t^r k(s_1, X_{s_1})ds_1} \nabla_x v(r, X_r)\sigma(r, X_r) dW_r \tag{4.4.6}$$

for any time $t \leq s_0 \leq s$. This formula is important and general enough for the applications we need.

▷ **Applications and localization procedure.** We apply the above formula to the function $v(s, y) = u(s, y)$, which is supposed to be smooth, provided that we restrict it to times $s < T$: by taking $s_0 = t$ and exploiting the equation (4.4.2) solved by u, we derive

$$u(s, X_s)e^{-\int_t^s k(s_1, X_{s_1})ds_1} = u(t, x) - \int_t^s e^{-\int_t^r k(s_1, X_{s_1})ds_1} g(r, X_r) dr$$
$$+ \int_t^s e^{-\int_t^r k(s_1, X_{s_1})ds_1} \nabla_x u(r, X_r)\sigma(r, X_r) dW_r \tag{4.4.7}$$

for any $s \in [t, T[$. We wish now to take the expectation to let the stochastic integral disappear but the lack of growth conditions on $\nabla_x u$ does not permit us to proceed directly. For this reason, for $n \geq 1$, we introduce

$$\tau_n = \inf\{s \geq t : |X_s - x| \geq n\} \wedge \left(T - \frac{(T-t)}{n}\right) \in [t, T[.$$

τ_n is a finite stopping time, which by definition has the key property

$$\mathbf{1}_{r < \tau_n}|\nabla_x u(r, X_r)\sigma(r, X_r)| \leq C_n \tag{4.4.8}$$

for a deterministic constant that does depend on n. Thus, for $s = \tau_n$ the equality (4.4.7) becomes

$$u(\tau_n, X_{\tau_n})e^{-\int_t^{\tau_n} k(s_1, X_{s_1})ds_1}$$

$$= u(t,x) - \int_t^{\tau_n} e^{-\int_t^r k(s_1, X_{s_1})ds_1} g(r, X_r) dr$$
$$+ \int_t^T \mathbf{1}_{r<\tau_n} e^{-\int_t^r k(s_1, X_{s_1})ds_1} \nabla_x u(r, X_r) \sigma(r, X_r) dW_r$$

with $\phi_{n,r} = \mathbf{1}_{r<\tau_n} e^{-\int_t^r k(s_1, X_{s_1})ds_1} \nabla_x u(r, X_r) \sigma(r, X_r)$ in $\mathbb{H}_T^2$ in view of (4.4.8) and of the uniform bound on k. It follows, for any $n \geq 1$,

$$u(t,x) = \mathbb{E}\bigg(u(\tau_n, X_{\tau_n}) e^{-\int_t^{\tau_n} k(s_1, X_{s_1})ds_1}$$
$$+ \int_t^{\tau_n} e^{-\int_t^r k(s_1, X_{s_1})ds_1} g(r, X_r) dr\bigg). \qquad (4.4.9)$$

▷ **Completion of proof.** Denote by U_{τ_n} the term inside in the above expectation: it is bounded by $Ce^{(T-t)|k|_\infty}(1+\sup_{s\leq T}|X_s|^2)(1+T)$ where the constant C is related to the growth conditions of u and g: by Theorem 4.3.1, this upper bound is integrable. Additionally, we easily check that τ_n converges a.s. to T as $n \to +\infty$, and thus $U_{\tau_n} \xrightarrow{\text{a.s.}} U_T$ (by using that u is continuous up to T and not only on the open set). All the conditions are met to pass to the limit in (4.4.9) thanks to the dominated convergence theorem. This completes the proof. □

▷ *Some precise details on the assumptions.*

i) The previous theorem is a uniqueness result. There does not exist only one set of minimal assumptions on the data b, σ, f, g, k to ensure that a solution u exists, with the announced estimates. The reader can refer to these specialized monographs: [44, 100, 45, 46, 43]. To schematize, there exist two sets of different working assumptions.

  a) The coefficients b, σ, f, g, k are quite smooth in time and space, bounded with bounded derivatives. Then there exists a solution u that is very smooth, bounded with bounded derivatives.

  b) Under the same assumptions on b, σ, g, k but assuming that f is continuous only, it is not true in general that a smooth solution u exists. If the regularity cannot come from f, it may come from a smoothing effect of the operator[9] $\mathcal{L}$, provided that we suppose a *non-degeneracy condition*. A classic hypothesis is, for instance, the *uniform ellipticity*, which means that the smallest

---
[9] As in the Laplacian case for the heat equation.

eigenvalue of the symmetric non-negative matrix $\sigma\sigma^\top(t,x)$ is bounded away from 0:

$$\inf_{x\in\mathbb{R}^d,\ t\in[0,T],\ \xi\in\mathbb{R}^d \text{ with } |\xi|=1} \xi\cdot\sigma\sigma^\top(t,x)\xi > 0.$$

Another non-degeneracy hypothesis is hypoellipticity, but it goes far beyond our framework.

ii) In some cases (in particular under the assumption of uniform ellipticity), the Feynman-Kac representation holds without the regularity condition on f, which can be proved at the cost of extra technicalities. This is an interesting extension to handle the case of indicator functions (useful for computing the c.d.f. of X_T).

iii) The quadratic growth assumption related to u, f, g is made first to simplify the presentation and second, because we have introduced the stochastic differential equations in the space L_2 only. But results are still true by replacing 2 by $p > 0$.

4.4.3 Linear elliptic partial differential equation

Now we let the time horizon T tend to infinity: then the origin of times is meaningless and we thus suppose that the coefficients do not depend any longer on time. We obtain then a linear elliptic partial differential equation.

Theorem 4.4.4 *Assume that*

i) *the diffusion process X with coefficients b and σ satisfies the notations and assumptions of Theorem 4.3.1, with time-independent coefficients;*

ii) *the functions $g, k : \mathbb{R}^d \mapsto \mathbb{R}$ are two bounded continuous functions, and*

$$\alpha := \inf_{x\in\mathbb{R}^d} k(x) > 0;$$

iii) *there exists a function $u : \mathbb{R}^d \mapsto \mathbb{R}$ continuous, bounded, of class $\mathcal{C}^2$, satisfying*

$$\mathcal{L}u(x) - k(x)u(x) + g(x) = 0, \quad x \in \mathbb{R}^d. \qquad (4.4.10)$$

Then u is given by the probabilistic representation

$$u(x) = \mathbb{E}\left[\int_0^{+\infty} g(X_s^x) e^{-\int_0^s k(X_r^x)\,dr}\,ds\right], \qquad (4.4.11)$$

where X^x is the diffusion process starting from x at time 0.

PROOF:
The proof is similar to that of Theorem 4.4.3, with some simplifications since u does not depend on time. Let us define the stopping time

$$\tau_n = \inf\{s \geq 0 : |X_s - x| \geq n\} \wedge n;$$

thus, (4.4.9) becomes

$$u(x) = \mathbb{E}\left(u(X_{\tau_n}) e^{-\int_0^{\tau_n} k(X_{s_1})\,ds_1} + \int_0^{\tau_n} g(X_r) e^{-\int_0^r k(X_{s_1})\,ds_1}\,dr\right). \qquad (4.4.12)$$

Let n go to infinity: clearly $\tau_n \to +\infty$ a.s. since the solution X does not explode in finite time. Besides, the first term in the expectation is bounded by $|u|_\infty e^{-\alpha \tau_n} \xrightarrow[n\to+\infty]{\text{a.s.}} 0$, while being uniformly bounded by the integrable quantity $|u|_\infty$: therefore, its expectation tends to 0. The second term can be analyzed similarly by observing that $\int_0^\infty e^{-\int_0^r k(X_{s_1})\,ds_1} |g(X_r)|\,dr \leq \int_0^\infty e^{-\alpha r}|g|_\infty\,dr < +\infty$. The result is proved. □

There exist analogous results as $k \equiv 0$ ($\alpha = 0$), but it requires extra conditions on X and g, that are different from the current setting. In particular, X should be ergodic with a stationary distribution μ, and g should be centered for μ: in this case, $\mathbb{E}\left[\int_0^{+\infty} g(X_s^x)\,ds\right]$ can be meaningful since $\mathbb{E}(g(X_s^x)) \xrightarrow[s\to+\infty]{} \int_{\mathbb{R}^d} g(y)\mu(dy) = 0$. The interested reader can refer to [43, Chapter 1].

4.4.4 Linear parabolic partial differential equation with Cauchy-Dirichlet condition

A Dirichlet condition is translated in a probabilistic way by stopping the process X as it exits from a *domain* $D \subset \mathbb{R}^d$ (non-empty open connected set), see Figure 4.6. The boundary of D is denoted by ∂D.

Theorem 4.4.5 *Let $T > 0$ be fixed. Assume that*

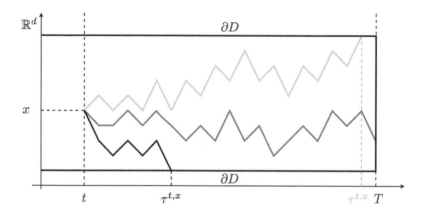

Figure 4.6 Cauchy-Dirichlet condition and several paths of stopped process.

i) the diffusion process X with coefficients b and σ fulfills the notations and assumptions of Theorem 4.3.1;

ii) D is a bounded domain of $\mathbb{R}^d$ and we define[10] $\tau^{t,x} = \inf\{s > t : X_s^{t,x} \notin D\}$ as the first exit time from D by X started at (t,x);

iii) the boundary ∂D is smooth for X in the sense that

$$\forall t \in [0,T], \forall x \in \partial D, \quad \mathbb{P}(\tau^{t,x} = t) = 1;$$

iv) the functions $f, g, k : [0,T] \times \overline{D} \mapsto \mathbb{R}$ are continuous;

v) there exists a continuous function $u : [0,T] \times \overline{D} \mapsto \mathbb{R}$, of class $\mathcal{C}^{1,2}$ on any open set $[0,T[\times D$, which fulfills

$$\begin{cases} \partial_t u(t,x) + \mathcal{L}u(t,x) - k(t,x)u(t,x) + g(t,x) = 0, & t < T, x \in D, \\ u(T,x) = f(T,x), & x \in \overline{D}, \\ u(t,x) = f(t,x), & (t,x) \in [0,T[\times \partial D. \end{cases}$$

(4.4.13)

Then u is given by the probabilistic representation

$$u(t,x) = \mathbb{E}\left[f(\tau^{t,x} \wedge T, X_{\tau^{t,x} \wedge T}^{t,x}) e^{-\int_t^{\tau^{t,x} \wedge T} k(r, X_r^{t,x}) dr}\right]$$

---

[10]With the usual convention $\inf \emptyset = +\infty$.

$$+ \int_t^{\tau^{t,x}\wedge T} g(s, X_s^{t,x})e^{-\int_t^s k(r, X_r^{t,x})\mathrm{d}r}\mathrm{d}s\Big]. \qquad (4.4.14)$$

PROOF:
The points at the boundary require specific analysis.

- For $t = T$, $\tau^{t,x} \wedge T = T$ and the formula (4.4.14) is obvious since it reduces to $u(T,x) = f(T,x)$.
- For $t < T$ and $x \in \partial D$, owing to hypothesis iii) we have $\tau^{t,x} = t$: it readily follows that $u(t,x) = f(t,x)$ as required.

Consider now the interior points $(t,x) \in [0,T[\times D$ and simply write $X_s = X_s^{t,x}$. The proof follows the same lines as Theorem 4.4.3, taking carefully into account that u is continuous up to boundary in space ∂D and in time T, but not necessarily its derivatives. Actually, it is appropriate to suitably localize by stopping at a small positive distance before hitting the boundary.

For $n \geq 1$, set

$$\tau_n = \inf\{s > t : d(X_s, \partial D) \leq \frac{1}{n}\} \wedge \left(T - \frac{(T-t)}{n}\right) \in [t, T[:$$

this is a stopping time such that for $r \leq \tau_n$, (r, X_r) stays in a compact of $[0,T[\times D$ on which the derivatives of u are uniformly bounded (by a constant depending on n). By applying (4.4.6) to $v(s,y) = u(s,y)$, $s_0 = t$ and $s = \tau_n$, we obtain for any $n \geq 1$

$$u(\tau_n, X_{\tau_n})e^{-\int_t^{\tau_n} k(s_1, X_{s_1})\mathrm{d}s_1} = u(t,x)$$
$$+ \int_t^{\tau_n} e^{-\int_t^r k(s_1, X_{s_1})\mathrm{d}s_1}\Big(-k(r, X_r)u(r, X_r) + [\partial_t + \mathcal{L}]u(r, X_r)\Big)\mathrm{d}r$$
$$+ \int_t^{\tau_n} e^{-\int_t^r k(s_1, X_{s_1})\mathrm{d}s_1}\nabla_x u(r, X_r)\sigma(r, X_r)\mathrm{d}W_r.$$

The term $-k(r, X_r)u(r, X_r) + [\partial_t + \mathcal{L}]u(r, X_r)$ equals $-g(r, X_r)$ because u satisfies the partial differential equation in $[0,T[\times D$ and $(r, X_r) \in [0,T[\times D$ for $r < \tau_n$. The last term is equal to $\int_t^T \mathbf{1}_{\tau_n < r} e^{-\int_t^r k(s_1, X_{s_1})\mathrm{d}s_1}\nabla_x u(r, X_r)\sigma(r, X_r)\mathrm{d}W_r$: the integrand is in $\mathbb{H}_T^2$ in view of the previous localization. Therefore, we can take the expectation in the above equality and obtain

$$u(t,x) = \mathbb{E}\Big(u(\tau_n, X_{\tau_n})e^{-\int_t^{\tau_n} k(s_1, X_{s_1})\mathrm{d}s_1}$$
$$+ \int_t^{\tau_n} e^{-\int_t^r k(s_1, X_{s_1})\mathrm{d}s_1}g(r, X_r)\mathrm{d}r\Big). \qquad (4.4.15)$$

Now let us pass to the limit in n: the sequence $(\hat{\tau}_n = \inf\{s > t :$

$d(X_s, \partial D) \leq \frac{1}{n}\})_n$ is increasing, bounded by $\inf\{s > t : d(X_s, \partial D) = 0\} = \tau^{t,x}$, and thus a.s. convergent. Its limit, denoted by $\hat{\tau}_\infty$ say, verifies by continuity of the trajectories of X

$$d(X_{\hat{\tau}_n}, \partial D) = \frac{1}{n} \to d(X_{\hat{\tau}_\infty}, \partial D) = 0.$$

Consequently, $X_{\hat{\tau}_\infty} \notin D$ (since D is open), i.e., $\hat{\tau}_\infty \geq \tau^{t,x}$: to sum up, we have shown $\hat{\tau}_\infty = \tau^{t,x}$ and therefore $\tau_n \xrightarrow[n \to +\infty]{a.s.} \tau^{t,x} \wedge T$.

Since the random variables in (4.4.15) are bounded and continuous in τ_n, it is easy to pass to the limit and obtain the advertised formula. □

▷ *Comments on the assumptions.* This is a very technical point and we give only the main features.

a) The assumption (iii) states that once on the boundary, with probability 1 the process will not go back in D during a small time interval $]t, t + h]$ ($h > 0$), but it will rather pass on both sides of the boundary infinitely often - like a Brownian motion at point 0. This condition is fulfilled if there is enough noise in the dynamics of X (the uniform ellipticity is enough) and if the domain satisfies, at each boundary point, an exterior cone condition,[11] which is automatically satisfied if the boundary is Lipschitz.

b) Similar to the partial differential equation with Cauchy condition, obtaining smooth solutions u requires that we have data b, σ, f, g, k smooth enough, and a domain with a smooth boundary. In the current case of the Dirichlet condition, an additional non-degeneracy condition for X is needed - at least on the boundary.

c) The domain is assumed to be bounded; in fact, it is enough for the boundary to be compact, so that it can be covered by a finite number of local chart changes, but this is not always necessary, as for instance in the case of half-space D. Let us mention that domains with corners - frequent in the applications - require a delicate and specific analysis.

---

[11] One can plot a cone that is locally exterior to D and that touches D only at the boundary.

4.4.5 Linear elliptic partial differential equation with Dirichlet condition

The extension to the elliptic problem is achieved in a similar way by letting $T \to +\infty$. We leave the proof to the reader. Here the coefficients are time-independent.

Theorem 4.4.6 *Assume that*

i) *the diffusion process X with coefficients b and σ satisfies the notations and assumptions of Theorem 4.3.1, with time-independent coefficients;*

ii) *D is a bounded domain of $\mathbb{R}^d$ and we define $\tau^x = \inf\{s > 0 : X_s^x \notin D\}$ as the first exit time from D by X initially started at x;*

iii) *the functions $f, g : \overline{D} \mapsto \mathbb{R}$ and $k : \overline{D} \mapsto \mathbb{R}^+$ are continuous;*

iv) *the boundary ∂D is smooth for X in the sense that*
$$\forall x \in \partial D, \quad \mathbb{P}(\tau^x = 0) = 1;$$

v) *for any $x \in D$, $\mathbb{E}(\tau^x) < +\infty$;*[12]

vi) *there exists a continuous function $u : \overline{D} \mapsto \mathbb{R}$, of class $\mathcal{C}^{1,2}$ for any open set of D, which fulfills*
$$\begin{cases} \mathcal{L}u(x) - k(x)u(x) + g(x) = 0, & x \in D, \\ u(x) = f(x), & x \in \partial D. \end{cases} \quad (4.4.16)$$

Then u is given by the probabilistic representation

$$u(x) = \mathbb{E}\left[f(X_{\tau^x}^x)e^{-\int_0^{\tau^x} k(X_r^x)dr} + \int_0^{\tau^x} g(X_s^x)e^{-\int_0^s k(X_r^x)dr}ds\right]. \quad (4.4.17)$$

4.5 PROBABILISTIC FORMULAS FOR THE GRADIENTS

We continue the discussion of the computation of sensitivity in Section 2.2.4, by specializing the presentation on the sensitivity of $\mathbb{E}(f(X_T^{0,x}))$ with respect to x (the initial condition of X at $t = 0$) through the evaluation of $\partial_x \mathbb{E}(f(X_T^{0,x}))$. By comparing with Theorem 4.4.3 with $g \equiv 0$ and $k \equiv 0$, we seek to represent in a probabilistic way the gradient $\nabla_x u(0, x)$ of the partial differential equation with Cauchy condition.

---

[12] When $g \equiv 0$, this condition can be weakened to $\mathbb{P}(\tau^x < +\infty) = 1$.

4.5.1 Pathwise differentiation method

To follow the approach described in Section 2.2.4, we have to define the a.s.-derivative of $x \mapsto X_t^{0,x}$, as we would for an ordinary differential equation. In the stochastic case, the usual framework to simultaneously define these derivatives a.s. for all t and x is to suppose that the coefficients are a little more regular than $\mathcal{C}^1$ in space: to avoid too technical considerations, we suppose directly in the continuation of the chapter that

$$\boxed{b \text{ and } \sigma \text{ are of class } \mathcal{C}^{0,2} \text{ with bounded derivatives,}}$$

in addition to notations and assumptions of Theorem 4.3.1. We state a differentiability result without proof; see [97] for details.

Proposition 4.5.1 (a.s. differentiability with respect to the initial condition) *Under the previous assumptions, it is possible to define the derivative of $X_t^{0,x}$ with respect to x, which we denote by $\nabla X_t^{0,x}$, and whose dynamics are obtained by formally differentiating (4.3.1) with respect to x: thus $(\nabla X_t^{0,x})_t$ is a matrix-valued process[13] with size $d \times d$ solution of*

$$\nabla X_t^{0,x} := \mathrm{Id} + \int_0^t \nabla_x b(s, X_s^{0,x}) \, \nabla X_s^{0,x} \, ds$$
$$+ \sum_{k=1}^d \int_0^t \nabla_x \sigma_k(s, X_s^{0,x}) \, \nabla X_s^{0,x} \, dW_{k,s} \quad (4.5.1)$$

where σ_k is the k-th column of the matrix σ.
Moreover, for any $T > 0$ we have $\mathbb{E}(\sup_{0 \leq t \leq T} |\nabla X_t^{0,x}|^2) < +\infty$.

Regarding simulation issues, it is worth noting that $(X_t^{0,x}, \nabla X_t^{0,x})_t$ is a stochastic differential equation (in dimension $d + d^2$), to which we can apply the simulation methods of the next Chapter 5.

Combining this with Proposition 2.2.7, we derive the following result.

Corollary 4.5.2 (pathwise differentiation method) *Assume that $f: \mathbb{R}^d \mapsto \mathbb{R}$ is a function $\mathcal{C}^1$ with bounded derivative. Then*

$$\nabla_x \mathbb{E}(f(X_T^{0,x})) = \mathbb{E}(\nabla_x f(X_T^{0,x}) \nabla X_T^{0,x}).$$

---
[13] Also called the tangent process.

4.5.2 Likelihood method

It is not possible to apply the likelihood method of Proposition 2.2.9 because in general the distribution of $X_T^{0,x}$ is not explicit. Instead, we make use of stochastic calculus in a smart way to work out a representation of the gradient in the form of expectation, without letting the derivative of f appear. The following formula, which is quite remarkable, is due to Bismut, Elworthy, and Li; it is the starting point to evaluate by Monte-Carlo method the gradient of solution of partial differential equation. For extensions and ramifications, we refer to [61].

Theorem 4.5.3 (Bismut-Elworthy-Li formula) *Let $T > 0$ be fixed. Assume that*

i) *b and σ satisfy the previous regularity assumptions;*

ii) *there exists a solution $u : [0, T] \times \mathbb{R}^d \mapsto \mathbb{R}$ of class $\mathcal{C}^{1,2}$ with bounded derivatives, and the solution of*

$$\begin{cases} \partial_t u(t,x) + \mathcal{L}u(t,x) = 0, & t < T, x \in \mathbb{R}^d, \\ u(T,x) = f(x), & x \in \mathbb{R}^d, \end{cases}$$

so that $u(t,x) = \mathbb{E}(f(X_T^{t,x}))$;

iii) *$\sigma(.)$ is invertible and its inverse is uniformly bounded.*

Then

$$\nabla_x u(0,x) = \mathbb{E}\Big(\frac{f(X_T^{0,x})}{T}\Big[\int_0^T [\sigma^{-1}(s, X_s^{0,x})\nabla X_s^{0,x}]^\mathsf{T} \mathrm{d}W_s\Big]^\mathsf{T}\Big).$$

Note that if $X = x + W$, then $\sigma^{-1}(s, X_s^{0,x})\nabla X_s^{0,x} = \mathrm{Id}$ and the above formula coincides with that obtained in Example 2.2.11 using computations based on the explicit Gaussian distribution.

PROOF:
Start from the decomposition (4.4.7): as the derivatives of u are uniformly bounded, the preliminary localizations are not necessary and we directly obtain, for any $s \in [0, T]$,

$$u(s, X_s^{0,x}) = u(0,x) + \int_0^s \nabla_x u(r, X_r^{0,x})\sigma(r, X_r^{0,x})\mathrm{d}W_r. \qquad (4.5.2)$$

The term in the stochastic integral is in $\mathbb{H}_T^2$; thus taking the expectation gives $u(0,x) = \mathbb{E}(u(s, X_s^{0,x}))$ for any $s \in [0, T]$, which reads as a time-invariant identity. By applying Corollary 4.5.2 to the function $y \mapsto u(s, y)$, we can deduce a second time-invariant identity

$$\nabla_x u(0,x) = \mathbb{E}(\nabla_x u(s, X_s^{0,x}) \nabla X_s^{0,x}), \quad \forall 0 \leq s \leq T.$$

Actually, the relation can also be obtained by using stochastic calculus (Itô formula) and showing that $(\nabla_x u(s, X_s^{0,x}) \nabla X_s^{0,x})_{0 \leq s \leq T}$ is martingale. Then, by leveraging the covariance equality of Theorem 4.2.3 — a bit complicated because of the vector notations — we obtain

$$\nabla_x u(0,x) = \mathbb{E}\left(\frac{1}{T} \int_0^T \nabla_x u(s, X_s^{0,x}) \nabla X_s^{0,x} \, ds\right)$$

$$= \mathbb{E}\left(\frac{1}{T} \left[\int_0^T \nabla_x u(s, X_s^{0,x}) \sigma(s, X_s^{0,x}) dW_s\right]\right.$$

$$\left. \times \left[\int_0^t [\sigma^{-1}(s, X_s^{0,x}) \nabla X_s^{0,x}]^\mathsf{T} dW_s\right]^\mathsf{T}\right).$$

Thanks to (4.5.2) applied at $s = T$, the first term in the bracket above is equal to $u(T, X_T^{0,x}) - u(0,x) = f(X_T^{0,x}) - u(0,x)$. As the stochastic integral is centered, the contribution with the factor $u(0,x)$ disappears and we obtain the announced formula. □

The above assumptions ensure implicitly that f is smooth (as $u(T,.)$). Nevertheless, extension to functions f without regularity is possible by being more careful in the use of stochastic calculus and since u is smooth for $t < T$ even if f is not.

4.6 EXERCISES

Exercise 4.1 (linear transformation of Brownian motion)

i) Let W be a standard d-dimensional Brownian motion and let U be an orthogonal matrix (i.e. $U^\mathsf{T} = U^{-1}$). Prove that UW defines a new standard d-dimensional Brownian motion.

ii) Application: let W_1 and W_2 be two independent Brownian motions. For any $\rho \in [-1, 1]$, justify that $\rho W_1 + \sqrt{1 - \rho^2} W_2$ and $-\sqrt{1 - \rho^2} W_1 + \rho W_2$ are two independent Brownian motions.

Exercise 4.2 (approximation of the integral of a stochastic

process) *For a standard Brownian motion, we study the convergence rate of the approximation*

$$\Delta I_n := \int_0^1 W_s \mathrm{d}s - \frac{1}{n}\sum_{i=0}^{n-1} W_{\frac{i}{n}}$$

as $n \to +\infty$.

i) *We start by a rough estimate. Prove that*

$$\mathbb{E}(|\Delta I_n|) \leq \sum_{i=0}^{n-1} \mathbb{E}\left(\int_{\frac{i}{n}}^{\frac{i+1}{n}} |W_s - W_{\frac{i}{n}}| \mathrm{d}s\right) = O(n^{-1/2}).$$

ii) *Using Lemma A.1.4, prove that ΔI_n is Gaussian distributed. Compute its parameters and conclude that*

$$\mathbb{E}(|\Delta I_n|) = O(n^{-1}).$$

iii) *A more generic proof of the above estimate consists of writing*

$$\Delta I_n := \sum_{i=0}^{n-1} \int_{\frac{i}{n}}^{\frac{i+1}{n}} (\frac{i+1}{n} - s) \mathrm{d}W_s$$

where we have applied the Itô formula to $s \mapsto (\frac{i+1}{n} - s)(W_s - W_{\frac{i}{n}})$ on each interval $[\frac{i}{n}, \frac{i+1}{n}]$. Using the Itô isometry, derive $\mathbb{E}(|\Delta I_n|^2) = O(n^{-2})$ and therefore the announced estimate.

iv) *Proceeding as in (iii), extend the previous estimate to*

$$\int_0^1 X_s \mathrm{d}s - \frac{1}{n}\sum_{i=0}^{n-1} X_{\frac{i}{n}}$$

where X is a scalar Itô process with bounded coefficients (Definition 4.2.4).

Exercise 4.3 (approximation of stochastic integral) *We consider the convergence rate of the approximation*

$$\Delta J_n := \int_0^1 Z_s \mathrm{d}W_s - \sum_{i=0}^{n-1} Z_{\frac{i}{n}}(W_{\frac{i+1}{n}} - W_{\frac{i}{n}})$$

where $Z_s := f(s, W_s)$ for some function f, such that $\mathbb{E}\int_0^1 |Z_s|^2 \mathrm{d}s + \sup_{i<n} \mathbb{E}|Z_{\frac{i}{n}}|^2 < +\infty$. We illustrate that the convergence order is, under mild conditions, equal to $1/2$ but it can be smaller for irregular f.

i) Show that $\mathbb{E}|\Delta J_n|^2 = \mathbb{E}\left(\sum_{i=0}^{n-1} \int_{\frac{i}{n}}^{\frac{i+1}{n}} |Z_s - Z_{\frac{i}{n}}|^2 ds\right)$.

ii) When $Z_s = W_s$, show that $\mathbb{E}|\Delta J_n|^2 \sim \operatorname{Cst} n^{-1}$ for some positive constant.

iii) Assuming that f is bounded, smooth with bounded derivatives, prove that $\mathbb{E}|\Delta J_n|^2 = O(n^{-1})$.

iv) Assume that Z is a square-integrable martingale. Show that $\mathbb{E}|Z_s - Z_{\frac{i}{n}}|^2 \leq \mathbb{E}|Z_{\frac{i+1}{n}}|^2 - \mathbb{E}|Z_{\frac{i}{n}}|^2$, and thus $\mathbb{E}|\Delta J_n|^2 \leq (\mathbb{E}|Z_1|^2 - \mathbb{E}|Z_0|^2)n^{-1}$.

v) Set $Z_s := \mathcal{N}'(W_s/\sqrt{1-s})/\sqrt{1-s}$. Establish that $n^{1/2}\mathbb{E}|\Delta J_n|^2$ is bounded away from 0, for n large enough.

Exercise 4.4 (exact simulation of Ornstein-Uhlenbeck process) Two processes $(X_t)_{t\geq 0}$ and $(Y_t)_{t\geq 0}$ have the same distribution if for any $n \in \mathbb{N}$ and any $0 \leq t_1 < \cdots < t_n$, the vectors $(X_{t_1}, \cdots, X_{t_n})$ and $(Y_{t_1}, \cdots, Y_{t_n})$ have the same distribution. Let us consider the Ornstein-Uhlenbeck process $(X_t)_{t\geq 0}$, solution of

$$X_t = x_0 - a\int_0^t X_s ds + \sigma W_t,$$

where $x_0 \in \mathbb{R}$, $\sigma > 0$, and $(W_t)_{t\geq 0}$ is a standard Brownian motion.

i) By applying the Itô formula to $e^{at}X_t$, give an explicit representation for X_t in terms of stochastic integrals.

ii) Deduce the explicit distribution of $(X_{t_1}, \cdots, X_{t_n})$.

iii) Find two functions $\alpha(t)$ and $\beta(t)$ such that $(X_t)_{t\geq 0}$ has the same distribution as $(Y_t)_{t\geq 0}$ with $Y_t = \alpha(t)(x_0 + W_{\beta(t)})$. Design a scheme for the exact simulation of the Ornstein-Uhlenbeck process.

Exercise 4.5 (Transformations of SDE and PDE) For any $t \in [0,T)$ and $x \in \mathbb{R}$, we denote by $(X_s^{t,x}, s \in [t,T])$ the solution to

$$X_s = x + \int_t^s b(X_r)dr + \int_t^s \sigma(X_r)dW_r, \qquad t \leq s \leq T$$

where the coefficients $b, \sigma : \mathbb{R} \to \mathbb{R}$ are smooth with bounded derivatives, and $\sigma(x) \geq c > 0$. For a given Borel set $A \subset \mathbb{R}$, we define $u(t,x) := \mathbb{P}(X_T^{t,x} \in A)$. We assume in the following $u(t,x) > 0$ for any $(t,x) \in [0,T) \times \mathbb{R}$, and that appropriate smoothness assumptions are satisfied (namely, $u \in C^{1,2}([0,T) \times \mathbb{R})$).

i) Let $x_0 \in \mathbb{R}$ and f be a bounded continuous function. Using the PDE satisfied by u on $[0,T) \times \mathbb{R}$, show that

$$\mathbb{E}[f(X_t)|X_T \in A] = \frac{\mathbb{E}[f(X_t)u(t, X_t)]}{u(0, x_0)}, \qquad \forall t < T,$$

where $X_t = X_t^{0, x_0}$ to simplify.

ii) We assume that for any $s \leq t < T$ the equation

$$\overline{X}_r = x + \int_s^r \left(b(\overline{X}_w) + \sigma^2(\overline{X}_w) \frac{\partial_x u}{u}(w, \overline{X}_w) \right) dw$$
$$+ \int_s^r \sigma(\overline{X}_w) dW_w, \qquad s \leq r \leq t$$

has a unique solution, denoted by $(\overline{X}_r^{s,x}, s \leq r \leq t)$. We set $v_t(s, x) := \mathbb{E}[f(\overline{X}_t^{s,x})]$.

  (a) What is the PDE solved by $(s, x) \mapsto v_t(s, x)$ on $[0, t) \times \mathbb{R}$?
  (b) Applying the Itô formula to $u(s, X_s)$ and $v_t(s, X_s)$, $0 \leq s \leq t$, and then to $u(s, X_s)v_t(s, X_s)$, show

$$\mathbb{E}[f(X_t)u(t, X_t)] = v_t(0, x_0)u(0, x_0), \qquad \forall t < T.$$

  (c) Conclude that for any $t < T$, the distribution of X_t given $\{X_T \in A\}$ is the distribution of $\overline{X}_t^{0, x_0}$.

iii) In the case $b = 0$, $\sigma(x) = 1$ and $A = (y - R, y + R)$, show that $\frac{\partial_x u(t,x)}{u(t,x)} \to -\frac{x-y}{T-t}$ for any (t, x) as $R \to 0$. Interpret the solution to the following equation in terms of a Brownian bridge:

$$\overline{X}_t = x_0 - \int_0^t \frac{\overline{X}_s - y}{T - s} ds + W_t.$$

Exercise 4.6 (Exit time from a domain) Consider the solution $(X_t^x)_t$ of a stochastic differential equation in $\mathbb{R}^d$, starting from x at time 0, with time-independent coefficients (b, σ) satisfying the usual Lipschitz conditions of Theorem 4.3.1. Let D be a non-empty open connected set of $\mathbb{R}^d$ and set

$$\tau_D^x = \inf\{t \geq 0 : X_t^x \notin D\}$$

for the first exit time from D.

i) *Assume first that* $\sup_{x \in D} \mathbb{E}[\tau_D^x] \leq c$, *for some constant* $c > 0$.

   (a) *Show that* $\mathbb{E}[(\tau_D^x)^k] \leq k! c^k$ *for any* $k \in \mathbb{N}$.
   Hint: *use the identity* $\frac{1}{k} T^k = \int_0^T (T-t)^{k-1} dt$ *and the Markov property of* X^x.

   (b) *Deduce that* $\sup_{x \in D} \mathbb{E}[e^{\lambda \tau_D^x}] < \infty$ *for any* $\lambda < c^{-1}$. *What are the consequences of this result on the simulation of the path of* X *up to* τ_D^x?

   (c) *Set* $\gamma(t) = \sup_{x \in D} \mathbb{P}(\tau_D^x > t)$. *Show* $\gamma(t+s) \leq \gamma(t) \gamma(s)$ *for any* $t, s \geq 0$.

   (d) *The previous question shows that the function* $t \mapsto \ln \gamma(t) \in [-\infty, 0]$ *is sub-additive: by the Fekete lemma, the limit*

   $$\lim_{t \to \infty} \frac{1}{t} \ln \gamma(t) = \inf_{t > 0} \frac{1}{t} \ln \gamma(t) =: -\alpha_D$$

   exists in $[-\infty, 0]$. *Show that* $\alpha_D \geq c^{-1}$.

ii) *Assume now that* D *is bounded. The infinitesimal generator of* X *is denoted by* $\mathcal{L}$.

   (a) *Suppose there exists* $f \in C^2(\mathbb{R}^d, \mathbb{R})$ *such that* $f(x) \geq 0$ *and* $\mathcal{L} f(x) \leq -1$ *for* $x \in D$. *Show that* $\sup_{x \in D} \mathbb{E}[\tau_D^x] \leq c :=$ $\sup_{y \in D} f(y)$.

   (b) *In the case* $\inf_{x \in D} [\sigma \sigma^T]_{i,i}(x) > 0$ *for some* i, *exhibit such a function* f.

Exercise 4.7 (Bismut-Elworthy-Li formula) *Let* X^x *be the Ornstein-Uhlenbeck process solution of*

$$X_t = x - a \int_0^t X_s ds + \sigma W_t,$$

with $x \in \mathbb{R}$ *and* $\sigma > 0$. *We aim at computing* $\partial_x \mathbb{E}(f(X_T^x))$ *for bounded smooth function* f.

i) *Using the sensitivity formula of Theorem 4.5.3, show that*

$$\partial_x \mathbb{E}(f(X_T^x)) = \mathbb{E}\left(f(X_T^x) \int_0^T \frac{e^{-at}}{\sigma T} dW_t\right).$$

ii) Show that the sensitivity formula is still valid by replacing $\int_0^T \frac{e^{-at}}{\sigma T} \mathrm{d}W_t$ by its conditional expectation given X_T^x. Compute this conditional expectation explicitly.

iii) Prove that the new formula coincides with that given by the likelihood ratio method using the Gaussian distribution of X_T^x (Proposition 2.2.9).

iv) Which representation among that *(i)* or *(iii)* has the smallest variance?

CHAPTER 5

Euler scheme for stochastic differential equations

Stochastic differential equations $(X_t : t \geq 0)$ provide flexible models for stochastic modeling in continuous time and space (see the examples in Section 4.3.3), and also powerful probabilistic tools for the numerical solution of partial differential equations via the expectation of functionals of the corresponding processes.

To implement a Monte-Carlo evaluation of $\mathbb{E}(f(X_T))$ or of more complicated functionals, the simulation of X is necessary, either at a fixed date, or at multiple dates, or ideally even at a continuum of dates to account for path dependency (for instance, for problems involving exit time). Unfortunately, there exist few diffusion models for which the simulation is explicit and simple: among them, let us mention arithmetic and geometric Brownian motion, the Ornstein-Uhlenbeck process. In almost all other cases, approximations are necessary and the purpose of the chapter is to study these approximations.

We will concentrate our efforts on a usual approximation method based on time discretization, which is called the Euler scheme: this method has the advantage of being simple, relatively efficient, and adaptable to any dimension of the space. This is a natural extension of the scheme for ordinary differential equations. The version for stochastic differential equations was first proposed by Maruyama [110] and sometimes is called the Euler-Maruyama scheme. For references on this subject, see [92].

164 ■ Euler scheme for stochastic differential equations

In this chapter, we consider a stochastic differential equation of the form

$$X_t = x + \int_0^t b(s, X_s)\mathrm{d}s + \int_0^t \sigma(s, X_s)\mathrm{d}W_s, \quad t \geq 0, \quad (5.0.1)$$

where X and W are d-dimensional. The dependence on the initial condition x is not important and we omit it in our notations. The hypotheses on the drift and diffusion coefficients (b, σ) are given along with the statement of the results.

5.1 DEFINITION AND SIMULATION

The Euler scheme for stochastic differential equations is a natural extension of the Euler scheme for an ordinary differential equation: it consists of locally freezing the coefficients on small time intervals.

5.1.1 Definition as an Itô process, quadratic moments

Definition 5.1.1 *The Euler scheme $X^{(h)}$ associated with the stochastic differential equation with coefficients (b, σ) and with a time step h is defined by*

$$\begin{cases} X_0^{(h)} = x; \\ X_t^{(h)} = X_{ih}^{(h)} + b(ih, X_{ih}^{(h)})(t - ih) + \sigma(ih, X_{ih}^{(h)})(W_t - W_{ih}), \\ \quad \text{for} \quad i \geq 0, \quad t \in (ih, (i+1)h]. \end{cases}$$

$$(5.1.1)$$

In other terms, $X^{(h)}$ is a piecewise arithmetic Brownian motion, for which the coefficients on the interval $(ih, (i+1)h]$ are calculated using the functions (b, σ) evaluated at $(ih, X_{ih}^{(h)})$. In general, the distribution of X_t^h is not explicitly known: at most, we can give an expression in an iterative form over different discretization intervals, but it is not very exploitable.

To easily apply stochastic calculus to the Euler scheme, it is useful to represent it as an Itô process (see Definition 4.2.4). For this, we use an appropriate notation for the discretization time before t:

$$\varphi_t^h := ih \quad \text{if } ih < t \leq (i+1)h.$$

With this notation in hand, we obtain the following representation and we derive quadratic moment estimates, bounded uniformly in time step.

Proposition 5.1.2 (Itô process) *The Euler scheme is an Itô process given by*

$$X_t^{(h)} = x + \int_0^t b(\varphi_s^h, X_{\varphi_s^h}^{(h)}) ds + \int_0^t \sigma(\varphi_s^h, X_{\varphi_s^h}^{(h)}) dW_s, \qquad t \geq 0. \quad (5.1.2)$$

Supposing that the coefficients (b, σ) satisfy

i) $|b(t, x) - b(t, y)| + |\sigma(t, x) - \sigma(t, y)| \leq C_{b,\sigma} |x - y|$ *for all* $(t, x, y) \in [0, T] \times \mathbb{R}^d \times \mathbb{R}^d$,

ii) $\sup_{0 \leq t \leq T} (|b(t, 0)| + |\sigma(t, 0)|) \leq C_{b,\sigma}$.

Then we have, for all $T > 0$,

$$\mathbb{E}(\sup_{0 \leq t \leq T} |X_t^{(h)}|^2) \leq C(1 + |x|^2),$$

for a constant C depending on T and $C_{b,\sigma}$ (and independent of h).

PROOF:
    The arguments used below will appear again several times in the following and this proof serves as a warm-up.
    ▷ The equality (5.1.2) is immediate.
    ▷ It remains to show that $X^{(h)}$ satisfies the same moment estimate as X. Set

$$M(s) = \sup_{0 \leq t \leq s} |X_t^{(h)}|^2, \qquad m(s) = \mathbb{E}(M(s))$$

and let us show first that this expectation is finite, proving by induction on i that this is true for $m(ih)$. For $i = 0$, there is nothing to prove.

Now suppose that $m(ih)$ is finite. We recall classic convexity inequalities, often used in the following:

$$(a_1 + \cdots + a_n)^2 \leq n(a_1^2 + \cdots + a_n^2), \quad \forall n \in \mathbb{N}^*, \forall (a_1, \ldots, a_n) \in \mathbb{R}^n, \tag{5.1.3}$$

$$\left(\int_s^t a_r dr\right)^2 \leq (t - s) \int_s^t |a_r|^2 dr, \quad \forall s \leq t, \forall (a_r)_r \text{ real square integrable}. \tag{5.1.4}$$

From this we deduce that for any s, we have

$$|b(\varphi_s^h, X_{\varphi_s^h}^{(h)})|^2 = |b(\varphi_s^h, 0) + b(\varphi_s^h, X_{\varphi_s^h}^{(h)}) - b(\varphi_s^h, 0)|^2$$
$$\leq 2C_{b,\sigma}^2 (1 + |X_{\varphi_s^h}^{(h)}|^2),$$

166 ■ Euler scheme for stochastic differential equations

$$|\sigma(\varphi_s^h, X_{\varphi_s^h}^{(h)})|^2 \leq 2C_{b,\sigma}^2(1 + |X_{\varphi_s^h}^{(h)}|^2). \tag{5.1.5}$$

So, using (5.1.1) and the previous bounds, we deduce

$$\mathbb{E}\Big(\sup_{ih \leq s \leq (i+1)h} |X_s^{(h)}|^2 \big| X_{ih}^{(h)}\Big) \leq 3|X_{ih}^{(h)}|^2 + 3|b(ih, X_{ih}^{(h)})|^2 h^2$$

$$+ 3|\sigma(ih, X_{ih}^{(h)})|^2 \underbrace{\mathbb{E}\Big(\sup_{ih \leq s \leq (i+1)h} |W_s - W_{ih}|^2 \big| X_{ih}^{(h)}\Big)}_{\leq c_2 h \text{ (Doob maximal inequality, Theorem 4.2.3)}}$$

$$\leq 3\big(|X_{ih}^{(h)}|^2 + 2C_{b,\sigma}^2(1 + |X_{ih}^{(h)}|^2)h^2 + 2C_{b,\sigma}^2(1 + |X_{ih}^{(h)}|^2)c_2 h\big);$$

then by taking the expectation in the above inequality, we get that $m((i+1)h) \leq m(ih) + \mathbb{E}\big(\sup_{ih \leq s \leq (i+1)h} |X_s^{(h)}|^2\big) \leq C_1(1 + m(ih)) < +\infty$, for a certain constant C_1: the result "m finite" is proved.

▷ A little more work is needed to get an explicit upper bound, uniform in h, but the ideas are the same, i.e., by combining the convexity inequalities (5.1.3–5.1.4) and the Doob maximal inequality applied to $\int_0^t \sigma(\varphi_s^h, X_{\varphi_s^h}^{(h)})dW_s$ (with an integrand in $\mathbb{H}_T^2$ by (5.1.5) and the fact that m is finite). Performing the calculations with fewer details this time, starting from (5.1.2) we can write for $t \leq T$

$$m(t) \leq 3\Big(|x|^2 + t\int_0^t \mathbb{E}\big(|b(\varphi_s^h, X_{\varphi_s^h}^{(h)})|^2\big)ds + c_2\int_0^t \mathbb{E}\big(|\sigma(\varphi_s^h, X_{\varphi_s^h}^{(h)})|^2\big)ds\Big)$$

$$\leq 3\Big(|x|^2 + 2TC_{b,\sigma}^2\int_0^t (1 + m(s))ds + 2c_2 C_{b,\sigma}^2 \int_0^t (1 + m(s))ds\Big)$$

$$\leq 3|x|^2 + KT + K\int_0^t m(s)ds \tag{5.1.6}$$

where we set $K = 6C_{b,\sigma}^2(T + c_2)$. An application of the Gronwall lemma[1] finally gives $m(T) \leq (3|x|^2 + KT)e^{KT}$. □

5.1.2 Simulation

Observe that to simulate a complete trajectory for $X^{(h)}$ is as difficult as for Brownian motion, thus impossible.

▷ Most often, we need only to sample $X^{(h)}$ at the discretization times ih; this is done iteratively:

1. Initialization with $X_0^{(h)} = x$;

2. Then iteration with $i = 0, \ldots$:

---

[1] If $f : [0, T] \mapsto \mathbb{R}^+$ and $f(t) \leq a + b\int_0^t f(s)ds < +\infty$, then $f(t) \leq ae^{bt}$.

(a) $X^{(h)}_{ih}$ is already generated;

(b) simulation of d random variables $W_{k,(i+1)h} - W_{k,ih}$, with centered Gaussian distribution and variance h;

(c) calculation of $X^{(h)}_{(i+1)h}$ with the help of (5.1.1).

The procedure is thus very simple and has a computational cost equal to $C(d)n$, if n intervals of discretization are covered, and where $C(d)$ comes from vector and matrix calculations in dimension d.

▷ If we need to generate $X^{(h)}$ outside the set of discretization times, say $X^{(h)}_t$ for $t \in (ih, (i+1)h]$, two cases appear depending on whether $X^{(h)}_{(i+1)h}$ has been already simulated or not.

a) If only the random variables $(X^{(h)}_{jh})_{j \leq i}$ has been simulated, it is enough to *extend the trajectory*, by simulating another Brownian increment $W_t - W_{ih}$ independent of the rest and calculating $X^{(h)}_t$ by (5.1.1).

b) If $(X^{(h)}_{jh})_{0 \leq j \leq n}$ (with $n \geq i+1$) has been simulated, we are rather in the situation of trajectory *refining* between ih and $(i+1)h$: we can then use the technique of the Brownian bridge seen in Section 4.1.3, considering only the values at ih and $(i+1)h$. This is justified by the following result.

Lemma 5.1.3 *Let $i \geq 0$.*

- *The distribution of $(X^{(h)}_t)_{t \in [ih,(i+1)h]}$ conditionally on $(X^{(h)}_t : t \leq ih, t \geq (i+1)h)$ coincides with the distribution of $(X^{(h)}_t)_{t \in [ih,(i+1)h]}$ conditionally on $(X^{(h)}_{ih}, X^{(h)}_{(i+1)h})$.*

- *If $\sigma(ih, X^{(h)}_{ih})$ is invertible, the distribution of*

$$\left(\sigma^{-1}(ih, X^{(h)}_{ih}) X^{(h)}_t\right)_{t \in [ih,(i+1)h]}$$

conditionally on $(X^{(h)}_{ih}, X^{(h)}_{(i+1)h})$ coincides with that of a d-dimensional Brownian bridge (with independent components) between the points

$$\left(ih, \sigma^{-1}(ih, X^{(h)}_{ih}) X^{(h)}_{ih}\right)$$

and

$$\left((i+1)h, \sigma^{-1}(ih, X^{(h)}_{ih}) X^{(h)}_{(i+1)h}\right).$$

168 ■ Euler scheme for stochastic differential equations

In particular, it doesn't depend on $b(ih, X_{ih}^{(h)})$.

PROOF:
The first statement comes from the Markov property of $X^{(h)}$ at times $(ih)_i$, as in the proof of Lemma 4.1.7. The second property follows from Lemma 4.1.7, applied component by component. □

5.1.3 Application to computation of diffusion expectation: discretization error and statistical error

For the computation of the expectation of a functional of the diffusion process X — motivated directly by a probabilistic application or indirectly by the numerical solution of a partial differential equation — a Monte-Carlo method can be naturally designed, using a simulation of the discretized trajectories of $X^{(h)}$. Let us discuss, for example, the case of the functional

$$\mathcal{E}(f, g, k, X) := f(X_T) e^{-\int_0^T k(r, X_r) dr} + \int_0^T g(s, X_s) e^{-\int_0^s k(r, X_r) dr} ds$$

appearing in Theorem 4.4.3. To simplify, let us divide the interval $[0, T]$ into N regular sub-intervals with equal size in order to use the Euler scheme with a step $h = T/N$: the functional for the Euler scheme is then approximated by

$$\mathcal{E}(f, g, k, X^{(h)}) = f(X_T^{(h)}) e^{-h \sum_{j=0}^{N-1} k(jh, X_{jh}^{(h)})}$$
$$+ h \sum_{i=0}^{N-1} g(ih, X_{ih}^{(h)}) e^{-h \sum_{j=0}^{i-1} k(jh, X_{jh}^{(h)})}. \quad (5.1.7)$$

Denote by $(X_{ih}^{(h,m)})_{0 \leq i \leq N}$ the m-th simulation of the Euler scheme along the discretization grid: we generate M independent copies.

Therefore, the Monte-Carlo evaluation of $\mathbb{E}(\mathcal{E}(f, g, k, X))$ is written as

$$\mathbb{E}(\mathcal{E}(f, g, k, X)) \approx \frac{1}{M} \sum_{m=1}^{M} \mathcal{E}(f, g, k, X^{(h,m)}). \quad (5.1.8)$$

Moreover, the error of the method can be decomposed in the superposition of two errors with a different nature:

$$\frac{1}{M} \sum_{m=1}^{M} \mathcal{E}(f, g, k, X^{(h,m)}) - \mathbb{E}(\mathcal{E}(f, g, k, X))$$

```
1  M, N, m, i: int ;         /* numbers of simulations and time
   steps, variable indices */
2  T, h: double ;            /* for the terminal time and the time
   step */
3  x0, x: vector ; /* for $X_0^{(h)}$ and the current value of $X_{ih}^{(h)}$
   */
4  disc: double ;            /* for the discount factor
   $\exp(-h\sum_{j=0}^{i-1} k(jh, X_{jh}^{(h)}))$ */
5  efgk, sum, sum2: double ; /* for $\mathcal{E}(f, g, k, X^{(h)})$, its sum
   and sum of its squares */
6  $sum \leftarrow 0$; $sum2 \leftarrow 0$; $h \leftarrow T/N$;
7  for $m = 1$ to $M$ do
8  |   $x \leftarrow x0$; $efgk \leftarrow 0$; $disc \leftarrow 1$;
9  |   for $i = 0$ to $N - 1$ do
10 |   |   $efgk \leftarrow efgk + h \times g(i \times h, x) \times disc$;
11 |   |   $disc \leftarrow disc \times \exp(-h \times k(i \times h, x))$;
12 |   |   $x \leftarrow x + b(i \times h, x) \times h + \sigma(i \times h, x) \times \sqrt{h} \times \mathcal{N}(0, \text{Id})$;
13 |   $efgk \leftarrow efgk + f(x) \times disc$;
14 |   $sum \leftarrow sum + efgk$;
15 |   $sum2 \leftarrow sum2 + efgk \times efgk$;
16 $sum \leftarrow sum/M$; $sum2 \leftarrow sum2/M$ ;    /* to obtain the
   averages */
17 moyMC $\leftarrow sum$ ;   /* 1st result: empirical mean */
18 IC $\leftarrow \frac{1.96}{\sqrt{M}} \times \sqrt{\frac{M}{(M-1)} \times (sum2 - sum \times sum)}$ ;   /* 2nd
   result: half Conf. Inter. */
```

Algorithm 5.1: Computation by Monte-Carlo method of $\mathbb{E}(f(X_T)e^{-\int_0^T k(r, X_r)dr} + \int_0^T g(s, X_s)e^{-\int_0^s k(r, X_r)dr}ds)$, with calculation of the empirical mean moyMC and confidence intervals at 95% (IC for the half-width).

$$= \underbrace{\frac{1}{M}\sum_{m=1}^{M}\mathcal{E}(f,g,k,X^{(h,m)}) - \mathbb{E}(\mathcal{E}(f,g,k,X^{(h,m)}))}_{\text{statistical error}}$$

$$+ \underbrace{\mathbb{E}(\mathcal{E}(f,g,k,X^{(h,m)})) - \mathbb{E}(\mathcal{E}(f,g,k,X))}_{\text{discretization error}}. \qquad (5.1.9)$$

- The statistical error is due to the finite number of simulations: by the law of large numbers, it converges to 0 as $M \to +\infty$. It depends also on h but its impact is minor. In the next chapter we analyze in more detail this statistical error, quantifying it precisely with non-asymptotic confidence intervals as in Chapter 2, or reducing it with variance reduction methods.

- The second error is due to the effect of time discretization only: we expect that the smaller the value of h is, the better the accuracy. The objective in the rest of this chapter is to give estimates of the convergence rate as a function of h.

The optimal tuning between $M \to +\infty$ and $h \to 0$ is discussed in the next chapter.

5.2 STRONG CONVERGENCE

We justify the convergence of the Euler approximation to the corresponding stochastic differential equation, in the sense of L_p-norm: this reads as a certain closeness of the trajectories of two processes, called *strong convergence*. The time regularity hypotheses on the coefficients (b, σ) are slightly strengthened.

Theorem 5.2.1 (strong convergence at order $\frac{1}{2}$) *Suppose that the coefficients (b, σ) from the equation (5.0.1) for X satisfy*

i) $|b(t,x) - b(s,y)| + |\sigma(t,x) - \sigma(s,y)| \leq C_{b,\sigma}(|x-y| + |t-s|^{\frac{1}{2}})$ for all $(t,s,x,y) \in [0,T] \times [0,T] \times \mathbb{R}^d \times \mathbb{R}^d$,

ii) $|b(0,0)| + |\sigma(0,0)| \leq C_{b,\sigma}$,

for a certain finite constant $C_{b,\sigma}$.

For any $p > 0$, there exists a constant C — depending on T, x, $C_{b,\sigma}$ and p — such that

$$\mathbb{E}(\sup_{0 \leq t \leq T} |X_t^{(h)} - X_t|^p) \leq Ch^{\frac{p}{2}}. \qquad (5.2.1)$$

We say that the strong convergence is of *order* $\frac{1}{2}$ because the L_p-norm of the error is of magnitude $h^{\frac{1}{2}}$.

PROOF:

We show the upper bound only for $p = 2$; the case $p \neq 2$ is similar but it requires some complementary results from stochastic calculus that are not presented in this book.

Set $\varepsilon_t = \sup_{0 \leq s \leq t} |X_s^{(h)} - X_s|^2$ and decompose the error as

$$X_t^{(h)} - X_t$$
$$= \int_0^t (b(\varphi_s^h, X_{\varphi_s^h}^{(h)}) - b(s, X_s^{(h)}))\mathrm{d}s + \int_0^t (b(s, X_s^{(h)}) - b(s, X_s))\mathrm{d}s$$
$$+ \int_0^t (\sigma(\varphi_s^h, X_{\varphi_s^h}^{(h)}) - \sigma(s, X_s^{(h)}))\mathrm{d}W_s + \int_0^t (\sigma(s, X_s^{(h)}) - \sigma(s, X_s))\mathrm{d}W_s.$$

Proceeding as for the inequality (5.1.6), we obtain (for $t \leq T$)

$$\mathbb{E}(\varepsilon_t) \leq 4\Big(t \int_0^t \mathbb{E}|b(\varphi_s^h, X_{\varphi_s^h}^{(h)}) - b(s, X_s^{(h)})|^2 \mathrm{d}s$$
$$+ t \int_0^t \mathbb{E}|b(s, X_s^{(h)}) - b(s, X_s)|^2 \mathrm{d}s \quad (5.2.2)$$
$$+ c_2 \int_0^t \mathbb{E}|\sigma(\varphi_s^h, X_{\varphi_s^h}^{(h)}) - \sigma(s, X_s^{(h)})|^2 \mathrm{d}s$$
$$+ c_2 \int_0^t \mathbb{E}|\sigma(s, X_s^{(h)}) - \sigma(s, X_s)|^2 \mathrm{d}s\Big) \quad (5.2.3)$$
$$\leq 4\Big(2(T + c_2)C_{b,\sigma}^2 \int_0^t (\mathbb{E}|X_s^{(h)} - X_{\varphi_s^h}^{(h)}|^2 + (s - \varphi_s^h))\mathrm{d}s$$
$$+ (T + c_2)C_{b,\sigma}^2 \int_0^t \mathbb{E}|X_s^{(h)} - X_s|^2 \mathrm{d}s\Big). \quad (5.2.4)$$

An exact calculation for the Euler scheme increment gives, for $s \in [0, T]$,

$$\mathbb{E}|X_s^{(h)} - X_{\varphi_s^h}^{(h)}|^2$$
$$= \mathbb{E}\Big(\mathbb{E}(|X_s^{(h)} - X_{\varphi_s^h}^{(h)}|^2 | X_{\varphi_s^h}^{(h)})\Big)$$
$$= (s - \varphi_s^h)^2 \mathbb{E}|b(\varphi_s^h, X_{\varphi_s^h}^{(h)})|^2 + (s - \varphi_s^h)\underbrace{\mathbb{E}(\mathrm{Tr}(\sigma\sigma^\mathsf{T}(\varphi_s^h, X_{\varphi_s^h}^{(h)})))}_{\leq d\ \mathbb{E}|\sigma(\varphi_s^h, X_{\varphi_s^h}^{(h)})|^2}$$
$$\leq 2C_{b,\sigma}^2(1 + m(\varphi_s^h))((s - \varphi_s^h)^2 + d\ (s - \varphi_s^h))$$
$$\leq 2C_{b,\sigma}^2(1 + m(T))(h + d)h$$

using (5.1.5) and the notation $m(t) = \mathbb{E} \sup_{0 \leq s \leq t} |X_s^{(h)}|^2$. Plugging this upper bound into (5.2.4) yields

$$\mathbb{E}(\varepsilon_t) \leq 8(T+c_2)C_{b,\sigma}^2 T\Big(2C_{b,\sigma}^2(1+m(T))(h+d)h + h\Big)$$
$$+ 4(T+c_2)C_{b,\sigma}^2 \int_0^t \mathbb{E}(\varepsilon_s)\,\mathrm{d}s.$$

An application of the Gronwall lemma allows us finally to get

$$\mathbb{E}(\varepsilon_t) \leq 8(T+c_2)C_{b,\sigma}^2 T\Big(2C_{b,\sigma}^2(1+m(T))(h+d) + 1\Big)e^{4(T+c_2)C_{b,\sigma}^2 T}h.$$

$\square$

▷ Remarks. The estimate of the convergence rate in Theorem 5.2.1 is optimal and cannot be improved in full generality. Observe that the global Lipschitz hypothesis for the coefficients is important in the proof to obtain the error estimate: it is very delicate in general to weaken this hypothesis while keeping the same convergence order.

– For example, if the coefficients are Hölder continuous, there exist very few results in this generalized context, see [68].

– Recently, in [70, 81] the authors have nicely demonstrated that the linear growth condition may be crucial in the derivation of error estimates: namely, for any given (arbitrary slow) convergence rate, they provide examples of stochastic differential equations with smooth coefficients but with superlinear growth, for which the Euler scheme converges in a strong sense more slowly than the given rate.

If the diffusion coefficient σ does not depend on x, it is possible to show — adapting cleverly the previous calculations and supposing that the coefficients are Lipschitz in time — that the order of convergence becomes 1 with respect to the time step (as for an ordinary differential equation); see Exercise 5.1. This shows that the largest contribution to the discretization error comes from the approximation of the stochastic integral. Convergence schemes with higher orders exist — such as the Milshtein scheme — but in general, they cannot be easily implemented when X is a multidimensional diffusion, which certainly restricts their practical interest.

▷ **Pathwise approximation.** From Theorem 5.2.1, we deduce the convergence rate in the pathwise sense. Indeed, for all $\eta \in]0, \frac{1}{2}[$ and setting $p = 2/\eta$, we remark that $e_N = N^{\frac{1}{2}-\eta} \sup_{0 \le t \le T} |X_t^{T/N} - X_t|$ is such that $\mathbb{E}(\sum_{N \ge 1} e_N^p) = \sum_{N \ge 1} \mathbb{E}(e_N^p) < +\infty$: hence, the series with the general term $(e_N^p)_{N \ge 1}$ is a.s. convergent, and thus bounded.

Corollary 5.2.2 (pathwise convergence of order $\frac{1}{2}^-$) *Under the hypotheses of Theorem 5.2.1, the Euler scheme converges in the pathwise sense at order $\frac{1}{2} - \eta$ (for all $\eta \in]0, \frac{1}{2}[$):*

$$\sup_{0 \le t \le T} |X_t^{T/N} - X_t| < CN^{-\frac{1}{2}+\eta} \quad \text{a.s.,} \tag{5.2.5}$$

with a finite random variable C (depending on η).

5.3 WEAK CONVERGENCE

If the objective is to evaluate $\mathbb{E}(f(X_T))$ by the Monte-Carlo method using the Euler scheme, the final discretization error $\mathbb{E}(f(X_T^{(h)})) - \mathbb{E}(f(X_T))$ is called the *weak error* because this is the error between the *distributions* of $X_T^{(h)}$ and X_T and not between their trajectories (strong error).

However, strong convergence provides convergence rates for the weak error when f is regular. For example, if f is Lipschitz, we get immediately an upper bound on the discretization error

$$\left| \mathbb{E}(f(X_T^{(h)})) - \mathbb{E}(f(X_T)) \right| \le C_f \mathbb{E} |X_T^{(h)} - X_T| = O(\sqrt{h})$$

using the previous upper bound for the strong convergence. For such functions f, the weak error $\mathbb{E}(f(X_T^{(h)})) - \mathbb{E}(f(X_T))$ converges at order $\frac{1}{2}$ as a function of h. Actually, the estimation is very rough because it neglects possible averaging and cancellation effects in the expectation: indeed, we will show that the convergence rate is h if f is regular.

5.3.1 Convergence at order 1

Theorem 5.3.1 (weak convergence at order 1) *Let $T > 0$ and suppose that*

i) *the functions b, σ, f, g, k are continuous bounded, of class $C^{1,4}([0,T] \times \mathbb{R}^d)$ with bounded derivatives; and*

174 ■ Euler scheme for stochastic differential equations

ii) *the function* $u : [0, T] \times \mathbb{R}^d \mapsto \mathbb{R}$ *defined by*

$$u(t,x) = \mathbb{E}\bigg[f(X_T^{t,x})e^{-\int_t^T k(r, X_r^{t,x})dr} + \int_t^T g(s, X_s^{t,x})e^{-\int_t^s k(r, X_r^{t,x})dr} ds\bigg]$$

is[2] bounded continuous, of class $\mathcal{C}^{1,4}$ *with bounded derivatives, and satisfies*

$$\begin{cases} \partial_t u(t,x) + \mathcal{L}u(t,x) - k(t,x)u(t,x) + g(t,x) = 0, & t < T, x \in \mathbb{R}^d, \\ u(T, x) = f(x), \end{cases}$$

where $\mathcal{L}$ *is the infinitesimal generator of* X.

Then, if $X^{(h)}$ is the Euler scheme associated with X with a step $h = T/N$ (with $N \in \mathbb{N}^*$), we have

$$\mathbb{E}\bigg[f(X_T^{(h)})e^{-h\sum_{j=0}^{N-1} k(jh, X_{jh}^{(h)})} + h\sum_{i=0}^{N-1} g(ih, X_{ih}^{(h)})e^{-h\sum_{j=0}^{i-1} k(jh, X_{jh}^{(h)})}\bigg]$$
$$- \mathbb{E}\bigg[f(X_T)e^{-\int_0^T k(r, X_r)dr} + \int_0^T g(s, X_s)e^{-\int_0^s k(r, X_r)dr} ds\bigg] = O(h).$$

PROOF:

Denote by $\text{Err.}_{\text{disc.}}(h)$ the above discretization error. In view of the forthcoming stochastic calculus questions, it is more convenient to represent it using Itô processes:

$$\text{Err.}_{\text{disc.}}(h)$$
$$= \mathbb{E}\bigg[f(X_T^{(h)})e^{-\int_0^T k(\varphi_r^h, X_{\varphi_r^h}^{(h)})dr} + \int_0^T g(\varphi_s^h, X_{\varphi_s^h}^{(h)})e^{-\int_0^{\varphi_s^h} k(\varphi_r^h, X_{\varphi_r^h}^{(h)})dr} ds\bigg]$$
$$- \mathbb{E}\bigg[f(X_T)e^{-\int_0^T k(r, X_r)dr} + \int_0^T g(s, X_s)e^{-\int_0^s k(r, X_r)dr} ds\bigg].$$

The technique of proof follows a general principle: use the limit solution (i.e., the function u) to represent the error. Observe that the second expectation is $u(0, x) = u(0, X_0^{(h)})$ and that the terminal function is $f(\cdot) = u(T, \cdot)$, thus the discretization error can be written as

$$\mathbb{E}\bigg[u(T, X_T^{(h)})e^{-\int_0^T k(\varphi_r^h, X_{\varphi_r^h}^{(h)})dr} + \int_0^T g(\varphi_s^h, X_{\varphi_s^h}^{(h)})e^{-\int_0^{\varphi_s^h} k(\varphi_r^h, X_{\varphi_r^h}^{(h)})dr} ds\bigg]$$

---
[2]Actually, the required properties of u come from the hypothesis i), but we will not prove it.

$- u(0, X_0^{(h)})$.

We can decompose this difference using stochastic calculus, in particular the Itô formula applied to $u(t, X_t^{(h)})e^{-\int_0^t k(\varphi_r^h, X_{\varphi_r^h}^{(h)}) dr}$ (as for the formulas (4.4.1) and (4.4.6)): this gives

$$u(t, X_t^{(h)})e^{-\int_0^t k(\varphi_r^h, X_{\varphi_r^h}^{(h)}) dr} - u(0, X_0^{(h)})$$

$$= \int_0^t e^{-\int_0^s k(\varphi_r^h, X_{\varphi_r^h}^{(h)}) dr} \left(\partial_t u(s, X_s^{(h)}) - k(\varphi_s^h, X_{\varphi_s^h}^{(h)}) u(s, X_s^{(h)}) \right.$$

$$+ \sum_{l_1=1}^d u'_{x_{l_1}}(s, X_s^{(h)}) b_{l_1}(\varphi_s^h, X_{\varphi_s^h}^{(h)})$$

$$\left. + \frac{1}{2} \sum_{l_1, l_2=1}^d u''_{x_{l_1}, x_{l_2}}(s, X_s^{(h)}) [\sigma \sigma^\mathsf{T}]_{l_1, l_2}(\varphi_s^h, X_{\varphi_s^h}^{(h)}) \right) ds \qquad (5.3.1)$$

$$+ \int_0^t e^{-\int_0^s k(\varphi_r^h, X_{\varphi_r^h}^{(h)}) dr} \nabla_x u(s, X_s^{(h)}) \sigma(\varphi_s^h, X_{\varphi_s^h}^{(h)}) dW_s. \qquad (5.3.2)$$

The above stochastic integral has a zero expectation under the hypothesis of the theorem. By taking the expectation in the above equality, setting $t = T$ and using the expression of $\partial_t u$ (see the partial differential equation satisfied by u at the point $(s, X_s^{(h)})$), we get a new representation of the error:

Err.$_{\text{disc.}}(h)$

$$= \mathbb{E}\left[\int_0^T \left(e^{-\int_0^{\varphi_s^h} k(\varphi_r^h, X_{\varphi_r^h}^{(h)}) dr} - e^{-\int_0^s k(\varphi_r^h, X_{\varphi_r^h}^{(h)}) dr} \right) g(\varphi_s^h, X_{\varphi_s^h}^{(h)}) ds \right]$$

$$+ \mathbb{E}\left[\int_0^T e^{-\int_0^s k(\varphi_r^h, X_{\varphi_r^h}^{(h)}) dr} \left([k(s, X_s^{(h)}) - k(\varphi_s^h, X_{\varphi_s^h}^{(h)})] u(s, X_s^{(h)}) \right. \right.$$

$$+ [g(\varphi_s^h, X_{\varphi_s^h}^{(h)}) - g(s, X_s^{(h)})]$$

$$+ \sum_{l_1=1}^d u'_{x_{l_1}}(s, X_s^{(h)}) [b_{l_1}(\varphi_s^h, X_{\varphi_s^h}^{(h)}) - b_{l_1}(s, X_s^{(h)})]$$

$$+ \frac{1}{2} \sum_{l_1, l_2=1}^d u''_{x_{l_1}, x_{l_2}}(s, X_s^{(h)})$$

$$\left. \left. \times [[\sigma \sigma^\mathsf{T}]_{l_1, l_2}(\varphi_s^h, X_{\varphi_s^h}^{(h)}) - [\sigma \sigma^\mathsf{T}]_{l_1, l_2}(s, X_s^{(h)})] \right) ds \right].$$

This shows that the global error is an average of local errors coming from the fact that the coefficients k, g, b, σ have been frozen on each time interval.

The first expectation can be easily bounded by $T|g|_\infty |k|_\infty e^{|k|_\infty T} \sup_s (s-$

176 ■ Euler scheme for stochastic differential equations

$\varphi_s^h) = O(h)$. As for the term $e^{-\int_0^s k(\varphi_r^h, X_{\varphi_r^h}^{(h)})\mathrm{d}r}[k(s, X_s^{(h)}) - k(\varphi_s^h, X_{\varphi_s^h}^{(h)})]u(s, X_s^{(h)})$, let us again apply the Itô formula between φ_s^h and s to get that this term equals

$$\int_{\varphi_s^h}^s \left[u(r, X_r^{h,x})a_{k,0}(r) + \sum_{l_1=1}^d u'_{x_{l_1}}(r, X_r^{h,x})a_{k,1,l_1}(r) \right.$$
$$\left. + \sum_{l_1,l_2=1}^d u''_{x_{l_1},x_{l_2}}(r, X_r^{h,x})a_{k,2,l_1,l_2}(r) \right] \mathrm{d}r$$

plus a stochastic integral, which can be shown to have zero expectation. The processes $a_{k,0}, a_{k,1,l_1}, a_{k,2,l_1,l_2}$ are expressed as functions of the exponential factor, of k and its derivatives: it is not important to have their exact expression and it is enough to note that these processes are uniformly bounded in r. Taking into account the uniform bounds on u and its derivatives, we have shown that

$$\left| \mathbb{E}\left(e^{-\int_0^s k(\varphi_r^h, X_{\varphi_r^h}^{(h)})\mathrm{d}r}[k(s, X_s^{(h)}) - k(\varphi_s^h, X_{\varphi_s^h}^{(h)})]u(s, X_s^{(h)}) \right) \right|$$
$$\leq C(s - \varphi_s^h) \leq Ch,$$

for a uniform constant C. The same arguments can be applied for the terms with g, b_{l_1} and $[\sigma\sigma^\top]_{l_1,l_2}$. To sum up, we have justified that $|\mathrm{Err.}_{\mathrm{disc.}}(h)| \leq O(h) + \int_0^T C\, h\, \mathrm{d}s = O(h)$, as advertised. □

By strengthening the regularity hypothesis, we can continue the expansion and show that the error admits an asymptotic expansion at any order in powers of h, see [139].

5.3.2 Extensions

The regularity hypotheses of the previous theorem are sometimes too strong for applications. We can compensate for the lack of regularity of the functions f and g by supposing that the distribution of X_t has a smoothing density, under the condition of non-degeneracy (such as uniform ellipticity). The analysis is much more involved, because it is mixing the arguments from the partial differential equation and from stochastic analysis [9]. With a purely stochastic approach, we can even suppose that the non-degeneracy takes place only at the initial point $(0, x)$; see [61]. We state a result of this type, without its proof.

Theorem 5.3.2 (weak convergence for measurable f) *For coefficients b and σ of class $\mathcal{C}^\infty$ with bounded derivatives and supposing that $\sigma\sigma^\top(0,x)$ is invertible, we have*

$$\mathbb{E}[f(X_T^{(h)})] - \mathbb{E}[f(X_T)] = O(h)$$

for any measurable bounded function f.

Similar error estimates for the densities of $X_T^{(h)}$ and X_T are also available, with rather definitive results.

On the other hand, the results are much more incomplete if we want to reduce to minimum the regularity and boundedness hypotheses of b and σ; this is again an active research domain. See [89, Proposition 2.3] for examples of error estimates of type h^α with $\alpha \in [\frac{1}{2}, 1[$.

▷ **To learn more.** The convergence rate h can sometimes be extended to more complex functionals, as for example those appearing in the calculation of the sensitivity with respect to the initial condition, i.e., $\nabla_x \mathbb{E}(f(X_T^{0,x}))$. Taking the results and notation of Section 4.5, the additional process to simulate is $(\nabla X_t^{0,x})_t$, the solution of

$$\nabla X_t^{0,x} := \mathrm{Id} + \int_0^t \nabla_x b(s, X_s^{0,x})\, \nabla X_s^{0,x}\, \mathrm{d}s$$
$$+ \sum_{k=1}^d \int_0^t \nabla_x \sigma_k(s, X_s^{0,x})\, \nabla X_s^{0,x}\, \mathrm{d}W_{k,s}.$$

An approximation by the Euler scheme is natural[3] and gives, at the discretization time ih,

$$\begin{cases} \nabla X_0^{0,x,(h)} = \mathrm{Id}, \\ \nabla X_{(i+1)h}^{0,x,(h)} = \nabla X_{ih}^{0,x,(h)} + \nabla_x b(ih, X_{ih}^{0,x,(h)})\, \nabla X_{ih}^{0,x,(h)} h \\ \qquad + \sum_{k=1}^d \nabla_x \sigma_k(ih, X_{ih}^{0,x,(h)})\, \nabla X_{ih}^{0,x,(h)}(W_{k,(i+1)h} - W_{k,ih}), \end{cases}$$
(5.3.3)

or, equivalently in the form of an Itô process,

$$\nabla X_t^{0,x,(h)} = \mathrm{Id} + \int_0^t \nabla_x b(\varphi_s^h, X_{\varphi_s^h}^{0,x,(h)})\, \nabla X_{\varphi_s^h}^{0,x}\, \mathrm{d}s$$

---

[3] It is also coherent because $\nabla X^{0,x,(h)}$ defined like that is the derivative of $X^{0,x,(h)}$ with respect to x.

$$+ \sum_{k=1}^{d} \int_0^t \nabla_x \sigma_k(\varphi_s^h, X_{\varphi_s^h}^{0,x,(h)}) \, \nabla X_{\varphi_s^h}^{0,x,(h)} \, dW_{k,s}.$$

Thus, the simulation of $\nabla X^{0,x,(h)}$ at the discretization times is as easy as for $X^{0,x,(h)}$.

The sensitivity formulas given in Corollary 4.5.2 and Theorem 4.5.3 can be approximated at rate h (under appropriate hypotheses on b, σ, f, see [61]):

$$\mathbb{E}(\nabla_x f(X_T^{0,x,(h)}) \nabla X_T^{0,x,(h)}) - \mathbb{E}(\nabla_x f(X_T^{0,x}) \nabla X_T^{0,x}) = O(h),$$

$$\mathbb{E}\Big(\frac{f(X_T^{0,x,(h)})}{T} \Big[\sum_{i=0}^{N-1} [\sigma^{-1}(ih, X_{ih}^{0,x,(h)}) \nabla X_{ih}^{0,x,(h)}]^\mathsf{T} (W_{(i+1)h} - W_{ih})\Big]^\mathsf{T}\Big)$$

$$- \mathbb{E}\Big(\frac{f(X_T^{0,x})}{T} \Big[\int_0^T [\sigma^{-1}(s, X_s^{0,x}) \nabla X_s^{0,x}]^\mathsf{T} dW_s\Big]^\mathsf{T}\Big) = O(h).$$

The proof of these results is based on advanced stochastic analysis which goes beyond the scope of this book; in fact the arguments based on the partial differential equation cannot be applied because these functionals are not connected to the Feynman-Kac representation.

5.4 SIMULATION OF STOPPED PROCESSES

In this section, our attention is focused on the approximation by the Euler scheme of

$$\mathbb{E}(f(\tau \wedge T, X_{\tau \wedge T})), \quad \text{where} \quad \tau = \inf\{s > 0 : X_s \notin D\}$$

is the first exit time from D of X starting at $(0, x)$. We recall that this evaluation of the distribution of the couple "time/exit position" is connected to the probabilistic solution of the Cauchy-Dirichlet problem (see Theorem 4.4.5).

Contrary to the previous case of $\mathbb{E}(f(X_T))$, which requires us to know only the value of the diffusion at T, the current case requires us to have a lot of information on the trajectory to know the exit time with accuracy: this requirement comes from the significant irregularity of the trajectory of X (as that of Brownian motion). Several numerical schemes exist: their convergence analyses have been performed in detail but the proofs lead to delicate mathematical development combining differential geometry, stochastic analysis, and probabilistic limit theorems. To spare the reader and not forget the important ideas, we restrict to result statements about convergence rate without specifying hypotheses or giving proofs. The interested reader is referred to the given bibliography.

5.4.1 Discrete approximation of exit time

The simplest approach certainly consists of stopping the Euler scheme because one of its values at the discretization times is outside the domain D: this means to set

$$\tau_{\text{disc.}}^h := \inf\{ih > 0 : X_{ih}^{(h)} \notin D\}.$$

Calculation of $\tau_{\text{disc.}}^h$ does not require additional simulation. On the other hand, the approximation is rather rough.

▷ **Heuristic.** To easily understand the underlying phenomenon, consider the case of Brownian motion with $X = W = X^{(h)}$: the values at the points ih are generated without discretization, but the exit time is wrong, with a systematic bias, $\tau_{\text{disc.}}^h \geq \tau$. To get an idea of the order of the error, consider the case where $D =]-1, 1[$ with X Brownian motion in dimension 1, and let us evaluate [4] $\mathbb{E}(\tau_{\text{disc.}}^h - \tau)$. Because $(W_t^2 - t)_t$ is a martingale, the optional sampling theorem gives

$$\mathbb{E}(\tau_{\text{disc.}}^h - \tau) = \lim_{T \to \infty} \mathbb{E}(T \wedge \tau_{\text{disc.}}^h - T \wedge \tau) \quad \text{(monotone convergence)}$$

$$= \lim_{T \to \infty} \mathbb{E}(W_{T \wedge \tau_{\text{disc.}}^h}^2 - W_{T \wedge \tau}^2) \quad \text{(optional sampling theorem)}$$

$$\geq \mathbb{E}(W_{\tau_{\text{disc.}}^h}^2 - W_\tau^2) = \mathbb{E}(W_{\tau_{\text{disc.}}^h}^2) - 1$$

applying in the last inequality, the Fatou lemma to the first term and the dominated convergence theorem to the second. Now the heuristic is the following: at the discretization times before $\tau_{\text{disc.}}^h$, we have by definition $|W_{\tau_{\text{disc.}}^h - h}| < 1$, while at $\tau_{\text{disc.}}^h$, we have $|W_{\tau_{\text{disc.}}^h}| \geq 1$; as the increments of the Brownian motion are of order $\sqrt{h}$ on average, we can reasonably conjecture that

$$\mathbb{E}(\tau_{\text{disc.}}^h - \tau) \geq \mathbb{E}\Big((|W_{\tau_{\text{disc.}}^h}| - 1)(|W_{\tau_{\text{disc.}}^h}| + 1)\Big)$$

$$\geq \mathbb{E}\Big(|W_{\tau_{\text{disc.}}^h}| - |W_{\tau_{\text{disc.}}^h - h}|\Big)$$

---

[4] We leave to the reader the (nontrivial) proof of the fact that $\tau_{\text{disc.}}^h$ and τ are a.s. finite and with finite expectation.

$$\geq c\sqrt{h}$$

for a constant $c > 0$, proving that the convergence order cannot be more than $\frac{1}{2}$.

```
1  N, i: int ;          /* number of time steps and time index */
2  T, h: double ;       /* for the terminal time and the time
                           step */
3  x0, x: vector ;      /* for X_0^(h) and the current value of X_{ih}^(h)
                           */
4  ExitTime: double ;                  /* for the exit time */
5  h ← T/N ;
6  x ← x0; ExitTime ← T ;
7  for i = 0 to N − 1 do
8   │  x ← x + b(i × h, x) × h + σ(i × h, x) × √h × 𝒩(0, Id) ;
9   │  if x ∉ D then
10  │   │  ExitTime ← (i + 1) × h ;     /* the exit from the
                                            domain is detected */
11  │   └  exit ;                       /* we exit from the loop for */
12 Return f(ExitTime, x);
```

Algorithm 5.2: Approximated simulation of $f(\tau \wedge T, X_{\tau \wedge T})$ by Euler scheme and discrete exit time.

▷ **Exact order of convergence.** The order of convergence is exactly $\frac{1}{2}$ because there exists an expansion of the error in $\sqrt{h}$ (see [60]), in full generality on the domain D, on the diffusion X, on the function f[5], under certain regularity and non-degeneracy hypotheses: for some constant C_1, we have

$$\mathbb{E}(f(\tau_{\text{disc.}}^h \wedge T, X_{\tau_{\text{disc.}}^h \wedge T}^{(h)})) - \mathbb{E}(f(\tau \wedge T, X_{\tau \wedge T})) = C_1 \sqrt{h} + o(\sqrt{h}).$$

The rate $\sqrt{h}$ is not due to the error approximation $X^{(h)} - X$ at the discretization times, but due to the discrete approximation of the exit time (see the discussion at the beginning about Brownian motion).

---

[5] A discount factor and a source term could be added without modifying the nature of the result.

▷ **Conclusion.** This method is very simple to use but it leads to *significant over-estimations of the exit time*, with an error of order $\frac{1}{2}$ with respect to the time step, thus significantly larger than in the case of $\mathbb{E}(f(X_T))$.

5.4.2 Brownian bridge method

To account for the possible exit of $X^{(h)}$ from the domain between two consecutive discretization times, one may try to improve the approximation of the exit time by continuous observation of the trajectory

$$\tau^h_{\text{cont.}} := \inf\{s > 0 : X^{(h)}_s \notin D\},$$

knowing the values $(X^{(h)}_{ih})_{i \geq 0}$, which have been already simulated.

▷ **Brownian bridge.** The objective is to improve the order of convergence $\frac{1}{2}$, and to achieve the order 1 as in the absence of the spatial boundary. It is not necessary to simulate exactly $\tau^h_{\text{cont.}}$: it is enough to detect in which interval $[ih, (i+1)h]$ the time $\tau^h_{\text{cont.}}$ is, and then proceed to the approximation

$$\tau^h_{\text{cont.}} \approx (i+1)h.$$

This last approximation introduces an additional error of order 1 in h. To achieve this exit detection, set

$$\mathbb{P}\Big(\exists t \in [ih, (i+1)h] : X^{(h)}_t \notin D \big| X^{(h)}_{jh} : j \leq N\Big)$$
$$= \mathbb{P}\Big(\exists t \in [ih, (i+1)h] : X^{(h)}_t \notin D \big| X^{(h)}_{ih}, X^{(h)}_{(i+1)h}\Big)$$
$$:= p\big(D; ih, (i+1)h; X^{(h)}_{ih}, X^{(h)}_{(i+1)h}\big);$$

the first equality comes from Lemma 5.1.3 on Brownian bridges. The function p is the exit probability of a certain *Brownian bridge*. Then by simulating a sequence $(U_i)_{i \geq 0}$ of independent random variables with uniform distribution on $[0,1]$, the first index i such that

$$U_i \leq p\big(D; ih, (i+1)h; X^{(h)}_{ih}, X^{(h)}_{(i+1)h}\big)$$

simulates the index of the interval in which $\tau^h_{\text{cont.}}$ is located.

▷ **Computation of the exit probability for the Brownian bridge.** Most of the difficulties of this method are in the explicit calculation of the function $p(D; s, t; x, y)$.

a) **(Exterior points)** If $x \notin D$ or $y \notin D$, then obviously $p(D; s, t; x, y) = 1$.

b) **(Interior points far from the boundary)** If
$$\min(d(x, D), d(y, D)) \gg (t - s)^{\frac{1}{2}},$$
then $p(D; s, t; x, y)$ is extremely small with respect to $t - s$ and we can approximate it by 0 without getting any significant error in the procedure.

c) **(Half-space)** In the case where $D = \{w \in \mathbb{R}^d : (w - z) \cdot \vec{v} \geq 0\}$ is a half-space with a boundary passing through z and orthogonal to the unit vector $\vec{v}$, p is explicit:
$$p(D; s, t; x, y) = \exp\left(-2\frac{[(x - z) \cdot \vec{v}][(y - z) \cdot \vec{v}]}{[\vec{v} \cdot \sigma\sigma^\mathsf{T}(s, x)\vec{v}](t - s)}\right), \quad \forall (x, y) \in D^2.$$
This formula can be established combining Lemma 5.1.3 and the equality (4.1.2); see also (4.1.5).

In most other cases, there is no explicit expression of p. When the domain D is smooth, we can replace in the calculation of p the domain D by its local approximation by a half-space: we project[6] x on the boundary ∂D, which gives $\pi(x)$; we denote by $\vec{n}(x)$ the unit inward normal vector at $\pi(x) \in \partial D$. Finally we approximate D by $\{w \in \mathbb{R}^d : (w - \pi(x)) \cdot \vec{n}(x) \geq 0\}$; see Figure 5.1.

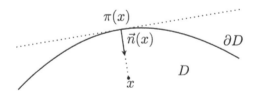

Figure 5.1 Local approximation by half-space.

Then, by taking
$$\tilde{p}(D; s, t; x, y) = \exp\left(-2\frac{[(x - \pi(x)) \cdot \vec{n}(x)][(y - \pi(x)) \cdot \vec{n}(x)]}{[\vec{n}(x) \cdot \sigma\sigma^\mathsf{T}(s, x)\vec{n}(x)](t - s)}\right)$$
instead of p, the global algorithm has an explicit implementation and achieves the order of convergence 1 with respect to the time step h [58].

---
[6]The projected point $\pi(x)$ is unique if x is close enough to ∂D.

```
1  N, i: int ;         /* number of time steps and time index */
2  T, h: double ;      /* for the terminal time and the time
                          step */
3  x0, xl, xr: vector ; /* for $X_0^{(h)}$ and the current values of
                          $X_{ih}^{(h)}$ and $X_{(i+1)h}^{(h)}$ */
4  ExitTime: double ;                /* for the exit time */
5  h ← T/N ;
6  x ← x0; ExitTime ← T ;
7  for i = 0 to N − 1 do
8  |   xr ← xl + b(i × h, xl) × h + σ(i × h, xl) × $\sqrt{h}$ × $\mathcal{N}$(0, Id) ;
9  |   if rand ≤ p(D, i × h, (i + 1) × h, xr, xl) then
10 |   |   ExitTime ← (i + 1) × h ;   /* the exit of the
                                         Brownian bridge is detected */
11 |   |   exit ;                     /* we exit from the loop for */
12 |   xl ← xr ;
13 Return f(ExitTime, xr);
```

Algorithm 5.3: Approximated simulation of $f(\tau \wedge T, X_{\tau \wedge T})$ by Euler scheme and Brownian bridge.

5.4.3 Boundary shifting method

The discrete exit time over-estimates, on average, the continuous exit time: the bias is of order $\sqrt{h}$. To compensate *exactly* this bias, a natural idea consists of properly shrinking the domain to generate more frequent exits (see Figure 5.2): intuitively, the size of inward shift should be of order $\sqrt{h}$; actually, there exists a precise universal tuning.

This procedure proposed first for the Brownian motion in dimension 1 in [21] was then generalized to the case of a general multidimensional diffusion in [60]. Surprisingly, the tuning of the boundary shifting is rather simple and universal, as a function of the following constant[7] which is expressed with the help of the Riemann zeta function:

$$c_0 = -\frac{\zeta(\frac{1}{2})}{\sqrt{2\pi}} = 0.5826\ldots.$$

Denote by $\vec{n}(x)$ the unit inward normal vector at the point on the boundary ∂D closest to x and define as follows the new exit time from the domain with a shifted boundary:

$$\tau^h_{\text{disc. shift.}} := \inf\left\{ih > 0 : X^{(h)}_{ih} \notin D \right.$$
$$\left. \text{or } d(X^{(h)}_{ih}, \partial D) \leq c_0 \sqrt{h} |\sigma^\mathsf{T} \vec{n}|(ih, X^{(h)}_{ih})\right\}.$$

Then, under appropriate regularity hypotheses, we have

$$\mathbb{E}[f(\tau^h_{\text{disc. shift.}} \wedge T, X^{(h)}_{\tau^h_{\text{disc. shift.}} \wedge T})] - \mathbb{E}[f(\tau \wedge T, X_{\tau \wedge T})] = o(\sqrt{h}).$$

The order of convergence is thus strictly better than $\frac{1}{2}$; from numerical tests, it is conjectured that it is equal to 1, but the mathematical proof remains an open question.

Note that this algorithm does not require additional simulation and its computational cost is the same as for the algorithm with discrete exit times, but it has better accuracy.

In Table 5.1, we present the numerical values from [59, Table 4.3] comparing the three previous methods on a 2-dimensional example, for different N; the benchmark value is $\mathbb{E}[f(\tau \wedge T, X_{\tau \wedge T})] \approx 1.727$. The number of samples is large enough for the statistical error of the Monte-Carlo method to be less than 0.002 with probability 95%. The methods of Brownian bridge and boundary shifting appear to be equally accurate; on the other hand, the one by discrete exit times turns out to be rather imprecise, in agreement with the theoretical estimates.

---

[7] This constant appears in asymptotic theorems for Gaussian random walks.

```
1  N, i: int ;        /* number of time steps and time index */
2  T, h: double ;     /* for the terminal time and the time
                         step */
3  c0 = 0.5826: double ;  /* for the universal constant */
4  x0, x: vector ; /* for $X_0^{(h)}$ and the current value of $X_{ih}^{(h)}$
                         */
5  ExitTime: double ;              /* for the exit time */
6  h ← T/N ;
7  x ← x0; ExitTime ← T ;
8  for i = 0 to N − 1 do
9  |  x ← x + b(i × h, x) × h + σ(i × h, x) × $\sqrt{h}$ × $\mathcal{N}$(0, Id) ;
10 |  if x ∉ D then
11 |  |  ExitTime ← (i + 1) × h ;
12 |  |  exit ;              /* we exit from the loop for */
13 |  if d(x, ∂D) ≤ c₀ × $\sqrt{h}$ × |σᵀn|(i × h, x) then
14 |  |  ExitTime ← (i + 1) × h ;
15 |  |  exit ;              /* we exit from the loop for */
16 Return f(ExitTime, x);
```

Algorithm 5.4: Approximated simulation of $f(\tau \wedge T, X_{\tau \wedge T})$ by the Euler scheme and discrete exit time with a shifted boundary.

Methods	$N = 13$	$N = 26$	$N = 52$	$N = 126$
Discrete exit times	2.244	2.110	2.008	1.913
Brownian bridge	1.730	1.728	1.727	1.727
Boundary shifting	1.752	1.726	1.725	1.727

Table 5.1 Comparison of the numerical evaluation of $\mathbb{E}[f(\tau \wedge T, X_{\tau \wedge T})]$.

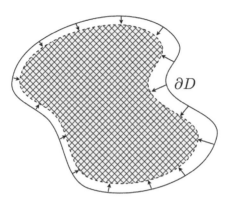

Figure 5.2 The modified (grid pattern) domain is obtained by shifting locally the initial boundary in the unit inward normal direction, by a quantity $c_0\sqrt{h}|\sigma^\top(t,x)\vec{n}(x)|$.

5.5 EXERCISES

Exercise 5.1 (strong convergence) *Show that in Theorem 5.2.1, the convergence rate is of order 1 if σ is constant and b is $\mathcal{C}^2$ in space and $\mathcal{C}^1$ in time.*

Exercise 5.2 (Milshtein scheme) *Denote by $(X_t)_{t\geq 0}$ the solution of the stochastic differential equation*

$$X_t = x + \int_0^t \sigma(X_s)\mathrm{d}W_s + \int_0^t b(X_s)\mathrm{d}s$$

where $\sigma, b : \mathbb{R} \to \mathbb{R}$ are bounded $\mathcal{C}^2$-functions with bounded derivatives.

1. *Show the short time L_2-approximation*

$$\mathbb{E}\left((X_t - [x + b(x)t + \sigma(x)W_t])^2\right) = \frac{(\sigma\sigma'(x))^2}{2}t^2 + o(t^2).$$

The above estimate is instrumental to show the strong convergence of the Euler scheme (5.1.1) at order $1/2$: essentially, the global quadratic error (of size N^{-1}) is the summation of the above local estimates (of order N^{-2} with $t = h$) over N times, to finally get the bound (5.2.1) for $p = 2$.

2. *Similarly, show*

$$\mathbb{E}\left(\left(X_t - [x + b(x)t + \sigma(x)W_t + \frac{1}{2}\sigma\sigma'(x)(W_t^2 - t)]\right)^2\right) = \mathcal{O}(t^3).$$
(5.5.1)

This estimate leads to a high-order scheme, called the Milshtein scheme, which is written

$$\begin{cases} X_0^{(h,M)} = x, \\ X_{(i+1)h}^{(h,M)} = X_{ih}^{(h,M)} + b(X_{ih}^{(h,M)})h + \sigma(X_{ih}^{(h,M)})(W_{(i+1)h} - W_{ih}) \\ \qquad\qquad + \frac{1}{2}\sigma\sigma'(X_{ih}^{(h,M)})[(W_{(i+1)h} - W_{ih})^2 - h]. \end{cases}$$

3. *Using the above estimate (5.5.1), prove that*

$$\sup_{0 \le i \le N} \mathbb{E}(|X_{ih}^{(h,M)} - X_{ih}|^2) = \mathcal{O}(h^2), \qquad (5.5.2)$$

i.e. a convergence rate at order 1 for the strong convergence.

This Milshtein scheme is effective, mainly for one-dimensional situations, because in the multidimensional case, it is written using the iterated stochastic integrals $\int_0^t W_s^i dW_s^j$ for $1 \le i \ne j \le d$, which cannot be easily simulated (in dimension 1, we simply have $\int_0^t W_s dW_s = \frac{1}{2}(W_t^2 - t)$).

Exercise 5.3 (convergence rate of weak convergence) *We consider the model of geometric Brownian motion:*

$$X_t = x + \int_0^t \sigma X_s dW_s + \int_0^t \mu X_s ds$$

with $x > 0$.

1. *Compute $\mathbb{E}(X_T^2)$.*

2. *Let $X^{(h)}$ be the related Euler scheme with time step h. Set $y_i = \mathbb{E}((X_{ih}^{(h)})^2)$. Find a relation between y_{i+1} and y_i.*

3. *Deduce that $\mathbb{E}((X_T^{(h)})^2) = \mathbb{E}(X_T^2) + \mathcal{O}(h)$. Compare with Theorem 5.3.1.*

Exercise 5.4 (solving SDE using change of variables) *Consider an SDE of the form*

$$X_t = x_0 + \int_0^t b(X_s)\mathrm{d}s + \int_0^t \sigma(X_s)\mathrm{d}W_s. \tag{5.5.3}$$

We study two transformations that lead to simpler equations and may help in the resolution of the initial SDE.

i) *(Lamperti transformation). We assume that the equation (5.5.3) is defined on $\mathbb{R}$, and we assume that the coefficients $b : \mathbb{R} \to \mathbb{R}$ and $\sigma : \mathbb{R} \to (0, \infty)$ are of class C^1 with bounded derivative, that the function $\frac{1}{\sigma(x)}$ is not integrable at $\pm\infty$, and that the function $b/\sigma - \sigma'/2$ is Lipschitz.*

   (a) *Verify that the function $f(x) = \int_{x_0}^x \frac{\mathrm{d}y}{\sigma(y)}$ is a bijection from $\mathbb{R}$ into $\mathbb{R}$.*

   (b) *Show that the solution to the equation (5.5.3) can be put in the form $X_t = f^{-1}(Y_t)$, where the process Y solves*

$$Y_t = \int_0^t \tilde{b}(Y_s)\mathrm{d}s + W_t,$$

   for some function $\tilde{b}$ to explicit.

   Comment: *The interest in Lamperti transformation is to retrieve an SDE with unit-diffusion coefficient; the approximation by Euler scheme of this SDE becomes more accurate (Exercise 5.1).*

ii) *(Doss-Sussmann transformation). We consider now the case where the coefficients $b, \sigma : \mathbb{R}^d \to \mathbb{R}^d$ are Lipschitz, σ is of class C_b^2 (bounded with bounded derivatives), and W_t is still a scalar Brownian motion. Denote by $F(\theta, x)$ the flow of the differential equation $x'_\theta = \sigma(x_\theta)$ in $\mathbb{R}^d$, i.e.*

$$\partial_\theta F(\theta, x) = \sigma(F(\theta, x)), \qquad F(0, x) = x, \qquad \forall\, (\theta, x) \in \mathbb{R} \times \mathbb{R}^d,$$

   and by $\partial_x F(\theta, x)$ the Jacobian matrix of F. Using that F is of class C^2 on $\mathbb{R} \times \mathbb{R}^d$ and that $\partial_x F(\theta, x)$ is invertible for any (θ, x), show that the solution X_t of (5.5.3) is written in the form

$$X_t(\omega) = F(W_t(\omega), Z_t(\omega))$$

for any t, where the differentiable process Z_t solves — ω by ω — the Ordinary Differential Equation (ODE)

$$Z_t = x_0 + \int_0^t (\partial_x F(W_s, Z_s))^{-1}\left(b - \frac{1}{2}(\partial_x \sigma)\sigma\right)(F(W_s, Z_s))\mathrm{d}s,$$

with the matrix $(\partial_x F(\theta, x))^{-1}$ being the inverse of $\partial_x F(\theta, x)$.

Exercise 5.5 (simulation of exit time) *Write a simulation program to compute $\mathbb{E}(1_{\sup_{0 \le t \le T} S_t \le U}(K - S_T)_+)$ where S is a geometric Brownian motion. We will compare the three schemes (discrete time approximation, Brownian bridge method, boundary shifting method).*

Exercise 5.6 (exit time, nonconvergence) *In the theoretical results of Section 5.4, the assumption of non-degeneracy (i.e. $\sigma(x)$ invertible) plays an important role in the validity of the convergence results. Otherwise, we may face some pathological situations, as illustrated below.*

i) *In dimension 1, consider the model with $b(y) = y$, $\sigma(y) = 0$, $x_0 = 1$ and $D = (-\infty, \exp(1))$, $T = 1$. Prove that $\tau_{\mathrm{disc.}}^h > 1$ and $\tau = 1$, so that*

$$\mathbb{P}(\tau_{\mathrm{disc.}}^h > T) - \mathbb{P}(\tau > T) = 1$$

does not converge to 0 as $h \to 0$.

ii) *In dimension 1, consider the model $b(x) = \cos(x)$, $\sigma(x) = \sin(x)$, $x_0 = \pi/2$ with $D = (-\pi, 2\pi)$. Prove that $\tau = +\infty$ a.s. and $\tau_{\mathrm{disc.}}^h < +\infty$ a.s..*

CHAPTER 6

Statistical error in the simulation of stochastic differential equations

In this chapter we discuss the statistical error that appears in the simulation of stochastic differential equations, in connection with the discretization step that is chosen for the process approximation. The main goal of the chapter is to provide guidance for a proper choice of the number of trajectories and the time step.

6.1 ASYMPTOTIC ANALYSIS: NUMBER OF SIMULATIONS AND TIME STEP

In Section 5.1.3 of the previous chapter, while calculating the expectation of

$$\mathcal{E}(f,g,k,X) = f(X_T)e^{-\int_0^T k(r,X_r)dr} + \int_0^T g(s,X_s)e^{-\int_0^s k(r,X_r)dr}ds,$$

we decomposed the error

$$\text{Error}_{h,M} := \frac{1}{M}\sum_{m=1}^{M}\mathcal{E}(f,g,k,X^{(h,m)}) - \mathbb{E}(\mathcal{E}(f,g,k,X))$$

of the Monte-Carlo method in the superposition of two errors (see (5.1.9)):

1. the discretization error $\mathbb{E}(\mathcal{E}(f,g,k,X^{(h)})) - \mathbb{E}(\mathcal{E}(f,g,k,X))$,

which depends only on h, has been mostly analyzed in Chapter 5; and

2. the statistical error $\dfrac{1}{M}\sum_{m=1}^{M}\mathcal{E}(f,g,k,X^{(h,m)})-\mathbb{E}(\mathcal{E}(f,g,k,X^{(h)}))$, which depends on h and M.

The computational cost $\mathcal{C}_{\text{cost}}$ of such an algorithm is, in general, equal to $M \times [C(d)N] = C(d)MTh^{-1}$, where $C(d)$ stands for the computational cost of the elementary vector/matrix operations made within a time interval. Everything else being equal, the global cost increases with the dimension and the horizon of the simulation, but only in the form of a factor, independently of M and h. Thus they have minor influence, i.e. a much smaller impact compared to that of M and h^{-1}, and for this reason we do not take them into account in what follows. We simply write

$$\mathcal{C}_{\text{cost}} \sim_c Mh^{-1}.$$

The amount of memory used is minor because it is enough to save only the statistics of the realizations (empirical mean and standard deviation), without keeping all the simulation history: see Algorithm 5.1. For non-linear processes from Part **C**, the algorithms are quite different and the required amount of memory is much more significant; it is even a limiting factor.

▷ **Question:** Given a computational cost $\mathcal{C}_{\text{cost}}$, what is the best strategy to allocate the effort between the discretization (i.e., $h \to 0$) and the independent simulations (i.e., $M \to +\infty$)?

We will answer this question by examining the asymptotic convergence rate. The square of the quadratic error can be easily calculated:

$$\mathbb{E}\left(\text{Error}_{h,M}^2\right) = \frac{\mathrm{Var}(\mathcal{E}(f,g,k,X^{(h)}))}{M}$$
$$+ \left[\mathbb{E}(\mathcal{E}(f,g,k,X^{(h)})) - \mathbb{E}(\mathcal{E}(f,g,k,X))\right]^2$$
$$\approx \frac{\mathrm{Var}(\mathcal{E}(f,g,k,X))}{M} + [Ch]^2.$$

This approximation comes, on the one hand, from the (expected) convergence of $X^{(h)}$ to X, and, on the other hand, from the weak convergence analysis in Theorem[1] 5.3.1. The following discussion could be

---

[1] Supposing that the estimate $O(h)$ that we have proved is rather an equivalent of the form Ch.

Asymptotic analysis: number of simulations and time step ■ 193

easily adapted if the order of the weak convergence were equal to $\frac{1}{2}$ or any other value. So, three asymptotic regimes appear, with respect to the relative values of M^{-1} and h^2.

1. If $M \gg h^{-2}$, the statistical error becomes negligible and $\text{Error}_{h,M} \approx Ch$. The computational effort is $\mathcal{C}_{\text{cost}} \gg h^{-3}$ and the quadratic error is thus

$$\sqrt{\mathbb{E}(\text{Error}_{h,M}^2)} \gg \mathcal{C}_{\text{cost}}^{-1/3}.$$

Computing confidence intervals as in Chapter 2 gives no information on the sought value $\mathbb{E}(\mathcal{E}(f,g,k,X))$, because the discretization error is predominant.

2. If $M \ll h^{-2}$, the discretization error becomes negligible and we can show, as in Chapter 2, a central limit theorem: the distribution of $\sqrt{M}\left(\frac{1}{M}\sum_{m=1}^{M}\mathcal{E}(f,g,k,X^{(h,m)}) - \mathbb{E}(\mathcal{E}(f,g,k,X))\right)$ converges to that of a centered Gaussian random variable with variance $\text{Var}(\mathcal{E}(f,g,k,X))$ (which can be asymptotically calculated with a sample of M simulations). Hence, we can produce confidence intervals for the sought value: denoting by $\sigma_{h,M}^2$ the empirical variance of $\mathcal{E}(f,g,k,X^{(h)})$, we have that with a probability of approximately 95%

$$\overline{\mathcal{E}(f,g,k,X^{(h)})}_M - 1.96\frac{\sigma_{h,M}}{\sqrt{M}}$$
$$\leq \mathcal{E}(f,g,k,X) \leq \overline{\mathcal{E}(f,g,k,X^{(h)})}_M + 1.96\frac{\sigma_{h,M}}{\sqrt{M}}, \quad (6.1.1)$$

where $\overline{\mathcal{E}(f,g,k,X^{(h)})}_M := \frac{1}{M}\sum_{m=1}^{M}\mathcal{E}(f,g,k,X^{(h,m)})$. As for the computational effort, we have $\mathcal{C}_{\text{cost}} \gg M^{3/2}$ and thus

$$\sqrt{\mathbb{E}(\text{Error}_{h,M}^2)} \gg \mathcal{C}_{\text{cost}}^{-1/3}.$$

3. If $M \sim h^{-2}$, the statistical and the discretization errors have the same magnitude; we can write confidence intervals, but they are not any more asymptotically centered in $\mathcal{E}(f,g,k,X)$ (as in the case $M \ll h^{-2}$) and, unfortunately, the bias (of the discretization) cannot be estimated *on the fly*.[2] As the bias is of the

---
[2] That is, simultaneously, with the same simulation sample.

same magnitude as the confidence interval, the latter is of less interesting because we cannot be sure with high probability that the sought value is inside the confidence interval. In this case, $\sqrt{\mathbb{E}(\text{Error}_{h,M}^2)} = O(\mathcal{C}_{\text{cost}}^{-1/3})$.

To conclude, by considering the availability of having a posteriori error estimates and by optimizing the final error with respect to the computational cost, we find that the second case $M = h^{-2+\varepsilon}$ (for a small $\varepsilon > 0$) seems to be the most attractive because

- it (almost) achieves the best accuracy for a given computational cost, and

- it gives asymptotically centered confidence intervals, and thus it is meaningful for a posteriori error estimates.

6.2 NON-ASYMPTOTIC ANALYSIS OF THE STATISTICAL ERROR IN THE EULER SCHEME

In this section, we show how the concentration inequalities (see Section 2.4.4) for the Gaussian random variables can be applied to the Euler scheme, in order to provide non-asymptotic control of the statistical fluctuations $\frac{1}{M}\sum_{m=1}^{M}\mathcal{E}(f,g,k,X^{(h,m)}) - \mathbb{E}(\mathcal{E}(f,g,k,X^{(h)}))$. To make the analysis clear, we consider a simplified case, where $\mathcal{E}(f,g,k,X^{(h)}) = f(X_T^{(h)})$, corresponding to $g = k \equiv 0$. This choice is inspired by [47].

Theorem 6.2.1 (concentration inequality for Euler scheme) *Let $T > 0$ and suppose that*

i) $|b(t,x) - b(t,y)| + |\sigma(t,x) - \sigma(t,y)| \leq C_{b,\sigma}|x-y|$ *for all* $(t,x,y) \in [0,T] \times \mathbb{R}^d \times \mathbb{R}^d$;

ii) $\sup_{0 \leq t \leq T} (|b(t,0)| + |\sigma(t,0)|) \leq C_{b,\sigma}$ *and* $\sigma(\cdot)$ *are uniformly bounded; and*

iii) f *is Lipschitz:* $|f(x) - f(y)| \leq |f|_{\text{Lip.}}|x-y|$ *for any* $(x,y) \in \mathbb{R}^d \times \mathbb{R}^d$.

Then, if $(X_T^{(h,m)})_{1 \leq m \leq M}$ *is a sequence of independent identically distributed random variables, derived from the Euler scheme with step*

Non-asymptotic analysis of the statistical error in the Euler scheme   ■ 195

$h = T/N$ $(N \in \mathbb{N}^*)$, we have

$$\mathbb{P}\Big(\big|\frac{1}{M}\sum_{m=1}^{M} f(X_T^{(h,m)}) - \mathbb{E}(f(X_T^{(h)}))\big| > \varepsilon\Big)$$
$$\leq 2\exp\Big(-\frac{M\varepsilon^2}{2C_{b,\sigma,T}|f|^2_{\text{Lip.}}}\Big), \quad \forall \varepsilon \geq 0,$$

for an explicit constant $C_{b,\sigma,T}$ depending only on b,σ,d and T.

Since the constant $C_{b,\sigma,T}$ does not depend on h, it is possible to take the limit[3] as $h \to 0$ in the above inequality and also obtain deviation bounds for the stochastic differential equation itself, but in practice it is less useful for the underlying Monte-Carlo method.

With this result in hand, we are able to obtain non-asymptotic confidence intervals, by applying Proposition 2.4.1 with $\lambda = 5\%$; with probability of at least 95%, we have

$$-2.72\frac{\sqrt{C_{b,\sigma,T}}|f|_{\text{Lip.}}}{\sqrt{M}}$$
$$\leq \frac{1}{M}\sum_{m=1}^{M} f(X_T^{(h,m)}) - \mathbb{E}(f(X_T^{(h)})) \leq 2.72\frac{\sqrt{C_{b,\sigma,T}}|f|_{\text{Lip.}}}{\sqrt{M}}.$$

PROOF:

▷ **Preliminary exponential inequality.** Let Φ be a function with compact support and Lipschitz with constant $|\Phi|_{\text{Lip.}} > 0$: Theorem 2.4.12 applied to the function $f = \Phi/|\Phi|_{\text{Lip.}}$ and the d-dimensional centered Gaussian distribution with a covariance Id_d (satisfying the logarithmic Sobolev inequality with a constant $C_{\gamma_d} = 2$ from Corollary 2.4.16) gives

$$\mathbb{E}\Big(e^{\lambda[\Phi(Y)-\mathbb{E}(\Phi(Y))]}\Big) \leq e^{\frac{1}{2}\lambda^2|\Phi|^2_{\text{Lip.}}}, \quad \forall \lambda \in \mathbb{R}$$

where $Y \stackrel{d}{=} \mathcal{N}(0,\text{Id}_d)$. This inequality is again true if Φ is not with compact support: indeed, as in the proof of Corollary 2.4.13, there exists a sequence of functions $(\Phi_n)_n$ with compact support, that are Lipschitz with the same Lipschitz constant $|\Phi|_{\text{Lip.}}$, such that $\Phi_n(Y) \stackrel{L_1,\text{a.s.}}{\longrightarrow} \Phi(Y)$. So, by the Fatou lemma, we deduce that

$$\mathbb{E}\big(e^{\lambda\Phi(Y)}\big) = \mathbb{E}\big(\liminf_{n\to+\infty} e^{\lambda\Phi_n(Y)}\big)$$

---
[3] As in the proof of Corollary 2.4.13.

$$\leq \liminf_{n \to +\infty} \mathbb{E}\left(e^{\lambda \Phi_n(Y)}\right)$$

$$\leq \liminf_{n \to +\infty} e^{\lambda \mathbb{E}(\Phi_n(Y)) + \frac{1}{2}\lambda^2 |\Phi|^2_{\text{Lip.}}} = e^{\lambda \mathbb{E}(\Phi(Y)) + \frac{1}{2}\lambda^2 |\Phi|^2_{\text{Lip.}}}.$$

▷ **Application to the deviations of** $f(X_T^{(h)}) - \mathbb{E}(f(X_T^{(h)}))$. Set

$$F_i := \mathbb{E}(f(X_T^{(h)}) | X_{ih}^{(h)}) := f_i(X_{ih}^{(h)})$$

and assume for a moment that each function f_i is Lipschitz with a constant $|f_i|_{\text{Lip.}}$. Observe that

$$f(X_T^{(h)}) - \mathbb{E}(f(X_T^{(h)})) = \sum_{i=0}^{N-1}(F_{i+1} - F_i) = \sum_{i=0}^{N-1}(F_{i+1} - \mathbb{E}(F_{i+1}|X_{ih}^{(h)})).$$

Conditionally on $X_{(N-1)h}^{(h)}$,

$$X_{Nh}^{(h)} \stackrel{d}{=} X_{(N-1)h}^{(h)} + b((N-1)h, X_{(N-1)h}^{(h)})h + \sqrt{h}\sigma((N-1)h, X_{(N-1)h}^{(h)})Y$$

where $Y \stackrel{d}{=} \mathcal{N}(0, \text{Id}_d)$; thus

$$\mathbb{E}\left(e^{\lambda[f(X_T^{(h)}) - \mathbb{E}(f(X_T^{(h)}))]}\right) = \mathbb{E}\left(e^{\lambda \sum_{i=0}^{N-1}(F_{i+1} - \mathbb{E}(F_{i+1}|X_{ih}^{(h)}))}\right)$$

$$= \mathbb{E}\left(\mathbb{E}\left(e^{\lambda \sum_{i=0}^{N-1}(F_{i+1} - \mathbb{E}(F_{i+1}|X_{ih}^{(h)}))} | X_{jh}^{(h)} : j \leq N-1\right)\right)$$

$$= \mathbb{E}\left(e^{\lambda \sum_{i=0}^{N-2}(F_{i+1} - \mathbb{E}(F_{i+1}|X_{ih}^{(h)}))} \mathbb{E}\left(e^{\lambda[F_N - \mathbb{E}(F_N|X_{(N-1)h}^{(h)})]} | X_{(N-1)h}^{(h)}\right)\right)$$

$$\leq \mathbb{E}\left(e^{\lambda \sum_{i=0}^{N-2}(F_{i+1} - \mathbb{E}(F_{i+1}|X_{ih}^{(h)}))}\right) e^{\frac{1}{2}\lambda^2 |f_N|^2_{\text{Lip.}} \|\sigma\|^2_\infty h}$$

using that $y \mapsto f_N(X_{(N-1)h}^{(h)} + b((N-1)h, X_{(N-1)h}^{(h)})h + \sqrt{h}\sigma((N-1)h, X_{(N-1)h}^{(h)})y)$ is Lipschitz with a constant $\sqrt{h}\|f_N\|_{\text{Lip.}}\|\sigma((N-1)h, X_{(N-1)h}^{(h)})\|_2$. Iterating the previous argument $N-2$ times, we obtain

$$\mathbb{E}\left(e^{\lambda[f(X_T^{(h)}) - \mathbb{E}(f(X_T^{(h)}))]}\right) \leq e^{\frac{1}{2}\lambda^2 \sum_{i=1}^{N} h|f_i|^2_{\text{Lip.}} \|\sigma\|^2_\infty}$$

for any $\lambda \in \mathbb{R}$. Considering the empirical mean for M simulations, we deduce from their independence that

$$\mathbb{E}\left(e^{\lambda[\frac{1}{M}\sum_{m=1}^{M} f(X_T^{(h,m)}) - \mathbb{E}(f(X_T^{(h)}))]}\right) \leq e^{\frac{1}{2}\frac{\lambda^2}{M} \sum_{i=1}^{N} h|f_i|^2_{\text{Lip.}} \|\sigma\|^2_\infty}.$$

Then applying the Chebyshev exponential inequality as in (2.4.12), we easily obtain

$$\mathbb{P}\left(\left|\frac{1}{M}\sum_{m=1}^{M} f(X_T^{(h,m)}) - \mathbb{E}(f(X_T^{(h)}))\right| > \varepsilon\right)$$

$$\leq 2\exp\left(-\frac{M\varepsilon^2}{2\sum_{i=1}^{N}h|f_i|_{\text{Lip.}}^2|||\sigma||_2|_\infty^2}\right).$$

▷ **Upper bound for Lipschitz constants** $|f_i|_{\text{Lip.}}$. We will show that

$$|f_i|_{\text{Lip.}} \leq |f|_{\text{Lip.}}(1 + C_{b,\sigma}^2 h^2 + 2C_{b,\sigma}h + dC_{b,\sigma}^2 h)^{\frac{N-i}{2}}. \quad (6.2.1)$$

Let us consider the case $i = 0$ in details, the other cases are similar. For x and y in $\mathbb{R}^d$, set $\delta_i = \mathbb{E}|X_{ih}^{x,(h)} - X_{ih}^{y,(h)}|^2$ for $i \geq 0$. By Gaussian-based computations, we directly find that

$$\mathbb{E}\big(|X_{(i+1)h}^{x,(h)} - X_{(i+1)h}^{y,(h)}|^2|W_{jh}: j \leq i\big)$$
$$= |X_{ih}^{x,(h)} + b(ih, X_{ih}^{x,(h)})h - X_{ih}^{y,(h)} - b(ih, X_{ih}^{y,(h)})h|^2$$
$$+ \text{Tr}\big([\sigma(ih, X_{ih}^{x,(h)}) - \sigma(ih, X_{ih}^{y,(h)})][\sigma(ih, X_{ih}^{x,(h)}) - \sigma(ih, X_{ih}^{y,(h)})]^\mathsf{T}\big)h$$
$$\leq |X_{ih}^{x,(h)} - X_{ih}^{y,(h)}|^2 + |b(ih, \cdot)|_{\text{Lip.}}^2 |X_{ih}^{x,(h)} - X_{ih}^{y,(h)}|^2 h^2$$
$$+ 2|b(ih, \cdot)|_{\text{Lip.}}|X_{ih}^{x,(h)} - X_{ih}^{y,(h)}|^2 h + d|\sigma(ih, \cdot)|_{\text{Lip.}}^2 |X_{ih}^{x,(h)} - X_{ih}^{y,(h)}|^2 h.$$

Passing to the expectation and using that $\delta_0 = |x - y|^2$, we get

$$\delta_i \leq \delta_{i-1}(1 + C_{b,\sigma}^2 h^2 + 2C_{b,\sigma}h + dC_{b,\sigma}^2 h)$$
$$\leq |x - y|^2 (1 + C_{b,\sigma}^2 h^2 + 2C_{b,\sigma}h + dC_{b,\sigma}^2 h)^i.$$

Finally, $|f_0(x) - f_0(y)| \leq \mathbb{E}\big|f(X_{Nh}^{x,(h)}) - f(X_{Nh}^{y,(h)})\big| \leq |f|_{\text{Lip.}}\sqrt{\delta_N}$, which leads to (6.2.1).

▷ **End of the proof.** From (6.2.1) and $1 + x \leq \exp(x)$, we obtain that

$$\sum_{i=1}^N h|f_i|_{\text{Lip.}}^2 \leq |f|_{\text{Lip.}}^2 \sum_{i=1}^N h\exp(N(C_{b,\sigma}^2 h^2 + 2C_{b,\sigma}h + dC_{b,\sigma}^2 h))$$
$$\leq |f|_{\text{Lip.}}^2 T\exp(C_{b,\sigma}^2 T^2 + 2C_{b,\sigma}T + dC_{b,\sigma}^2 T).$$

By setting

$$C_{b,\sigma,T} := T\exp(C_{b,\sigma}^2 T^2 + 2C_{b,\sigma}T + dC_{b,\sigma}^2 T)|||\sigma||_2|_\infty^2, \quad (6.2.2)$$

we complete the proof. □

6.3 MULTI-LEVEL METHOD

Here we present another approach to the computation of expectation for functionals of a diffusion process. Instead of allocating all the effort to the calculation by the Euler scheme with a small time step (to get a small discretization error) and with a large number of simulations (to get at the same time a small statistical error) (see the asymptotic analysis of Section 6.1) the multi-level approach will rather progressively spread the effort over the Euler schemes with a geometrically decreasing step, for example, twice smaller at each next level. The main idea is to take advantage of the fact that at coarse levels the error is large but the computational cost is small, and the error gradually decreases while the computational cost increases.

This principle is quite standard in deterministic numerical analysis (see [145]); a general approach in the stochastic case has been developed by Heinrich [75]. In the sequel, we rather follow [54], which analysis is based on the multi-level ideas applied to the process discretization.

▷ **Description of the method.** To fix the ideas, consider the problem of the evaluation of $\mathbb{E}(f(X_T))$ and take a sequence of discretization steps $h_0, \ldots, h_l, \ldots, h_L$, which decreases geometrically. The index l stands for the *level* number, associated with the time step

$$h_l = h_0 2^{-l}.$$

With this choice, the step is divided by 2 from the level l to the level $l+1$. The coarse level corresponds to $l = 0$, while the fine level, fixing the accuracy to be achieved, is $l = L$.

Instead of directly generating M simulations of the Euler scheme with the step h_L, we simulate the Euler schemes with the intermediate steps and write a telescopic sum

$$\mathbb{E}(f(X_T^{(h_L)})) = \mathbb{E}(f(X_T^{(h_0)})) + \sum_{l=1}^{L} \mathbb{E}(f(X_T^{(h_l)}) - f(X_T^{(h_{l-1})})).$$

a) The term of level 0 is evaluated with the help of M_0 independent simulations of $X_T^{(h_0)}$: this is standard. The corresponding computational cost is

$$\mathcal{C}_{\text{cost}}^{(0)} \sim_c M_0 h_0^{-1},$$

as discussed in Section 6.1. We have

$$\mathbb{E}(f(X_T^{(h_0)})) \approx \frac{1}{M_0} \sum_{m=1}^{M_0} f(X_T^{(h_0,0,m)}).$$

b) The term $\mathbb{E}(f(X_T^{(h_l)})-f(X_T^{(h_{l-1})}))$ of the increment from the level $l-1$ to the level l is generated using M_l independent simulations of the difference $f(X_T^{(h_l)}) - f(X_T^{(h_{l-1})})$: for this step, it is crucial that *the same Brownian increments are used to construct the two Euler schemes with the steps h_{l-1} and h_l*. In practice, it can be easily performed because it is enough

- to simulate the increments for a finer grid with time step h_l and re-sum them to get the increments for the time step h_{l-1}, or
- to simulate the increments with time step h_{l-1}, and then to refine them to a finer time step using the Brownian bridge technique (Lemma 4.1.7).

We obtain

$$\mathbb{E}(f(X_T^{(h_l)})-f(X_T^{(h_{l-1})})) \approx \frac{1}{M_l} \sum_{m=1}^{M_l} \left(f(X_T^{(h_l,l,m)})-f(X_T^{(h_{l-1},l,m)})\right).$$

Except for using the same Brownian motion for $X_T^{(h_l,l,m)}$ and $X_T^{(h_{l-1},l,m)}$, the simulations are independent, within the same level, and across the different levels. The computational cost at a level l is

$$\mathcal{C}_{\text{cost}}^{(l)} \sim_c M_l(h_l^{-1} + h_{l-1}^{-1}) \sim_c M_l h_l^{-1}$$

because $h_{l-1} = 2h_l$.

The multi-level Monte-Carlo estimator is then defined by

$$\overline{f(X_T)}_{M_0,\ldots,M_L}^{h_0,\ldots,h_L} := \frac{1}{M_0} \sum_{m=1}^{M_0} f(X_T^{(h_0,0,m)})$$

$$+ \sum_{l=1}^{L} \frac{1}{M_l} \sum_{m=1}^{M_l} \left(f(X_T^{(h_l,l,m)}) - f(X_T^{(h_{l-1},l,m)})\right).$$

▷ **Mathematical analysis.** To analyze precisely the statistical fluctuations of the differences $f(X_T^{(h_l,l,m)}) - f(X_T^{(h_{l-1},l,m)})$, suppose that

$\boxed{f \text{ is Lipschitz.}}$

The expectation of the multi-level Monte-Carlo estimator is equal to

$$\mathbb{E}(\overline{f(X_T)}_{M_0,\ldots,M_L}^{h_0,\ldots,h_L}) = \mathbb{E}(f(X_T^{(h_0)})) + \sum_{l=1}^{L} \mathbb{E}(f(X_T^{(h_l)}) - f(X_T^{(h_{l-1})}))$$

$$= \mathbb{E}(f(X_T^{(h_L)})) = \mathbb{E}(f(X_T)) + O(h_L) \qquad (6.3.1)$$

by assuming that the weak error is of order 1 with respect to h. Using the independence of the simulations within the levels and between the levels, its variance writes

$$\mathrm{Var}(\overline{f(X_T)}_{M_0,\ldots,M_L}^{h_0,\ldots,h_L}) = \frac{\mathrm{Var}(f(X_T^{(h_0)}))}{M_0}$$

$$+ \sum_{l=1}^{L} \frac{\mathrm{Var}(f(X_T^{(h_l)}) - f(X_T^{(h_{l-1})}))}{M_l}. \qquad (6.3.2)$$

Because $X_T^{(h_l)}$ and $X_T^{(h_{l-1})}$ are generated from the same Brownian motion W, we can advantageously insert $\pm f(X_T)$ in the variance term, where X_T is the solution at time T of the stochastic differential equation, constructed using W: then using the *strong convergence* of order $\frac{1}{2}$ of the Euler scheme (Section 5.2) and because f is Lipschitz, we obtain

$$\mathrm{Var}(f(X_T^{(h_l)}) - f(X_T^{(h_{l-1})})) \leq \mathbb{E}|f(X_T^{(h_l)}) - f(X_T^{(h_{l-1})})|^2$$
$$\leq 2\mathbb{E}|f(X_T^{(h_l)}) - f(X_T)|^2 + 2\mathbb{E}|f(X_T^{(h_{l-1})}) - f(X_T)|^2$$
$$= O(h_l) + O(h_{l-1})$$
$$= O(h_l) \quad (\text{because } h_{l-1} = 2h_l).$$

Combining these two estimates on the expectation and the variance, we get an upper bound on the squared quadratic error (up to a constant):

$$\mathbb{E}\left(\mathrm{Error}_{h,M}^2\right) := \mathbb{E}\left(|\overline{f(X_T)}_{M_0,\ldots,M_L}^{h_0,\ldots,h_L} - \mathbb{E}(f(X_T))|^2\right)$$
$$\leq_c h_L^2 + \frac{\mathrm{Var}(f(X_T^{(h_0)}))}{M_0} + \sum_{l=1}^{L} \frac{h_l}{M_l}. \qquad (6.3.3)$$

On the other hand, the global cost of the algorithm is

$$\mathcal{C}_{\mathrm{cost}} = \sum_{l=0}^{L} \mathcal{C}_{\mathrm{cost}}^{(l)} \sim_c \sum_{l=0}^{L} \frac{M_l}{h_l}. \qquad (6.3.4)$$

▷ **Optimal asymptotic choice.** Let us determine the computational effort required to achieve the accuracy $\varepsilon \to 0$ — meaning that $\sqrt{\mathbb{E}\left(\text{Error}_{h,M}^2\right)} \leq \varepsilon$ — letting the number of simulations tend to infinity, and the same for the number L of levels. Keep in mind that we have already chosen $h_l = h_0 2^{-l}$, and h_0 is fixed in the analysis below. Comparing (6.3.3) and (6.3.4), we see that the best strategy corresponds to taking a computational cost $\frac{M_l}{h_l}$ that is identical on all the levels: observe that this rule is similar to the optimal stratification (see Section 3.2.2). So, this leads to the following tuning

$$M_l = M_0 2^{-l}.$$

We neglect the rounding effects, which can lead to a non-integer M_l. To conclude, this multi-level method has a global cost $\mathcal{C}_{\text{cost}} \sim_c (L+1)M_0$ and achieves the accuracy $\sqrt{\mathbb{E}\left(\text{Error}_{h,M}^2\right)} \leq_c 2^{-L} + \sqrt{\frac{L+1}{M_0}}$. To ensure that the last expression is of order $\varepsilon \to 0$ with the best cost, it is necessary and sufficient to take $L = \frac{\lfloor \log(\varepsilon) \rfloor}{\log(2)}$ and $M_0 \sim_c \varepsilon^{-2} |\log(\varepsilon)|$, which gives the cost $\mathcal{C}_{\text{cost}} \sim_c \varepsilon^{-2} |\log(\varepsilon)|^2$. We have shown the following main result.

Theorem 6.3.1 (multi-level method) *Consider the evaluation of $\mathbb{E}(f(X_T))$ by the multi-level method. Suppose that f is Lipschitz, and that the Euler scheme has a strong convergence error of order $\frac{1}{2}$ and a weak convergence error of order 1 for f: to achieve accuracy (in quadratic mean) equal to $\varepsilon \to 0$, it is enough to take $L = \frac{\lfloor \log(\varepsilon) \rfloor}{\log(2)}$ levels, with a time step for each level equal to $h_l \sim_c 2^{-l}$ and a number of simulations $M_l \sim_c \varepsilon^{-2} |\log(\varepsilon)| 2^{-l}$. Then the computational cost is*

$$\mathcal{C}_{\text{cost}} \sim_c \varepsilon^{-2} |\log(\varepsilon)|^2.$$

Let us return to the discussion of Section 6.1 and compare this with the efficiency of the method that uses only one level: the latter achieves accuracy $\varepsilon \to 0$ with a cost $\mathcal{C}_{\text{cost}} \sim_c \varepsilon^{-3}$, corresponding to the time step $\sim_c \varepsilon$ and the number of simulations $\sim_c \varepsilon^{-2}$. The multi-level method with a cost almost equal to ε^{-2} is significantly more efficient.

In fact, the gain of efficiency is possible, under the condition that we know the orders of weak and strong convergences of the functional at hand (here $f(X_T)$). When the functional is non-continuous, the analysis of strong convergence is much more delicate, however this can be done in certain cases [7]. When the order of strong convergence is not $\frac{1}{2}$, as it is here, the previous analysis can obviously be adapted: we can then check that if the order is more than $\frac{1}{2}$, the computational effort is greater on the coarse levels and we have the opposite if the order is less than $\frac{1}{2}$.

The choice to divide the time step by 2 from one level to the next one is quite arbitrary: this factor 2 can be also optimized (see [54]), which suggests the factors 4 or 7 to get better results.

In addition to a careful adjustment of the level parameters, it is also important to provide confidence intervals, which can be done using the tools from Chapter 2.

Finally, let us mention that this algorithm can be efficiently implemented using parallel computing (each processor is responsible for the calculations at a level). In the framework of Theorem 6.3.1, the computational effort for each processor is of the same order, which means a good synchronization of the calculations between the processors.

6.4 UNBIASED SIMULATION USING A RANDOMIZED MULTI-LEVEL METHOD

We present another approach to construct estimators of expectations of path functionals associated with stochastic differential equations, say $\mathbb{E}(f(X_T))$, to simplify. The idea is closely related to the previous multi-level method, but the main difference is to allow the number of levels L to be random. This procedure is developed in [128] and related works.

- On the one hand, it results in a scheme that produces an unbiased estimator of the quantity of interest $\mathbb{E}(f(X_T))$.

- On the other hand, as seen below, the variance of the estimator may be infinite, which may rule out the application of the usual central limit theorem to get confidence intervals. If the estimator variance is finite, we retrieve a convergence of the global procedure as the inverse of the square root of the number of simulations.

To derive the unbiased estimator, we work under the assumption that X_T is approximated by different Euler schemes $(X^{(h_0)}, \ldots, X^{(h_l)}, \ldots)$ with associated time steps

$$h_l = h_0 2^{-l}$$

and built with the same Brownian motion. We also assume that the Euler scheme converges in the strong sense at a certain order $r > 0$: for any $p \geq 1$, for some constant C_p we have

$$\left[\mathbb{E}(|X_T^{(h_l)} - X_T|^p)\right]^{1/p} \leq C_p h_l^r = C_p h_0^r 2^{-lr}, \quad \forall l \geq 0. \tag{6.4.1}$$

This estimate is of the form (5.2.1) in the case $r = \frac{1}{2}$, but here we allow r to take larger values; this will be important for the subsequent discussion.

Let the number of levels L be a positive integer-valued random variable, such that

$$\mathbb{P}(L \geq l) > 0, \quad \forall l \geq 1. \tag{6.4.2}$$

We additionally assume that L is independent of the Brownian motion used to define X and the Euler schemes. The unbiased estimator of $\mathbb{E}(f(X_T))$ is designed as follows.

Theorem 6.4.1 *Assume that f is Lipschitz and set*

$$\Delta f_l := f(X_T^{(h_l)}) - f(X_T^{(h_{l-1})}), \quad \forall l \geq 1.$$

Then

$$Z := f(X_T^{(h_0)}) + \sum_{l=1}^{L} \frac{\Delta f_l}{\mathbb{P}(L \geq l)} \tag{6.4.3}$$

is integrable and is an unbiased estimator of $\mathbb{E}(f(X_T))$, i.e. $\mathbb{E}(Z) = \mathbb{E}(f(X_T))$.

PROOF:
We first check that the series

$$\sum_{l=1}^{L} \frac{|\Delta f_l|}{\mathbb{P}(L \geq l)} = \sum_{l \geq 1} 1_{L \geq l} \frac{|\Delta f_l|}{\mathbb{P}(L \geq l)}$$

converges in L_1. Using first the independence between L and the processes, and second, the Lipschitz property of f and the error estimate (6.4.1), we obtain

$$\mathbb{E}\left(\sum_{l \geq 1} 1_{L \geq l} \frac{|\Delta f_l|}{\mathbb{P}(L \geq l)}\right) = \sum_{l \geq 1} \mathbb{E}\left(|\Delta f_l|\right) \leq |f|_{\text{Lip.}} C_1 h_0^r \sum_{l \geq 1} 2^{-lr} < +\infty.$$

This allows us to write

$$\mathbb{E}(Z) = \mathbb{E}(f(X_T^{(h_0)})) + \sum_{l \geq 1} \mathbb{E}\left(1_{L \geq l} \frac{\Delta f_l}{\mathbb{P}(L \geq l)}\right)$$

$$= \mathbb{E}(f(X_T^{(h_0)})) + \sum_{l \geq 1} \mathbb{E}(\Delta f_l) = \mathbb{E}(f(X_T))$$

since the third term is a telescoping sum. □

Corollary 6.4.2 (Convergence of the randomized multi-level method) *Under the notations and assumptions of Theorem 6.4.1, let $(Z_1, \ldots, Z_M)$ be a sample of i.i.d. copies of Z and set*

$$\overline{Z}_M = \frac{1}{M} \sum_{m=1}^{M} Z_m. \tag{6.4.4}$$

Then $\overline{Z}_M$ is an unbiased estimator of $\mathbb{E}(f(X_T))$ and converges a.s. to $\mathbb{E}(f(X_T))$ as $M \to +\infty$.

PROOF:
The bias property follows from that of Z. The a.s. convergence of $\overline{Z}_M$ stems from the strong law of large numbers since Z is integrable. □

So far, the distribution of L is quite arbitrary (apart from the condition (6.4.2)). To analyze the cost of simulating Z, we need to specify some conditions on L. To simplify the analysis, we restrict to geometric distribution:

$$L \stackrel{d}{=} \mathcal{G}(2^{-\gamma}), \quad \text{with } \gamma > 1. \tag{6.4.5}$$

A more general discussion is derived in [128]. In the simulation of Z, the contribution Δf_l at level l requires $C(d)T/h_l + C(d)T/h_{l-1}$ elementary computations: therefore the expected cost is

$$\mathbb{E}(\mathcal{C}_{\text{cost}}) \leq \text{Cst } \mathbb{E}\left(\sum_{l=0}^{L} 2^l\right).$$

Since $\gamma > 1$, we easily check that $\mathbb{E}(\mathcal{C}_{\text{cost}})$ is finite. Therefore, with the choice (6.4.5), the computational cost of the unbiased estimator $\overline{Z}_M$ is proportional to the size M of the sample.

In order to derive confidence intervals on $\overline{Z}_M$, we investigate the integrability property of Z, namely, the p-th moment for $p \in (1, 2]$. The triangle inequality for the L_p-norm gives

$$(\mathbb{E}|Z|^p)^{1/p} \leq \left(\mathbb{E}|f(X_T^{(h_0)})|^p\right)^{1/p} + \sum_{l \geq 1} \frac{(\mathbb{E}(\mathbf{1}_{L \geq l}|\Delta f_l|^p))^{1/p}}{\mathbb{P}(L \geq l)}$$

$$= \left(\mathbb{E}|f(X_T^{(h_0)})|^p\right)^{1/p} + \sum_{l\geq 1}\frac{(\mathbb{E}(|\Delta f_l|^p))^{1/p}}{(\mathbb{P}(L\geq l))^{1-1/p}}$$

using the independence between L and the processes

$$\leq \text{Cst}\left(1 + \sum_{l\geq 1} 2^{-l[r-\gamma(1-1/p)]}\right)$$

using the geometric distribution of L and (6.4.1). Thus, the p-th moment is finite if
$$r - \gamma(1 - 1/p) > 0, \qquad (6.4.6)$$
i.e. if the strong convergence order is large enough. In particular, to get a standard central limit theorem on $\overline{Z}_M$, we need $p = 2$, thus $r > \gamma/2$ and since $\gamma > 1$ (to ensure a finite expected simulation cost), it requires $r > 1/2$, which is stronger than the standard estimate (5.2.1). In other words, two regimes arise.

- When the strong convergence order of the Euler scheme is $r = 1/2$, the unbiased estimator $\overline{Z}_M$ is not necessarily square integrable but only in L_p with $p < 2$, provided that we choose $\gamma \in (1, \frac{1}{2(1-1/p)})$.

- When the strong convergence order of the Euler scheme is larger, i.e. $r > 1/2$ (see Exercise 5.1 where $r = 1$, or Exercise 5.2 for the Milshtein scheme), we can choose $\gamma > 1$ and satisfying (6.4.6) with $p = 2$. Consequently, with a computational expected cost proportional to M, we obtain an unbiased estimator of $\mathbb{E}(f(X_T))$, with Gaussian-type confidence intervals of magnitude $1/\sqrt{M}$.

We briefly summarize these properties in a theorem.

Theorem 6.4.3 (unbiased estimator with randomized multi-level method) *Consider the evaluation of $\mathbb{E}(f(X_T))$ by the randomized multi-level method where the number of levels L is geometrically distributed as $\mathcal{G}(2^{-\gamma})$ with $\gamma > 1$.*

Suppose that f is Lipschitz, and that the Euler scheme has a strong convergence error of order $r > 0$.

Then, Z and $\overline{Z}_M$ are unbiased estimators of $\mathbb{E}(f(X_T))$, their expected simulation costs are finite. The Monte-Carlo estimator $\overline{Z}_M$ converges a.s. to $\mathbb{E}(f(X_T))$ as $M \to +\infty$. Its L_p-moment is finite provided that
$$r - \gamma(1 - 1/p) > 0.$$

In particular, if $r > 1/2$, we can appropriately choose the geometric distribution of L in order to get a square-integrable estimator and usual Gaussian confidence intervals for $\sqrt{M}(\overline{Z}_M - \mathbb{E}(f(X_T)))$.

More discussions and comparisons with non-randomized multi-level methods are provided in [128] and references therein.

6.5 VARIANCE REDUCTION METHODS

To reduce the size of confidence intervals as a result of decreasing the variance, the techniques seen in Chapter 3, in principle, can be applied to the present calculation of an expectation of a diffusion functional, with some additional specifications. We present several techniques, based on control variables and on importance sampling.

6.5.1 Control variates

We go back to Definition 3.3.1 of the control variate: it is a centered square integrable random variable Z, sufficiently correlated with

$$\mathcal{E}(f, g, k, X) = f(X_T)e^{-\int_0^T k(r, X_r)dr} + \int_0^T g(s, X_s)e^{-\int_0^s k(r, X_r)dr}ds,$$

the variable for which we evaluate the expectation, or with its discretized version

$$f(X_T^{(h)})e^{-h\sum_{j=0}^{N-1} k(jh, X_{jh}^{(h)})} + h\sum_{i=0}^{N-1} g(ih, X_{ih}^{(h)})e^{-h\sum_{j=0}^{i-1} k(jh, X_{jh}^{(h)})}.$$

The stochastic calculus naturally provides centered square integrable random variables: these are the stochastic integrals $Z_\phi = \int_0^T \phi_r dW_r$. Among the vast choice of $(\phi_s)_{0 \le s \le T}$, let us identify the optimal control variate. Going back to the proof of the Feynman-Kac formula (4.4.3)

$$u(t, x) = \mathbb{E}\left[f(X_T^{t,x})e^{-\int_t^T k(r, X_r^{t,x})dr} + \int_t^T g(s, X_s^{t,x})e^{-\int_t^s k(r, X_r^{t,x})dr}ds\right],$$

observe that by the Itô formula, we get (see (4.4.7) written between $t = 0$ and $s = T$)

$$f(X_T)e^{-\int_0^T k(s_1, X_{s_1})ds_1} + \int_0^T e^{-\int_0^r k(s_1, X_{s_1})ds_1} g(r, X_r)dr - Z = u(0, x),$$

with $Z := \int_0^T e^{-\int_0^r k(s_1, X_{s_1})ds_1} \nabla_x u(r, X_r) \sigma(r, X_r) dW_r.$ (6.5.1)

Consequently, the choice $\phi_r^* := e^{-\int_0^r k(s_1, X_{s_1})ds_1} \nabla_x u(r, X_r) \sigma(r, X_r)$ leads to a constant value of $\mathcal{E}(f, g, k, X) - Z_{\phi^*}$: only one simulation is thus enough to evaluate

$$u(0, x) = \mathcal{E}(f, g, k, X) - Z_{\phi^*} = \mathbb{E}(\mathcal{E}(f, g, k, X)).$$

In practice, as for the optimal importance sampling from Chapter 3, the simulation of Z_{ϕ^*} is out of reach:

- The first reason is that the simulation of Z_{ϕ^*} requires knowledge of $\nabla_x u$, the gradient of the unknown function u.

- The second reason is that the simulation of a stochastic integral can be done only via its time discretization, implying the loss of optimality of the control variate.

In practice, we search for an explicit approximation[4] of u — an approximation that we denote by v — and instead of Z_{ϕ^*}, we take the control variate

$$Z_{\phi^{(h)}} = \sum_{i=0}^{N-1} e^{-h\sum_{j=0}^{i-1} k(jh, X_{jh}^{(h)})} \nabla_x v(ih, X_{ih}^{(h)}) \sigma(ih, X_{ih}^{(h)})(W_{(i+1)h} - W_{ih}).$$

If $\nabla_x v$ and $\nabla_x u$ are close enough and if h is small enough, the variance of

$$f(X_T^{(h)}) e^{-h\sum_{j=0}^{N-1} k(jh, X_{jh}^{(h)})} + h \sum_{i=0}^{N-1} g(ih, X_{ih}^{(h)}) e^{-h\sum_{j=0}^{i-1} k(jh, X_{jh}^{(h)})} - Z_{\phi^{(h)}}$$

is significantly reduced.

6.5.2 Importance sampling

The tools of Chapter 3 were developed for the real-valued or vector-valued random variables: it is not a priori easy to adapt them to continuous time processes (infinite-dimensional case).

---

[4] In fact, this question depends very much on the problem to solve: either we use natural and intuitive approximations or we use asymptotic expansions to derive the first-order term of the function u.

- When the Euler scheme is used for the calculation of $\mathbb{E}(f(X_T)e^{-\int_0^T k(s_1,X_{s_1})\mathrm{d}s_1} + \int_0^T e^{-\int_0^r k(s_1,X_{s_1})\mathrm{d}s_1} g(r,X_r)\mathrm{d}r)$, the simulated random functional depends only on the Brownian increments, which form a $N\times d$ vector of independent Gaussian random variables: thus the problem is reduced to a finite-dimensional case. From this point of view, the adaptive method, presented in Section 3.4.4, can be implemented.

- In full generality, the changes of probability measure for the stochastic differential equations are described by the Girsanov theorem, which is an infinite-dimensional version of certain results from Chapter 3. This can be used as a guide for choosing a suitable change of probability measure, but here there is no miracle since at the simulation stage we finish by reducing the problem to the case of a discretized Brownian motion as before. We will not present any more of these techniques; the reader may consult [121] for examples.

6.6 EXERCISES

Exercise 6.1 (central limit theorem for varying h and M) *We study the CLT-type convergence of*

$$\mathrm{Error}_{h,M} = \frac{1}{M}\sum_{m=1}^M \mathcal{E}(f,g,k,X^{(h,m)}) - \mathbb{E}(\mathcal{E}(f,g,k,X))$$

by varying both the number of simulations M and the time step h. We consider the asymptotics $M\to+\infty$ and $h\to 0$, with different regimes on Mh^2.

i) *Assume $Mh^2 \to 0$, that f,g,k are bounded continuous functions, and that the weak error is of order 1 w.r.t. h. Show a central limit theorem on $\sqrt{M}\,\mathrm{Error}_{h,M}$ with a limit equal to a centered Gaussian random variable with variance $\mathrm{Var}(\mathcal{E}(f,g,k,X))$.*

   Deduce that (6.1.1) is an asymptotic confidence interval at level 95%.

ii) *Assume $Mh^2 = \mathrm{Cst} \neq 0$, and prove a central limit theorem but with a non-centered Gaussian random variable at the limit. For this, we assume that the weak error can be expanded at order 1 w.r.t. h.*

Exercise 6.2 (multi-level method with various strong convergence order) *Assume that the strong convergence of the Euler scheme is of order 1 w.r.t. h (as in the case of constant σ, see Exercise 5.1) and the weak convergence order is still 1 w.r.t. h.*

i) *By a similar analysis to that of Theorem 6.3.1, determine the optimal allocation of computational effort within the different levels (as a function of number of simulations).*

ii) *What is the global complexity $\mathcal{C}_{\text{cost}}$ as a function of the tolerance error ε?*

iii) *More generally, assume that the Euler scheme converges strongly at order $\alpha \in (0,1]$, and weakly at order $\beta \in (0,1]$ (observe that $\alpha \leq \beta$). Derive the complexity/accuracy analysis associated with a multi-level method. What are the configurations of (α, β) for which (after optimizing the effort within levels)*

$$\mathcal{C}_{\text{cost}} \sim_c \varepsilon^{-2},$$

i.e. we retrieve the standard Monte-Carlo convergence rate?

Exercise 6.3 (control variate for arithmetic mean and geometric mean) *The following example is inspired by the valuation of Asian options in financial engineering [90]. Let S be a geometric Brownian motion of the form $S_t = e^{\sigma W_t + (\mu - \frac{\sigma^2}{2})t}$ with $\mu \in \mathbb{R}$, $\sigma > 0$. We aim at computing by Monte-Carlo method the expectation*

$$\mathbb{E}(A^{\text{arith.}}) \quad \text{with} \quad A^{\text{arith.}} := (\int_0^1 S_t dt - K)_+$$

for some given $K > 0$.

i) *Justify why $A^{\text{geom.}} := (\exp(\int_0^1 \log(S_t)dt) - K)_+$ is a possible control variate for the computation of $\mathbb{E}(A^{\text{arith.}})$. We recall that if $Z \stackrel{d}{=} \mathcal{N}(m - \frac{V}{2}, V)$ with $V > 0$, then*

$$\mathbb{E}(e^Z - K)_+ = e^m \mathcal{N}\left[\frac{1}{\sqrt{V}}\ln(e^m/K) + \frac{\sqrt{V}}{2}\right]$$
$$- K\mathcal{N}\left[\frac{1}{\sqrt{V}}\ln(e^m/K) - \frac{\sqrt{V}}{2}\right].$$

ii) For which range of parameters (μ, σ, T) will this control variate be the most efficient?

iii) Now we approximate $A^{\text{arith.}}$ by

$$A_n^{\text{arith.}} := (\frac{1}{n} \sum_{i=0}^{n-1} S_{\frac{i}{n}} - K)_+.$$

Follow the arguments of Exercise 4.2 and show that

$$\mathbb{E}(A_n^{\text{arith.}}) - \mathbb{E}(A^{\text{arith.}}) = O(n^{-1}).$$

iv) What is the natural control variate $A_n^{\text{geom.}}$ associated with $A_n^{\text{arith.}}$?

v) Write a simulation program for illustrating the previous variance reduction.

PART C: SIMULATION OF NON-LINEAR PROCESSES

CHAPTER 7

Backward stochastic differential equations

In Section 4.4.2, we saw how stochastic differential equations can be naturally associated with linear partial differential equations via the expectations of functionals of the corresponding stochastic trajectories. These couples (PDE/process) are particular cases of more general models, possibly non-linear, to which we devote this last part. A phenomenon is non-linear if does not satisfy the superposition principle: the sum of two solutions is not equal to the solution of the problem where we sum the data. Non-linearities can be of very different nature and usually come from interactions in the studied phenomena. Far from being exhaustive, in Part **C**, we describe three types of typical interaction:

1. one leads to *backward stochastic differential equations* (Chapters 7 and 8);

2. another corresponds to *branching diffusion processes* (Section 7.5); and

3. the last one is associated with *stochastic differential equations with interaction* (*non-linear diffusion in the McKean sense*) (Chapter 9).

Definition 7.0.1 (informal definition of forward and backward stochastic differential equations)
▷ Forward SDE *was studied in Chapter 4: it has the dynamics of the type* $X_t = x_0 + \int_0^t \ldots \mathrm{d}s + \int_0^t \ldots \mathrm{d}W_s$, *with a specified* initial condition

x_0 and certain coefficients (...) *that impose the dynamics in time*.
▷ *On the contrary, a* backward SDE $(Y_t)_t$ *has a specified* terminal condition *at* T *(as a target to achieve), and the value* Y_t *at time t must be adapted to the information at t in order to not anticipate the future*.

Regarding backward stochastic differential equations and the associated PDE terminology, for the latter we refer to *semi-linear PDE*, which means that the function g in Theorem 4.4.3 depends (possibly non-linearly) on the solution u and its first derivative ∇u.

Further in this chapter, we start by giving several examples from very different fields, to stimulate the interest of the reader. A Feynman-Kac-type formula making the PDE/SDE connection will also be given. In fact the new stochastic differential equation is not conventional and must be solved starting from the end, which explains the terminology *backward*: this is a standard case of a *dynamic programming equation* appearing in optimization problems, where repeated decisions/commands have to be applied over time. We prove a theorem of existence and uniqueness for this type of equation, and we propose a *backward* time discretization inspired by the Euler scheme and analyze its order of convergence. However, its effective simulation is made by solving a discrete time dynamic programming equation, requiring the calculation of a large number of conditional expectations. This generic and delicate problem is treated independently in Chapter 8.

7.1 EXAMPLES

We describe several examples coming from either deterministic or stochastic equations: then we will see that it is easy to pass from one point of view to the other via the Feynman-Kac formulas. The point of view chosen in the model presentation is rather guided by its roots, either deterministic or stochastic.

7.1.1 Examples coming from reaction-diffusion equations

A large class of problems studied in this chapter relates to the family of reaction-diffusion systems, whose solutions $u = (u_1, \ldots, u_K) : (t, x) \in$

$[0, T] \times \mathbb{R}^d \mapsto \mathbb{R}^K$ satisfy the equations of the form

$$\begin{cases} \partial_t u_k(t, x) + \mathcal{L}^k u_k(t, x) + g_k(t, x, u(t, x)) = 0, \\ \qquad\qquad\qquad\qquad\qquad t < T, x \in \mathbb{R}^d, k = 1, \ldots, K, \\ u_k(T, \cdot) \text{ given}, k = 1, \ldots, K, \end{cases} \quad (7.1.1)$$

for certain functions $g_k : (t, x, v) \in [0, T] \times \mathbb{R}^d \times \mathbb{R}^K \mapsto \mathbb{R}$ which carry all the non-linearity of the problem. Here $\mathcal{L}^k$ stands for the infinitesimal generator of a certain stochastic differential equation X^k with coefficients (b^k, σ^k), i.e., from Definition 4.4.1

$$\mathcal{L}^k = \mathcal{L}^{X^k}_{b^k, [\sigma^k \sigma^k]^\mathsf{T}} = \frac{1}{2} \sum_{i,j=1}^d [[\sigma^k \sigma^k]^\mathsf{T}]_{i,j}(t, x) \partial^2_{x_i x_j} + \sum_{i=1}^d b_i^k(t, x) \partial_{x_i}.$$

As a consequence, the difference with Theorem 4.4.3 is essentially twofold.

- u is possibly vector-valued (system of K PDEs): in fact, this is an extension that is not mathematically complicated but it makes perfect sense for applications.

- Mainly, there is a coupling between the components of u (via $g = (g_1, \ldots, g_K)$). Even in the case $K = 1$, the non-linear effect remains in g. That last effect is the newest with respect to Part B.

Of course, as in Chapter 4 we could add boundary conditions (of Dirichlet type) or set the problem in infinite horizon (elliptic equation, without time derivative); adaptation of the following ideas and arguments is left to the reader. For complete references on these reaction-diffusion equations; see the books of Henry [76] and Smoller [136].

Example 7.1.1 (ecology) *In a region of the plane ($d = 2$), suppose the existence of K interacting species in the same environment and denote $u_k(t, x)$ the population density of the individuals of the species k at a point of the space x and at a given time t. The model of this type was introduced in the end of the 1930s in the works of Fisher [41] and Kolmogorov, Petrovsky, and Piskunov [94]; for more references, see the more recent book [133].*

216 ■ Backward stochastic differential equations

Let us rapidly discuss the case of two species $K = 2$:

$$\begin{cases} \partial_t u_1(t,x) = \alpha^1 \Delta u_1(t,x) + g_1(t,x,u_1,u_2), \\ \partial_t u_2(t,x) = \alpha^2 \Delta u_2(t,x) + g_2(t,x,u_1,u_2), \quad t > 0, x \in \mathbb{R}^2, \\ u_i(0,\cdot) \text{ given}, \end{cases}$$
(7.1.2)

where the two Laplacian operators model the diffusive movement (as a Brownian motion) of each species. With respect to the equation (7.1.1), the sign of $\partial_t u$ is changed, which is a simple time reversal $t \leftrightarrow T - t$ as the one seen in Chapter 4.

The function $g = (g_1, g_2)$ plays the role of a growth rate of the population: it describes the available local resources and interactions between the species. Namely:

1. *The case $\partial_{u_2} g_1 < 0$ and $\partial_{u_1} g_2 > 0$ corresponds to the predator-prey model, where the growth rate of the species 1 (prey) decreases in the case of high density of the species 2 (predator) and conversely for the growth rate of predators.*

2. *The case $\partial_{u_2} g_1 > 0$ and $\partial_{u_1} g_2 > 0$ corresponds to the symbiosis model, where each species benefits from the other.*

3. *Taking $\partial_{u_2} g_1 < 0$ and $\partial_{u_1} g_2 < 0$ describes a competition model between the species.*

Example 7.1.2 (neuroscience) *The famous model of Hodgkin and Huxley (Nobel Prize in medicine in 1963) is a set of equations describing the psychological phenomenon of signal transmission in the axone (nerve fiber), showing the dependences between electrical excitability and various chemical ion concentrations. The system is of size $K = 4$, the first unknown u represents the electric potential and the three other unknowns (v_1, v_2, v_3) are the chemical concentrations*[1]*:*

---

[1] In the historical experiment [78] on the giant squid, the electric currents were due to ions of potassium, sodium, and other residual currents.

the system writes for $(t,x) \in \mathbb{R}^+ \times \mathbb{R} \mapsto \mathbb{R}$ (linear neuron)

$$\begin{cases} c_0 \partial_t u = \frac{1}{R}\partial_{xx}^2 u + \kappa_1 v_1^3 v_2 (c_1 - u) + \kappa_2 v_3^4 (c_2 - u) + \kappa_3 (c_3 - u), \\ \partial_t v_1 = \varepsilon_1 \partial_{xx}^2 v_1 + g_1(u)(h_1(u) - v_1), \\ \partial_t v_2 = \varepsilon_2 \partial_{xx}^2 v_2 + g_2(u)(h_2(u) - v_2), \\ \partial_t v_3 = \varepsilon_3 \partial_{xx}^2 v_3 + g_3(u)(h_3(u) - v_3), \\ u(0,\cdot) \text{ and } v_i(0,\cdot) \text{ given}, \end{cases} \qquad (7.1.3)$$

for different positive constants $c_i, R, \kappa_i, \varepsilon_i$ and different functions g_i, h_i. There also exists a simplification of this model, known as the FitzHugh-Nagumo model.

Example 7.1.3 (chemistry) *Suppose that a container contains N chemical compounds taking part in R independent reactions. Denote c_i the concentration of the i-th compound and θ the temperature. Then their evolution follows the $K = N + 1$ equations (in $\mathbb{R}^+ \times \mathbb{R}^3$)*

$$\begin{cases} \varepsilon_p \partial_t c_i = D_i \Delta c_i + \sum_{j=1}^R \nu_{ij} g_j(c_1, \ldots, c_N, \theta), \quad i = 1, \ldots, N, \\ \rho c_p \partial_t \theta = k \Delta \theta - \sum_{j=1}^R \sum_{i=1}^N \nu_{ij} H_i g_j(c_1, \ldots, c_N, \theta), \\ c_i(0,\cdot) \text{ and } \theta(0,\cdot) \text{ given}, \end{cases} \qquad (7.1.4)$$

where g_j is the speed of the j-th reaction and H_i is the partial molar enthalpy of the i-th compound. For more details, see [50].

Example 7.1.4 (materials physics) *The Allen-Cahn equation [25] is a prototype model of phase transition with a diffusive interface, used to model, for example, a solid/liquid phase transition [142]. The system is one-dimensional, $K = 1$, and takes the form (in $\mathbb{R}^+ \times \mathbb{R}^3$)*

$$\begin{cases} \partial_t u = \varepsilon \Delta u + u(1 - u^2), \\ u(0,\cdot) \text{ given}. \end{cases} \qquad (7.1.5)$$

The solution u thus represents an order parameter defining the arrangement of atoms in a crystal lattice.

Other examples are given in [76, Chapter 2] and [136, Chapter 14].

7.1.2 Examples coming from stochastic modeling

Example 7.1.5 (finance) *Consider a simplified financial market including:*

- a stock, with a price evolution $(X_t)_{t\geq 0}$ following the geometric Brownian motion model (see Section 4.3.3):

$$X_t = x_0 \exp((b - \frac{1}{2}\sigma^2)t + \sigma W_t);$$

- a riskless asset with an interest rate r for both lending and borrowing.

Denote $(\pi_t)_t$ the amount invested in the stock over time. Under the constraint of being self-financing (neither exterior inputs and nor money withdrawals) the evolution the portfolio value obeys the conservation equation (see [119])

$$Y_t = Y_0 + \int_0^t \pi_s \frac{\mathrm{d}X_s}{X_s} + \int_0^t (Y_s - \pi_s)r\mathrm{d}s$$

$$= Y_0 + \int_0^t [b\,\pi_s + r(Y_s - \pi_s)]\mathrm{d}s + \int_0^t \sigma\pi_s \mathrm{d}W_s, \quad \forall t \geq 0.$$

Specifying a stochastic target at a future date $T > 0$ of the form $Y_T = f(X_T)$ (which is interpreted as a management objective[2]), we obtain a new equation written from the end, where the unknowns are the processes π and Y:

$$Y_t = f(X_T) - \int_t^T [b\,\pi_s + r(Y_s - \pi_s)]\mathrm{d}s - \int_t^T \sigma\pi_s \mathrm{d}W_s, \quad \forall t \in [0, T].$$

Let us now take into account a difference between the lending rate r and the borrowing rate $R > r$, each interest rate being applied according to $Y_s - \pi_s \geq 0$ (lending) or ≤ 0 (borrowing): then the previous equation becomes[3]

$$Y_t = f(X_T) - \int_t^T [b\,\pi_s + r(Y_s - \pi_s)_+ - R(Y_s - \pi_s)_-]\mathrm{d}s - \int_t^T \sigma\pi_s \mathrm{d}W_s.$$

Setting $Z_s = \sigma\pi_s$, we obtain a backward stochastic differential equation - this is backward because the condition at $t = T$ is known, instead of the condition at $t = 0$ - of the form

$$Y_t = f(X_T) + \int_t^T g(s, X_s, Y_s, Z_s)\mathrm{d}s - \int_t^T Z_s \mathrm{d}W_s \qquad (7.1.6)$$

---

[2] Options hedging, for example.
[3] Denoting $x_+ = \max(x, 0)$ and $x_- = \max(-x, 0)$.

where we recall that the unknowns are the stochastic processes Y and Z (which are implicitly adapted to the ambient Brownian filtration, and with omitted integrability conditions). Besides, this equation is coupled to a forward[4] stochastic differential equation for $(X_t)_{t\geq 0}$ like the one in Part **B**: thus (X, Y, Z) forms the prototype of a forward-backward stochastic differential equation, appearing in problems of stochastic control — here, find π that leads Y to the value $f(X_T)$ at time T. For other examples in finance, see the articles [38] and [37].

The following example develops a little more the *stochastic control* point of view — for more references, see [108]. This time, the process X has dynamics that depend also on (Y, Z), producing a stronger *forward-backward* coupling than in the previous example.

Example 7.1.6 (stochastic control) *Consider the problem of linear quadratic control: a controller, by his action $(c_t)_{t\geq 0}$ over time, manages a random system whose state at a time t is described by*

$$X_t^{(c)} = x_0 + \int_0^t (-aX_s^{(c)} + c_s)\mathrm{d}s + W_t \qquad (7.1.7)$$

(for a constant parameter a): this may be seen as an Ornstein-Uhlenbeck process driven linearly by the control c. This control must be optimized to minimize the quadratic *cost functional*

$$J(c) := \frac{1}{2}\mathbb{E}\left(\int_0^T ([X_t^{(c)}]^2 + c_t^2)\mathrm{d}t + [X_T^{(c)}]^2\right),$$

among all the adapted and square integrable controls. Let us apply the Pontryagin principle to determine the necessary form[5] of the optimum c^*: for any control c and any $\varepsilon \in \mathbb{R}$, $c^* + \varepsilon c$ is again an admissible control and by the definition of the optimum we have $J(c^*+\varepsilon c) \geq J(c^*)$. Precisely, we easily show that the process $\dot{X}^*$, which is a solution of the ordinary differential equation (with random coefficients);

$$\dot{X}_t^* = \int_0^t (-a\dot{X}_s^* + c_s)\mathrm{d}s \qquad (7.1.8)$$

---

[4]Forward because this time we specify the condition at $t = 0$.

[5]In this example, there exists a unique solution due to the convexity and the coercivity of $c \mapsto J(c)$.

220 ■ Backward stochastic differential equations

corresponds to the derivative $\dot{X}_t^* = \lim_{\varepsilon \to 0} (X_t^{(c^*+\varepsilon c)} - X_t^{(c^*)})/\varepsilon$. So, writing $[J(c^* + \varepsilon c) - J(c^*)]/\varepsilon \geq 0$ with $\varepsilon \to 0$, we obtain

$$0 \leq \mathbb{E}\left(\int_0^T (X_t^{(c^*)}\dot{X}_t^* + c_t c_t^*)dt + X_T^{(c^*)}\dot{X}_T^*\right). \tag{7.1.9}$$

To the Equation (7.1.7) satisfied by $X^{(c^*)}$, let us associate $(Y^{(c^*)}, Z^{(c^*)})$, the adapted solution of the following adjoint backward equation (admitting its existence and uniqueness)

$$\begin{cases} dY_t^{(c^*)} = (aY_t^{(c^*)} - X_t^{(c^*)})dt + Z_t^{(c^*)}dW_t, \\ Y_T^{(c^*)} = X_T^{(c^*)}. \end{cases} \tag{7.1.10}$$

By applying the Itô formula to $Y_t^{(c^*)}\dot{X}_t^*$, using (7.1.9) and (7.1.8–7.1.10) we deduce

$$0 \leq \mathbb{E}\left(\int_0^T (X_t^{(c^*)}\dot{X}_t^* + c_t c_t^*)dt + Y_T^{(c^*)}\dot{X}_T^*\right)$$

$$= \mathbb{E}\left(\int_0^T \left[(X_t^{(c^*)}\dot{X}_t^* + c_t c_t^*) + (aY_t^{(c^*)} - X_t^{(c^*)})\dot{X}_t^*\right.\right.$$

$$\left.\left. + Y_t^{(c^*)}(-a\dot{X}_t^* + c_t)\right] dt\right)$$

$$= \mathbb{E}\left(\int_0^T c_t[c_t^* + Y_t^{(c^*)}]dt\right).$$

As c is arbitrary, this inequality imposes $c_t^* = -Y_t^{(c^*)}$. Plugging this into (7.1.10), it follows that the solution (X, Y, Z) to the forward-backward stochastic differential equation

$$\begin{cases} dX_t = (-aX_t - Y_t)dt + dW_t, \\ X_0 = x_0, \end{cases} \quad \text{(SDE with initial condition)}$$

$$\begin{cases} dY_t = (aY_t - X_t)dt + Z_t dW_t, \\ Y_T = X_T, \end{cases} \quad \text{(SDE with terminal condition)}$$

solves the optimal stochastic control problem:

$$c_t^* = -Y_t \quad \text{and} \quad X_t^{(c^*)} = X_t.$$

The reader should notice that in this example, the numerical simulation of (X, Y, Z) should be much more complicated, despite the linearity of the coefficients. Indeed, in Example 7.1.5, we can simulate X in the forward time sense (Chapter 5) without knowing (Y, Z), and then simulate (Y, Z) in the backward sense. In this last example, the two simulation stages (forward and backward) are strongly dependent because X depends on Y, which depends on X.

7.2 FEYNMAN-KAC FORMULAS

7.2.1 A general result

The following result, which we state on purpose in a quite general form, gives a general connection between the PDE and the SDE points of view, and allows us to unify all the previous examples. This connection is obviously important for enlarging the possibilities of numerical resolution (by PDE discretization methods, or SDE simulation, or mixed) and diversifying the available theoretical tools. To avoid complicated notation, we suppose that the operators $\mathcal{L}^{[k]}$ in (7.1.1) are identical (one single process).

Theorem 7.2.1 *Let*

- *$d \in \mathbb{N}^*$ be the dimension of the given Euclidean space,*
- *$K \in \mathbb{N}^*$ be the size of the system of coupled equations, and*
- *$T > 0$ be a fixed terminal time.*

Consider the following:

- *The diffusion process X defined by coefficients $b : \mathbb{R}^d \mapsto \mathbb{R}^d$ and $\sigma : \mathbb{R}^d \mapsto \mathbb{R}^d \otimes \mathbb{R}^d$ satisfying the hypotheses of Theorem 4.3.1, and for which the solution is built with respect to a d-dimensional Brownian motion W:*

$$X_t = x_0 + \int_0^t b(s, X_s) \mathrm{d}s + \int_0^t \sigma(s, X_s) \mathrm{d}W_s. \qquad (7.2.1)$$

The infinitesimal generator $\mathcal{L}^X_{b,\sigma\sigma^\top}$ of X is simply denoted by $\mathcal{L}$. In addition, $\sigma(t, x)$ is invertible at any point (t, x) and its inverse $[\sigma(t, x)]^{-1}$ is uniformly bounded.

222 ■ Backward stochastic differential equations

- Two continuous functions $f = (f^{[1]}, \ldots, f^{[K]})^\mathsf{T} : \mathbb{R}^d \mapsto \mathbb{R}^K$ and $g = (g^{[1]}, \ldots, g^{[K]})^\mathsf{T} : [0,T] \times \mathbb{R}^d \times \mathbb{R}^K \times (\mathbb{R}^K \otimes \mathbb{R}^d) \mapsto \mathbb{R}^K$, with the growth conditions

$$|f(x)| + |g(t,x,u,v)| \leq C(1 + |x|^2 + |u| + |v|), \quad \forall (t,x,u,v),$$

for a constant $C \geq 0$.

- The system of K semi-linear partial differential equations for $u = (u^{[1]}, \ldots, u^{[K]})$

$$\begin{cases} \partial_t u^{[k]}(t,x) + \mathcal{L}u^{[k]}(t,x) + g^{[k]}(t,x,u(t,x), \nabla u(t,x)) = 0, & 1 \leq k \leq K, \\ u^{[k]}(T,x) = f^{[k]}(x). \end{cases}$$

(7.2.2)

Suppose that there exists a solution $u \in \mathcal{C}^{1,2}([0,T] \times \mathbb{R}^d, \mathbb{R}^K)$ to the previous system (with $\sup\limits_{(t,x) \in [0,T] \times \mathbb{R}^d} \dfrac{|u(t,x)| + |\nabla u(t,x)|^2}{1 + |x|^2} < +\infty$), then the processes (X, Y, Z) defined by (7.2.1),

$$Y_t := u(t, X_t) \text{ and } Z_t := \nabla u(t, X_t)\sigma(t, X_t),$$

satisfy the system of forward backward stochastic differential equations given by

$$\begin{cases} X_t = x_0 + \int_0^t b(s, X_s)\,\mathrm{d}s + \int_0^t \sigma(s, X_s)\,\mathrm{d}W_s, \\ Y_t = f(X_T) + \int_t^T g(s, X_s, Y_s, Z_s[\sigma(s, X_s)]^{-1})\,\mathrm{d}s - \int_t^T Z_s\,\mathrm{d}W_s. \end{cases}$$

(7.2.3)

PROOF:
Let us show the equality for the k-th component of (Y, Z), namely $Y_t^{[k]} = u^{[k]}(t, X_t)$ and $Z_t^{[k]} = \nabla u^{[k]}(t, X_t)\sigma(t, X_t)$. The Itô formula (4.4.4) applied to $u^{[k]}$ and X gives

$$\begin{aligned} \mathrm{d}Y_s^{[k]} &= \mathrm{d}u^{[k]}(s, X_s) \\ &= [\partial_t u^{[k]}(s, X_s) + \mathcal{L}u^{[k]}(s, X_s)]\,\mathrm{d}s + \nabla u^{[k]}(s, X_s)\sigma(s, X_s)\,\mathrm{d}W_s \\ &= -g^{[k]}(s, X_s, u(s, X_s), \nabla u(s, X_s))\,\mathrm{d}s + \nabla u^{[k]}(s, X_s)\sigma(s, X_s)\,\mathrm{d}W_s, \end{aligned}$$

which is written between $s = t$ and $s = T$ as:

$$u^{[k]}(T, X_T) = Y_t^{[k]} - \int_t^T g^{[k]}(s, X_s, u(s, X_s), \nabla u(s, X_s))\,\mathrm{d}s + \int_t^T Z_s^{[k]}\,\mathrm{d}W_s.$$

As $u^{[k]}(T, .) = f^{[k]}(.)$, the proof is finished. □

If the operator $\mathcal{L}$ were not the same for the K Equations (7.2.2), the probabilistic representation would be a little more delicate to write because it would involve multiple diffusion processes.

Writing (7.2.3) at $t = 0$ and taking the expectation (which removes the stochastic integral under the hypotheses of the theorem), we obtain a representation of u in the form of expectation.

Corollary 7.2.2 *Under the same assumptions as in Theorem 7.2.1, we have*

$$u(0, x_0) = \mathbb{E}\Big(f(X_T) + \int_0^T g\big(s, X_s, Y_s, Z_s[\sigma(s, X_s)]^{-1}\big)\mathrm{d}s\Big). \quad (7.2.4)$$

It is possible to write an analogous formula to represent $u(t, x)$ as expectation, letting the process X start at x at the time t instead of x_0 at the time 0. Among other possible extensions, let us mention that we could also write a representation formula for $\nabla u(0, x_0)$, using the Bismut-Elworthi-Li formula (Theorem 4.5.3). We will not further develop all these extensions, because the notation is quite cumbersome and may hide the fundamental ideas: we will focus on the Monte-Carlo simulation for problems with a simpler writing which will contain all the essential arguments used in more general problems. For the general case, see [63].

The careful reader would be interested in comparing Corollary 7.2.4 with Theorem 4.4.3, the latter containing a linear term in u: we see that it is possible either to treat it by including it in the source term g, or to consider it as a discount factor; both approaches are possible, being a little different numerically. Anyway, the second point of view is often used and is more direct.

To finish, let us return to the examples mentioned in the introduction.

> ▷ Examples 7.1.1 (Ecology), 7.1.2 (Neuroscience), 7.1.3 (Chemistry). These cases, for which we have, respectively, $K = 2$, $K = 4$, and $K =$ "number of chemical compounds plus 1," correspond to take Brownian motions (up to a constant), in order to give a probabilistic interpretation to the operator $\mathcal{L}$ under the Laplacian form.
>
> ▷ Example 7.1.4 (Materials physics). This case is similar, but we notice the appearance of a super-linear growth of $g(t, x, u, v) = u(1 - u^2)$, not covered by Theorem 7.2.2: this type of growth

is technically more delicate to deal with. Under a monotonicity condition (satisfied by $u \mapsto -u^3$), we can establish a probabilistic representation; see [123].

▷ **Example 7.1.5 (Finance).** Comparing (7.1.6) and (7.2.3), this amounts to taking $K = 1$, with a non-linear term for g, depending on u and its gradient ∇u.

▷ **Example 7.1.6 (Stochastic control).** Here again, $K = 1$ but the main difference is that the diffusion process depends (via its infinitesimal generator $\mathcal{L}$) on u (because of the term $-Y_t dt$ in the dynamics of X). Theorem 7.2.1 does not cover exactly this case but an extension of the result is again possible at the cost of additional technicalities; see [108].

7.2.2 Toy model

In the following, to really focus on the fundamental ideas underlying the simulation algorithms rather than on technicalities of the mathematical model, we will consider its most simplified form, taking

- the system reduced to one equation $K = 1$,
- the function g independent of ∇u.

By rewriting the result of Corollary 7.2.2 at the point (t, x) from which starts the diffusion $X^{t,x}$, our aim is now to calculate by Monte-Carlo method the function u given by

$$\begin{cases} u(t,x) = \mathbb{E}\left(f(X_T^{t,x}) + \int_t^T g(s, X_s^{t,x}, u(s, X_s^{t,x}))ds\right), \\ X_s^{t,x} = x + \int_t^s b(r, X_r^{t,x})dr + \int_t^s \sigma(r, X_r^{t,x})dW_r, \quad s \in [t,T]. \end{cases}$$

This is our toy model of non-linearity, sufficiently general to get all the flavor of the most important features and issues met in the design of a probabilistic scheme for solving such a non-linear equation.

We start by showing the existence and the uniqueness of a continuous function u that is a solution of the above equation. For this, for a given measurable function $v : [0,T] \times \mathbb{R}^d \mapsto \mathbb{R}$, set

$$Fv(t,x) = \mathbb{E}\left(f(X_T^{t,x}) + \int_t^T g(s, X_s^{t,x}, v(s, X_s^{t,x}))ds\right). \quad (7.2.5)$$

Lemma 7.2.3 *Suppose that b and σ satisfy the usual hypotheses of Theorem 4.3.1 and that*

- *$f : \mathbb{R}^d \mapsto \mathbb{R}$ is continuous and bounded;*
- *$g : [0,T] \times \mathbb{R}^d \times \mathbb{R} \mapsto \mathbb{R}$ is continuous, bounded, and Lipschitz in v:*

$$\sup_{(t,x) \in [0,T] \times \mathbb{R}^d} \sup_{v_1 \neq v_2} \left[\frac{|g(t,x,v_1) - g(t,x,v_2)|}{|v_1 - v_2|} + |g(t,x,v_1)| \right] := L_g < +\infty.$$

Then the map $v \mapsto Fv$ is a contraction in the Banach space of continuous bounded functions $v : [0,T] \times \mathbb{R}^d \mapsto \mathbb{R}$ with the uniform norm

$$\|v\|_\lambda := \sup_{(t,x) \in [0,T] \times \mathbb{R}^d} e^{\lambda t} |v(t,x)|,$$

for a λ large enough.

We can then apply the fixed point theorem to get the following result.

Theorem 7.2.4 (solution of the non-linear equation) *Under the previous hypotheses, there exists a unique continuous bounded solution u of $Fu = u$ and it satisfies the non-linear equation*

$$u(t,x) = \mathbb{E}\left(f(X_T^{t,x}) + \int_t^T g(s, X_s^{t,x}, u(s, X_s^{t,x})) ds \right). \quad (7.2.6)$$

PROOF:

(Proof of Lemma 7.2.3). We show first the contraction property. The quite strong boundedness hypotheses on g ensure that Fv is well defined and bounded by $|f|_\infty + T|g|_\infty$. Let v_1 and v_2 be two continuous functions: then

$$e^{\lambda t} |[Fv_1](t,x) - [Fv_2](t,x)| \leq \int_t^T e^{\lambda t} L_g \sup_{x' \in \mathbb{R}^d} |v_1(s,x') - v_2(s,x')| ds,$$

$$\|[Fv_1] - [Fv_2]\|_\lambda \leq L_g \int_t^T e^{-\lambda(s-t)} ds \|v_1 - v_2\|_\lambda$$

$$= \frac{L_g(1 - e^{-\lambda(T-t)})}{\lambda} \|v_1 - v_2\|_\lambda.$$

So, by taking $\lambda \geq 2L_g$, we obtain that F is contracting for the norm $\|.\|_\lambda$, with a constant less than $\frac{1}{2}$.

It remains to show the continuity of Fv given a continuous function v. With the help of stochastic calculus, we can prove that we can define with probability 1 the continuous function $(t, x, s) \mapsto X_s^{t,x}$; we skip the details [97]. Applying the dominated convergence theorem, it follows that for any continuous bounded function Ψ, $\mathbb{E}(\Psi(X_s^{t,x}))$ is continuous at (t, x) (and also at s): from this we easily deduce that $Fv(t, x) = \mathbb{E}(f(X_T^{t,x})) + \int_t^T \mathbb{E}(g(s, X_s^{t,x}, v(s, X_s^{t,x})))ds$ is continuous at (t, x). □

By adding several regularity assumptions on the problem data, it would be possible to show that conversely the obtained function u is a solution of a semi-linear partial differential equation of the type (7.2.2), which would complete the tour between the PDE and SDE approaches. For references in the PDE literature, see the specialized (quite technical) books [44], [100], [53], [105].

Using the flow composition property and the Markov property of stochastic differential equations (Proposition 4.3.2), we can also write an alternative expression in the form of a conditional expectation.

Theorem 7.2.5 *Under the hypotheses of Theorem 7.2.4, define for $t \in [0, T]$*

$$X_t = x_0 + \int_0^t b(r, X_r)dr + \int_0^t \sigma(r, X_r)dW_r, \qquad (7.2.7)$$

$$Y_t = u(t, X_t), \qquad (7.2.8)$$

without any specific reference to the starting point x_0 of the diffusion process X at the time 0. Then Y is a solution of the backward stochastic differential equation

$$Y_t = \mathbb{E}\left(f(X_T) + \int_t^T g(s, X_s, Y_s)ds \mid X_t \right). \qquad (7.2.9)$$

In fact, this representation is analogous to that of Theorem 7.2.1: indeed, it is enough to take the conditional expectation $\mathbb{E}(\cdot|\mathcal{F}_t)$ in (7.2.3) and to use the Markov property of X to replace the conditioning on $\mathcal{F}_t$[6] by the conditioning on X_t. We call it *backward SDE* because the value at t depends on the future values after t, which suggests a backward-in-time resolution.

---

[6] Which, as before, stands for the canonical augmented Brownian filtration.

The above representation highlights the non-linear equation satisfied by the process $(Y_t)_t$, while the presentation (7.2.6) is written as a non-linear equation on the function $u(t,x)$. The random value Y_t may be seen as the function $u(t,.)$ calculated at the random point X_t.

For the Monte-Carlo type algorithms, the *process* point of view with Y has the advantage of controlling the error in natural norms, and we return to it in Chapter 8. The *process* point of view has another advantage that we mention briefly, but without details: verifying the existence and the uniqueness of the adapted process Y satisfying $Y_t = \mathbb{E}(\xi + \int_t^T g(s, X_s, Y_s)\mathrm{d}s \mid \mathcal{F}_t)$ is possible without supposing that ξ is of the form $f(X_T)$ and that g depends in a Markovian way on X. This allows us to consider non-Markovian hypotheses, which are not attainable with PDE arguments. This efficient extension is based on smart techniques of stochastic calculus; see for example [37].

7.3 TIME DISCRETIZATION AND DYNAMIC PROGRAMMING EQUATION

To simulate the process Y given by (7.2.9), we will use the *dynamic programming* principle,[7] introduced by R.E. Bellman (1920–1984) in the 1940s: it consists of solving a global problem as a series of easier sub-problems. In the current setting, this means discretizing Y in time, writing the local evolution of the process, then simulating the discretized process. We analyze the first stage about time discretization.

7.3.1 Discretization of the problem

Take the notation of Chapter 5 concerning the Euler scheme: the time step is denoted by $h = T/N$, the discretization times are given by $t_i = ih$, the last discretization step before t is $\varphi_t^h := t_i$ if $t_i < t \leq t_{i+1}$, and the Euler scheme at the times $(t_i)_i$ is defined by

$$\begin{cases} X_0^{(h)} = x_0, \\ X_{(i+1)h}^{(h)} = X_{ih}^{(h)} + b(ih, X_{ih}^{(h)})h + \sigma(ih, X_{ih}^{(h)})(W_{(i+1)h} - W_{ih}). \end{cases} \quad (7.3.1)$$

---

[7]This principle is natural and standard for solving dynamic control problems.

From (7.2.9), the evolution of Y between two dates can be obtained by the tower property of the conditional expectations:

$$Y_{t_i} = \mathbb{E}\left(f(X_T) + \int_t^T g(s, X_s, Y_s)ds \mid \mathcal{F}_{t_i}\right)$$

$$= \mathbb{E}\left(\mathbb{E}(f(X_T) + \int_{t_{i+1}}^T g(s, X_s, Y_s)ds \mid \mathcal{F}_{t_{i+1}})\right.$$

$$\left. + \int_{t_i}^{t_{i+1}} g(s, X_s, Y_s)ds \mid \mathcal{F}_{t_i}\right)$$

$$= \mathbb{E}\left(Y_{t_{i+1}} + \int_{t_i}^{t_{i+1}} g(s, X_s, Y_s)ds \mid \mathcal{F}_{t_i}\right). \qquad (7.3.2)$$

By adding the terminal condition $Y_T = f(X_T)$, the above iterative equation is equivalent to (7.2.9) at the times $(t_i)_i$, but it is written in the form of an evolution local in time: this is the *dynamic programming principle*. A quite natural discretization scheme is the *backward Euler scheme*:

$$\begin{cases} Y_T^{(h)} = f(X_T^{(h)}), \\ Y_{ih}^{(h)} = \mathbb{E}\left(Y_{(i+1)h}^{(h)} + hg(t_i, X_{ih}^{(h)}, Y_{(i+1)h}^{(h)}) \mid \mathcal{F}_{t_i}\right), \quad i = N-1, \ldots, 0. \end{cases}$$
$$(7.3.3)$$

We call this recursive equation a *dynamic programming equation*. This scheme is *explicit* in the unknown process because in the term g, we use $Y_{(i+1)h}^{(h)}$; alternatively, it would be possible to use $g(t_i, X_{ih}^{(h)}, Y_{ih}^{(h)})$, which leads to an *implicit scheme*. This version does not fundamentally modify the analysis of the error hereafter; on the other hand, with respect to the explicit scheme, it requires additional implementation of a fixed-point algorithm. We have a preference for the explicit scheme (7.3.3).

7.3.2 Error analysis

We justify that the approximation (7.3.3) *converges in the strong sense at order* $\frac{1}{2}$. The result is stated under the Lipschitz regularity hypothesis in space and the Hölder-continuity hypothesis with exponent $\frac{1}{2}$ in time, the same as for the Euler scheme for X (see Theorem 5.2.1): to this purpose, for a function Ψ of multiple variables (t, x, z) we introduce the notation

$$|\Psi|_{1,x} = \sup_{t,z} \sup_{x_1 \neq x_2} \frac{|\Psi(t,x_1,z) - \Psi(t,x_2,z)|}{|x_1 - x_2|}, \quad (7.3.4)$$

$$|\Psi|_{1,z} = \sup_{t,x} \sup_{z_1 \neq z_2} \frac{|\Psi(t,x,z_1) - \Psi(t,x,z_2)|}{|z_1 - z_2|}, \quad (7.3.5)$$

$$|\Psi|_{\frac{1}{2},t} = \sup_{x,z} \sup_{t_1 \neq t_2} \frac{|\Psi(t_1,x,z) - \Psi(t_2,x,z)|}{|t_1 - t_2|^{\frac{1}{2}}}. \quad (7.3.6)$$

Theorem 7.3.1 (discretization of backward SDE) *Suppose that*

i) *the coefficients (b,σ) of the dynamics (7.2.7) of X are globally Lipschitz in x and Hölder-continuous with exponent $\frac{1}{2}$ in time:* $|b|_{1,x} + |\sigma|_{1,x} + |b|_{\frac{1}{2},t} + |\sigma|_{\frac{1}{2},t} < +\infty;$

ii) *the functions (f,g) from the non-linear equation in Theorem 7.2.4 are bounded, globally Lipschitz in (x,u), and Hölder-continuous with exponent $\frac{1}{2}$ in time:* $|f|_{1,x} + |g|_{1,x} + |g|_{1,u} + |g|_{\frac{1}{2},t} < +\infty;$

iii) *the value function u in the non-linear problem of Theorem 7.2.4 is Lipschitz in x and Hölder-continuous[8] with exponent $\frac{1}{2}$ in time:* $|u|_{1,x} + |u|_{\frac{1}{2},t} < +\infty.$

Then, there exists a constant $C > 0$ (independent of N) such that

$$\sup_{0 \leq i \leq N} \mathbb{E}|Y_{ih}^{(h)} - Y_{ih}|^2 \leq Ch.$$

PROOF:
From (7.3.2) and (7.3.3), we deduce, using triangular inequality, that

$$|Y_{ih}^{(h)} - Y_{ih}| \leq \mathbb{E}\left(|Y_{(i+1)h}^{(h)} - Y_{(i+1)h}| \mid \mathcal{F}_{t_i}\right)$$
$$+ \mathbb{E}\left(h|g(t_i, X_{ih}^{(h)}, Y_{(i+1)h}^{(h)}) - g(t_i, X_{t_i}, Y_{t_{i+1}})| \mid \mathcal{F}_{t_i}\right)$$
$$+ \mathbb{E}\left(\int_{t_i}^{t_{i+1}} |g(t_i, X_{t_i}, Y_{t_{i+1}}) - g(s, X_s, Y_s)| ds \mid \mathcal{F}_{t_i}\right).$$

By combining the convexity inequality

$$(\alpha + \beta + \gamma)^2 \leq (1+h)\alpha^2 + (1+h^{-1})(\beta + \gamma)^2$$

---

[8]The regularity required for u can be obtained either from PDE estimates, using Feynman-Kac representation, or directly by probabilistic arguments based on stochastic calculus.

230 ■ Backward stochastic differential equations

$$\leq (1+h)\alpha^2 + 2(1+h^{-1})(\beta^2 + \gamma^2)$$

with the Cauchy-Schwarz inequality, we deduce

$$|Y_{ih}^{(h)} - Y_{ih}|^2$$
$$\leq (1+h)\mathbb{E}\left(|Y_{(i+1)h}^{(h)} - Y_{(i+1)h}|^2 \mid \mathcal{F}_{t_i}\right)$$
$$+ 2(1+h^{-1})\mathbb{E}\left(h^2 |g(t_i, X_{ih}^{(h)}, Y_{(i+1)h}^{(h)}) - g(t_i, X_{t_i}, Y_{t_{i+1}})|^2 \mid \mathcal{F}_{t_i}\right)$$
$$+ 2(1+h^{-1})\mathbb{E}\left(h \int_{t_i}^{t_{i+1}} |g(t_i, X_{t_i}, Y_{t_{i+1}}) - g(s, X_s, Y_s)|^2 ds \mid \mathcal{F}_{t_i}\right).$$

Then, by using $Y_s = u(s, X_s)$, the time-space regularity properties for g and u, and the convexity inequalities of type (5.1.3), we get that

$$\mathbb{E}(|Y_{ih}^{(h)} - Y_{ih}|^2)$$
$$\leq (1+h)\mathbb{E}\left(|Y_{(i+1)h}^{(h)} - Y_{(i+1)h}|^2\right)$$
$$+ 4h^2(1+h^{-1})|g|_{1,x}^2 \mathbb{E}\left(|X_{ih}^{(h)} - X_{t_i}|^2\right)$$
$$+ 4h^2(1+h^{-1})|g|_{1,u}^2 \mathbb{E}\left(|Y_{(i+1)h}^{(h)} - Y_{t_{i+1}}|^2\right)$$
$$+ 6h(1+h^{-1}) \int_{t_i}^{t_{i+1}} \left[|g|_{\frac{1}{2},t}^2 (s-t_i) + \mathbb{E}\left(|g|_{1,x}^2 |X_s - X_{t_i}|^2\right.\right.$$
$$\left.\left. + |g|_{1,u}^2 |Y_s - Y_{t_{i+1}}|^2\right)\right] ds$$
$$\leq (1+Ch)\mathbb{E}(|Y_{(i+1)h}^{(h)} - Y_{(i+1)h}|^2) + Ch\mathbb{E}(|X_{ih}^{(h)} - X_{t_i}|^2)$$
$$+ Ch \sup_{s\in[t_i,t_{i+1}]} [(s-t_i) + \mathbb{E}(|X_s - X_{t_i}|^2)$$
$$+ (t_{i+1} - s) + \mathbb{E}(|X_{t_{i+1}} - X_s|^2)],$$

for some constant $C > 0$ depending on T, Lipschitz and Hölder constants of the functions, and some other universal constants. The term $\mathbb{E}(|X_{ih}^{(h)} - X_{t_i}|^2)$ is related to the strong convergence of the Euler scheme for a diffusion: it is of order $O(h)$, uniformly in i, owing to Theorem 5.2.1. It is easy to show (using the arguments of the proof of the latter theorem) that the increments of X are of order $\frac{1}{2}$ with respect to the time step (as for Brownian motion), i.e.,

$$\sup_{0\leq i\leq N} \sup_{s\in[t_i,t_{i+1}]} [\mathbb{E}(|X_s - X_{t_i}|^2) + \mathbb{E}(|X_{t_{i+1}} - X_s|^2)] = O(h).$$

Thus we have shown that the error propagates backward according to the inequality

$$\mathbb{E}(|Y_{ih}^{(h)} - Y_{ih}|^2) \leq (1+C'h)\mathbb{E}(|Y_{(i+1)h}^{(h)} - Y_{(i+1)h}|^2) + C'h^2 \quad (7.3.7)$$

for a new constant $C' > 0$. After multiplying this inequality by $(1+C'h)^i$ and summing between $N-1$ and i, a telescoping sum appears and it remains

$$(1+C'h)^i \mathbb{E}(|Y_{ih}^{(h)} - Y_{ih}|^2)$$
$$\leq (1+C'h)^N \mathbb{E}(|Y_{Nh}^{(h)} - Y_{Nh}|^2) + C'h^2 \sum_{j=i}^{N-1}(1+C'h)^j$$
$$\leq (1+C'h)^N \mathbb{E}(|Y_{Nh}^{(h)} - Y_{Nh}|^2) + C'h^2 \frac{(1+C'h)^N - 1}{C'h}.$$

Because $(1+C'h)^N \leq \exp(C'hN) = \exp(C'T)$, we get

$$\mathbb{E}(|Y_{ih}^{(h)} - Y_{ih}|^2) \leq \exp(C'T)\left[\mathbb{E}(|Y_{Nh}^{(h)} - Y_{Nh}|^2) + h\right].$$

Finally, we have $\mathbb{E}(|Y_T^{(h)} - Y_T|^2) = \mathbb{E}(|f(X_T^{(h)}) - f(X_T)|^2) = O(h)$ using again the strong convergence of order $\frac{1}{2}$ of the Euler scheme and the Lipschitz property of f. □

In the previous proof, the order of the strong approximation of X by $X^{(h)}$ plays an important role. In the case of order greater than $\frac{1}{2}$ (even exact simulation as in the case of Brownian motion or Ornstein-Uhlenbeck process), we can improve the order of the approximation of Y by $Y^{(h)}$. Without going into detail, this shows (without surprise) that a good approximation of the forward SDE X plays a determinant role in well approximating the backward SDE Y.

7.4 OTHER DYNAMIC PROGRAMMING EQUATIONS

▷ In the equality $Y_{t_i}^{(h)} = \mathbb{E}\left(Y_{t_{i+1}}^{(h)} + hg(t_i, X_{t_i}^{(h)}, Y_{t_{i+1}}^{(h)}) \mid \mathcal{F}_{t_i}\right)$, if we replace $Y_{t_{i+1}}^{(h)}$ by its formula, the tower property of the conditional expectation gives

$$Y_{t_i}^{(h)} = \mathbb{E}\left(Y_{t_{i+2}}^{(h)} + hg(t_i, X_{t_i}^{(h)}, Y_{t_{i+1}}^{(h)}) + hg(t_{i+1}, X_{t_{i+1}}^{(h)}, Y_{t_{i+2}}^{(h)}) \mid \mathcal{F}_{t_i}\right).$$

We can iterate this procedure until the terminal time where $Y_T^{(h)} = f(X_T^{(h)})$, which leads to a new dynamic programming equation

$$Y_{t_i}^{(h)} = \mathbb{E}\left(f(X_T^{(h)}) + h\sum_{j=i}^{N-1} g(t_j, X_{t_j}^{(h)}, Y_{t_{j+1}}^{(h)}) \mid \mathcal{F}_{t_i}\right), \quad i = N-1,\ldots,0.$$

(7.4.1)

Though the two equations are equivalent so far, differences appear between the dynamic programming equations (7.3.3) and (7.4.1) when the conditional expectations are calculated numerically, because the errors will propagate differently. The object of current research is to better understand which form leads to a better convergence order.

▷ As the Euler scheme $(X_{t_i}^{(h)})_i$ forms a Markov chain, it is not difficult to see that the conditional expectation (7.4.1) is a random variable that depends only on $X_{t_i}^{(h)}$ and writes as

$$Y_{t_i}^{(h)} = u^{(h)}(t_i, X_{t_i}^{(h)}) \qquad (7.4.2)$$

for some continuous bounded function $u^{(h)}(t_i, .) : \mathbb{R}^d \mapsto \mathbb{R}$, satisfying the *dynamic programming equation on functions* (instead of "on random variables"):

$$\begin{cases} u^{(h)}(t_i, x) = \mathbb{E}\Big(f(X_T^{(h),t_i,x}) \\ \qquad\qquad + h\sum_{j=i}^{N-1} g\big(t_j, X_{t_j}^{(h),t_i,x}, u^{(h)}(t_{j+1}, X_{t_{j+1}}^{(h),t_i,x})\big)\Big), \\ X_{t_i}^{(h),t_i,x} = x, \\ X_{t_{j+1}}^{(h),t_i,x} = X_{t_j}^{(h),t_i,x} + b(t_j, X_{t_j}^{(h),t_i,x})h \\ \qquad\qquad + \sigma(t_j, X_{t_j}^{(h),t_i,x})(W_{t_{j+1}} - W_{t_j}), \quad j \geq i. \end{cases} \qquad (7.4.3)$$

This is a discrete time analog of Theorem 7.2.4. The boundedness of $u^{(h)}(t_i, .)$ is obvious to show and its continuity is easier to prove than in the mentioned theorem: it is enough to iterate backward in time and to use the continuity (obvious) of $x \mapsto X_{t_j}^{(h),t_i,x}$. Using the flow property of the Euler scheme, we can also return to a *dynamic programming equation on functions* written between two successive dates:

$$u^{(h)}(t_i, x) = \mathbb{E}\Big(u^{(h)}(t_{i+1}, X_{t_{i+1}}^{(h),t_i,x}) + hg\big(t_i, x, u^{(h)}(t_{i+1}, X_{t_{i+1}}^{(h),t_i,x})\big)\Big). \qquad (7.4.4)$$

This point of view of the *dynamic programming equation on functions* shows clearly — once again — that the simulation of the process Y is not reduced (as in the case of the Euler scheme in Chapter 5) to the simulation of Brownian increments (and thus of $X^{(h)}$), but it also passes through the numerical approximation of the function $u^{(h)}$: this is a new problem of functions approximation, but its form is closely related to a computation on probability distributions, which guides the probabilistic tools studied in the next chapter.

▷ The dynamic programming equation (7.3.2) or (7.3.3) appears in a little different form while modeling optimal stopping problems; see [120, Section VI]. The problem consists of determining the random moment[9] $\tau \in \{t_0, \ldots, t_i, \ldots, T\}$ that is the most suitable to stop a random system, which we represent here as a game: the player receives a reward $f(X_{t_i})$ if he chooses to stop at time t_i. The expected gain at time t_i, denoted by Y_{t_i}, is then the maximum between his immediate gain (if he stops immediately) and his gain at t_{i+1} expected from time t_i (if he continues): thus

$$\begin{cases} Y_T = f(X_T), \\ Y_{t_i} = \max\left(f(X_{t_i}), \mathbb{E}(Y_{t_{i+1}} \mid \mathcal{F}_{t_i})\right), \quad i = N-1, \ldots, 0. \end{cases} \quad (7.4.5)$$

The process Y is known as the Snell envelope of $f(X_\cdot)$ and the above recursive equation is the dynamic programming equation that characterizes it. This type of problem is very popular in finance and is associated with the Bermudan (or American) options pricing.

The tools for the simulation of conditional expectation developed in the next chapter give effective schemes for solving these different dynamic programming equations.

7.5 ANOTHER PROBABILISTIC REPRESENTATION VIA BRANCHING PROCESSES

Some non-linear partial differential equations come from population dynamics modeling [94, 41], bringing into play the random genealogy phenomena. Thus we are not surprised to see that processes of birth and death coupled with diffusion processes give probabilistic representations of a certain class of reaction-diffusion partial differential equations. These ideas were developed in the middle of the 1960s by Skorokhod [135], and Itô and McKean [79] with the use of *branching diffusions*. See also the article [112], and the more recent [126].

---

[9]More precisely, the stopping time in the sense of Definition 4.2.2.

234 ■ Backward stochastic differential equations

Let us define the framework: we consider a diffusion process X with time-homogeneous coefficients (b, σ), possibly multidimensional, with infinitesimal generator $\mathcal{L} = \mathcal{L}^X_{b,[\sigma\sigma]^\mathsf{T}}$. A *branching diffusion* with branching intensity $\lambda > 0$ is built as follows.

- We start from $x \in \mathbb{R}^d$ at time t. During a random time τ_1 with distribution $\mathcal{E}\mathrm{xp}(\lambda)$, the process follows the dynamics of the stochastic differential equation of X (black trajectory in Figure 7.1).

- Then, the process dies and gives birth to k new (independent) processes, with probability $p_k \in [0,1]$, $\sum_{k \geq 0} p_k = 1$: these are the 3 trajectories, red, blue and green, in Figure 7.1.

- Each child process has a life duration with the distribution $\mathcal{E}\mathrm{xp}(\lambda)$ during which it evolves under the distribution of X, independently of the others, then gives in turn a birth to a random number of child processes, etc.

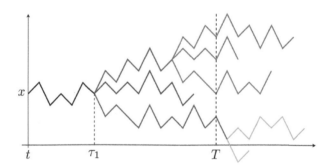

Figure 7.1 Branching diffusion.

Denote by N_T the number of children alive at time T and $(X_T^{t,x,i})_{1 \leq i \leq N_T}$ their respective positions. For a continuous function f with $|f|_\infty < 1^{10}$, define

$$u(t,x) := \mathbb{E}\Big(\prod_{i=1}^{N_T} f(X_T^{t,x,i})\Big), \qquad (7.5.1)$$

with the convention $\prod_{i=1}^{0} = 1$.

---

[10] Ensuring that u is well defined.

Theorem 7.5.1 *Suppose on the one hand the hypotheses of Theorem 4.3.1 about existence and uniqueness of X, and on the other hand, the hypotheses for the existence of a solution to the non-linear partial differential equation*

$$\begin{cases} \partial_t v + \mathcal{L}v + \lambda\Big(\sum_{k\geq 0} p_k v^k - v\Big) = 0, & 0 \leq t < T, x \in \mathbb{R}^d, \\ v(T, \cdot) = f(\cdot). \end{cases} \quad (7.5.2)$$

Then the solution is given by u defined in (7.5.1).

PROOF:

This follows from the strong Markov property applied to the times of birth. Let τ_1 be the first time of birth after t and n_{τ_1} be the corresponding number of children; for each child, labeled with an index k, born at τ_1, denote by N_T^k the number of its offspring alive at date T, as well as their respective positions by $(X_T^{t,x,k,i})_{1 \leq i \leq N_T^k}$. The use of the independence of the children and of their offspring yields

$$u(t,x) = \mathbb{E}\Big(\mathbf{1}_{\tau_1 > T} f(X_T^{t,x,1})\Big) + \mathbb{E}\Big(\mathbf{1}_{\tau_1 \leq T} \prod_{k=1}^{n_{\tau_1}} \prod_{i=1}^{N_T^k} f(X_T^{t,x,k,i})\Big)$$

$$= \mathbb{E}\Big(\mathbf{1}_{\tau_1 > T} f(X_T^{t,x,1})\Big) + \sum_{k \geq 0} p_k \mathbb{E}\Big(\mathbf{1}_{\tau_1 \leq T} [u(\tau_1, X_{\tau_1}^{t,x})]^k\Big)$$

$$= \mathbb{E}\Big(e^{-\lambda(T-t)} f(X_T^{t,x})\Big)$$

$$\quad + \sum_{k \geq 0} p_k \int_t^T \mathbb{E}\Big([u(s, X_s^{t,x})]^k\Big) \lambda e^{-\lambda(s-t)} ds,$$

where we use at the last inequality that $\tau_1 \stackrel{d}{=} \mathcal{E}xp(\lambda)$. We recover the probabilistic representation of Theorem 4.4.3 with $k(t,x) = \lambda$ and $g(t,x) = \lambda \sum_{k \geq 0} p_k u^k(t,x)$, which finishes the proof. □

For this type of non-linearity of polynomial form with positive coefficients, passing through the discrete time dynamic programming is not necessary to obtain a numerical solution, because the direct simulation of a branching diffusion process (with time discretization of X by Euler scheme if necessary) is enough to approximate $u(t,x)$ by the empirical mean of independent realizations of $\prod_{i=1}^{N_T} f(X_T^{t,x,i})$.

In [126], a similar interpretation is given when the branching rates and probabilities are allowed to depend on time and space.

7.6 EXERCISES

Exercise 7.1 (Pontryagin maximum principle) *We investigate furthermore the relation between stochastic control problems and backward stochastic differential equations (BSDE). The following exercise generalizes Example 7.1.6.*

Let $v : [0, T] \to \mathbb{R}$ be an adapted process, square integrable, and let $(X_t^v)_{t \in [0,T]}$ be the solution to the controlled stochastic differential equation

$$X_t^v = x_0 + \int_0^t b(X_s^v, v_s) \mathrm{d}s + \int_0^t \sigma(X_s^v, v_s) \mathrm{d}W_s,$$

where the scalar coefficients b, σ are Lipschitz in the two variables:

$$|b(x,v) - b(x',v')| + |\sigma(x,v) - \sigma(x',v')| \leq L(|x - x'| + |v - v'|).$$

We consider the optimal stochastic control problem related to the cost function

$$J(v) := \mathbb{E}\left(\int_0^T l(X_t^v, v_t) \mathrm{d}t + f(X_T^v)\right)$$

where l, f are C^2 functions with bounded second derivatives. The optimal control is the one minimizing $J(v)$ over all possible controls v: we assume here that it exists, we denote it by $v^\star$, and the related SDE is $X^\star$. Our aim is to characterize the optimal control via the following BSDE (Y, Z):

$$Y_t = f'(X_T^\star) + \int_t^T [Y_s b_x'(X_s^\star, v_s^\star) + Z_s \sigma_x'(X_s^\star, v_s^\star) + l_x'(X_s^\star, v_s^\star)] \, \mathrm{d}s$$
$$- \int_t^T Z_s \mathrm{d}W_s.$$

Solving this equation gives another method of optimal control.

In the following, consider the perturbed control $v_t^\varepsilon := v_t^\star + \varepsilon v_t$ (with $\varepsilon \to 0$) and for any quantity $H = H(v)$ depending on v, set $\Delta_\varepsilon H(v^\star) = H(v^\varepsilon) - H(v^\star)$. We assume that the perturbation v satisfies to the extra condition $\int_0^T \|v_t\|_{L^4(\mathbb{P})}^4 \mathrm{d}t < +\infty$ and that the estimation

$$\sup_{0 \leq t \leq T} \|\Delta_\varepsilon X_t^{v^\star}\|_{L^4(\mathbb{P})}^4 \leq C\varepsilon^4 \int_0^T \|v_t\|_{L^4(\mathbb{P})}^4 \mathrm{d}t \qquad (7.6.1)$$

holds (see v) for a proof).

i) Show that

$$\Delta_\varepsilon J = \mathbb{E}\left(\int_0^T (l'_x(X_t^{v^*}, v_t^*)\Delta_\varepsilon X_t^{v^*}\right.$$
$$\left. + l'_v(X_t^{v^*}, v_t^*)\varepsilon v_t)dt + f'(X_T^{v^*})\Delta_\varepsilon X_T^{v^*}\right) + O(\varepsilon^2).$$

ii) Write the dynamics of $(\Delta_\varepsilon X_t^{v^*})_{t\in[0,T]}$ and apply the Itô formula to $f'(X_T^{v^*})\Delta_\varepsilon X_T^{v^*} = Y_T \Delta_\varepsilon X_T^{v^*}$ in order to show

$$\Delta_\varepsilon J = \varepsilon \mathbb{E}\Big(\int_0^T (l'_v(X_t^{v^*}, v_t^*) + Y_t b'_v(X_t^{v^*}, v_t^*)$$
$$+ Z_t \sigma'_v(X_t^{v^*}, v_t^*))v_t dt\Big) + O(\varepsilon^2).$$

iii) Deduce that necessarily we have

$$Y_t b'_v(X_t^*, v_t^*) + Z_t \sigma'_v(X_t^*, v_t^*) + l'_v(X_t^*, v_t^*) = 0,$$

a.s. in (ω, t).

iv) Revisit Example 7.1.6 with the above result.

v) Establish the bound (7.6.1).
  Hint: use the Itô formula for decomposing $A_t = \|\Delta_\varepsilon X_t^{v^*}\|^4_{L^4(\mathbb{P})}$, apply the Young inequality $ab \le a^p/p + b^q/q$ (for any $a, b \ge 0$ and $1/p + 1/q = 1$) and the Gronwall lemma to conclude.

Exercise 7.2 (linear backward stochastic differential equation and comparison) *We study the solution to the backward stochastic differential equation (BSDE)*

$$Y_t = \xi + \int_t^T g(X_s, Y_s)ds - \int_t^T Z_s dW_s \quad (7.6.2)$$

where X is a standard SDE on $\mathbb{R}$. Let $(\mathcal{F}_t)_{t\ge 0}$ be the associated Brownian filtration.

i) Let $(\beta_t)_{t\in[0,T]}$ and $(\varphi_t)_{t\in[0,T]}$ be two adapted and bounded processes, and let ξ be a bounded $\mathcal{F}_T$-measurable random variable. We assume the existence of an adapted and square integrable solution $(Y_t, Z_t)_{t\in[0,T]}$ to the linear BSDE

$$Y_t = \xi + \int_t^T (\varphi_s + \beta_s Y_s)ds - \int_t^T Z_s dW_s.$$

Show that
$$\Gamma_t Y_t = \mathbb{E}\left(\Gamma_T \xi + \int_t^T \Gamma_s \varphi_s ds \mid \mathcal{F}_t\right)$$
where $\Gamma_s = \exp\left(\int_0^s \beta_r dr\right)$ *for* $s \in [0, T]$.

ii) *Deduce that if* $g(x, 0) = 0$ *for any* x *and if* $\xi \geq 0$, *then*
$$0 \leq Y_t \leq \hat{Y}_t$$
where $\hat{Y}$ *is the solution to the BSDE with driver* $\hat{g}(x, y) = Ly$.
Hint: *write the difference of two BSDEs as a linear BSDE and apply i).*

Exercise 7.3 (discretization of BSDE) *We study a variant of the discretization scheme of Theorem 7.3.1, and we consider the same notations and assumptions.*

i) *Assume the existence of a* $u \in \mathcal{C}_b^{1,2}([0, T] \times \mathbb{R}; \mathbb{R})$ *solution to the non-linear PDE*
$$\begin{cases} \frac{\partial u(t,x)}{\partial t} + b(t,x)\frac{\partial u(t,x)}{\partial x} + \frac{\sigma(t,x)^2}{2}\frac{\partial^2 u(t,x)}{\partial x^2} + g(x, u(t,x)) = 0, \\ u(T, x) = f(x). \end{cases}$$
Show that the couple $Y_t = u(t, X_t)$ *and* $Z_t = \sigma(t, X_t)\partial_x u(t, X_t)$ *solves* (7.6.2) *with* $\xi = f(X_T)$.

ii) *Instead of replacing* X *by its Euler scheme in* (7.3.3), *we consider a scheme without approximation of* X, *i.e.* $Y_T^{(h)} = f(X_T)$ *and*
$$Y_{kh}^{(h)} = \mathbb{E}(Y_{(k+1)h}^{(h)} \mid \mathcal{F}_{kh}) + hg(X_{kh}, Y_{kh}^{(h)}).$$
Using the above PDE, prove
$$|Y_{kh} - [\mathbb{E}(Y_{(k+1)h} \mid \mathcal{F}_{kh}) + hg(X_{kh}, Y_{kh})]| \leq Ch^2.$$

iii) *Deduce* $|Y_{kh}^{(h)} - Y_{kh}| \leq (1 + Ch)\mathbb{E}(|Y_{(k+1)h}^{(h)} - Y_{(k+1)h}| \mid \mathcal{F}_{kh}) + Ch^2$ *for some constant* C.

iv) *Conclude that the convergence order of the new scheme is* 1 *w.r.t.* h.

Comment: *this proves that the time discretization of the integral of* g *is responsible for a global error of order* h *while the* $\sqrt{h}$-*error of Theorem 7.3.1 is mainly due to the strong error of the Euler scheme.*

Exercise 7.4 (Optimal stopping problem) *We consider the optimal stopping problem in discrete time and finite horizon $N > 0$, associated with a Markov chain $(X_n)_{n \geq 0}$ with state space $\mathcal{E}$, starting from a given point x_0, and to the reward at time n given by $(Z_n)_{n \geq 0}$ where Z_n is integrable for any $n \geq 0$. Let $(\mathcal{F}_n)_{n \geq 0}$ be the filtration generated by $(X_n)_{n \geq 0}$ and $\mathcal{T}_N$ the set of stopping times of $(\mathcal{F}_n)_{n \geq 0}$ bounded by N: $\tau \in \mathcal{T}_N$ if and only if $\tau : \Omega \to \{0, \ldots, N\}$ and $\{\tau \leq n\} \in \mathcal{F}_n$. The reward process $(Z_n)_n$ is adapted to $(\mathcal{F}_n)_n$. We aim at characterizing the optimal expected reward*

$$J = \sup_{\tau \in \mathcal{T}_N} \mathbb{E}(Z_\tau),$$

using different representations, together with an optimal stopping time $\tau^\star \in \mathcal{T}_N$ such that $J = \mathbb{E}(Z_{\tau^\star})$. In the following, we assume that the state space $\mathcal{E}$ is finite.[11]

i) Let us define the maximal expected reward Y_n at time $n \in \{0, \ldots, N\}$ by

$$Y_n = \sup_{\tau \in \mathcal{T}_N : \tau \geq n} \mathbb{E}(Z_\tau \mid \mathcal{F}_n).$$

Show that $(Y_n)_{n \geq 0}$ solves the dynamic programming equation

$$Y_n = \max(Z_n, \mathbb{E}(Y_{n+1} \mid \mathcal{F}_n)), \qquad Y_N = Z_N \qquad (7.6.3)$$

and that $J = \mathbb{E}(Y_0)$. The process $(Y_n)_{0 \leq n \leq N}$ is called the Snell envelope of $(Z_n)_{0 \leq n \leq N}$. A particular case is described in (7.4.5). From the algorithmic point of view, (7.6.3) leads to numerical schemes based on iterations on value functions.

ii) Let $\tau^\star = \min\{n \geq 0 : Y_n = Z_n\}$ and define the stopped process $Y_n^\star = Y_{n \wedge \tau^\star}$. Justify that $\tau^\star \in \mathcal{T}_N$ and that

$$Y_n^\star = \mathbb{E}(Y_{n+1}^\star \mid \mathcal{F}_n).$$

iii) Show that $\tau^\star$ is an optimal stopping time.

---

[11] So that $\Omega = \mathcal{E}^N$ - the canonical space of $(X_n)_{0 \leq n \leq N}$ - is finite as well. It implies that the family $\mathcal{T}_N$ is finite, thus countable, so that we can easily define the supremum of the random variables $\mathbb{E}(Z_\tau \mid \mathcal{F}_n)$; in the general case of an uncountable family, the right notion is *essential supremum*; see [120, Section VI].

iv) Set $\tau_n^\star = \min\{j \geq n : Y_j = Z_j\}$ for the optimal stopping time after time n, i.e. $Y_n = \mathbb{E}(Z_{\tau_n^\star} \mid \mathcal{F}_n)$. Prove that the sequence $(\tau_n^\star)_n$ solves the following iteration on policies:

$$\tau_n^\star = \begin{cases} n & \text{if } \mathbb{E}(Z_{\tau_{n+1}^\star} \mid \mathcal{F}_n) \leq Z_n, \\ n+1 & \text{otherwise.} \end{cases}$$

v) In the case of Markovian reward $(Z_n := f_n(X_n))$ for $f_n : \mathcal{E} \to \mathbb{R}$, show that $Y_n = y_n(X_n)$ for some function $y_n : \mathcal{E} \to \mathbb{R}$ and give the dynamic programming equation solved by $(y_n)_{0 \leq n \leq N}$. Justify that $\tau^\star = \min\{n \geq 0 : X_n \in D_n\}$ for a sequence of domain $(D_n)_{0 \leq n \leq N}$.

These representations involving various dynamic programming equations constitute the basis for developing numerical schemes for computing J and the optimal stopping strategies.

CHAPTER 8

Simulation by empirical regression

In this chapter we develop and analyze tools for efficient numerical solution of the dynamic programming equation (7.3.3)

$$\begin{cases} Y_T^{(h)} = f(X_T^{(h)}), \\ Y_{t_i}^{(h)} = \mathbb{E}\left(Y_{t_{i+1}}^{(h)} + h\, g(t_i, X_{t_i}^{(h)}, Y_{t_{i+1}}^{(h)}) \mid X_{t_i}^{(h)}\right), \quad i = N-1, \ldots, 0. \end{cases} \quad (8.0.1)$$

The discussion at the end of the previous chapter shows (see equation (7.4.2)) that the discrete time process $(Y_{t_i}^{(h)})_{0 \le i \le N}$ can also be represented as

$$Y_{t_i}^{(h)} = u^{(h)}(t_i, X_{t_i}^{(h)})$$

for a sequence of functions $(u^{(h)}(t_i, .))_{0 \le i \le N}$, each of which is written as an expectation that depends on the others (see (7.4.4)):

$$u^{(h)}(t_i, x) = \mathbb{E}\left(u^{(h)}(t_{i+1}, X_{t_{i+1}}^{(h),t_i,x}) + h\, g(t_i, x, u^{(h)}(t_{i+1}, X_{t_{i+1}}^{(h),t_i,x}))\right). \quad (8.0.2)$$

8.1 THE DIFFICULTIES OF A NAIVE APPROACH

Let us quickly discuss some possible ideas for the numerical calculation of $u^{(h)}$, focusing our attention on the evaluation of the value $u^{(h)}(0, x_0)$.

- The distribution of $X_h^{(h),0,x_0}$ is Gaussian, which allows us to write $u^{(h)}(0, x_0)$ as a d-dimensional integral of the function

241

$x \mapsto u^{(h)}(h,x) + hg(0, x_0, u^{(h)}(h,x))$ with respect to the corresponding Gaussian probability measure.

1. The first difficulty is connected to the computation of the d-dimensional integral.

2. The second difficulty arises with the function to integrate: it is necessary to have the unknown values $u^{(h)}(h,x)$ at each point x (or at many points), which are themselves expressed as integrals of the functions $u^{(h)}(2h, \cdot)$, etc. This creates a cumbersome recursive procedure.

- To overcome the first difficulty, we can discretize the integral by properly choosing M points $(x_{1,1}, \ldots, x_{1,M})$ either in a deterministic way (quadrature method), or randomly (Monte-Carlo simulation), that represent well the distribution $X_h^{(h),0,x_0}$. But the necessity to know $u^{(h)}(h,x)$ at each point $x \in \{x_{1,1}, \ldots, x_{1,M}\}$ requires that we iterate the procedure at time t_1.

- So it turns out to be necessary at each point $x_{1,m}$ to regenerate M candidates representing the distribution of $X_{2h}^{(h),h,x_{1,m}}$: starting from M points at time t_1, we obtain M^2 space points at time t_2, ... and finally M^N at time $t_N = T$.

- To sum up, we get a stochastic tree whose size grows exponentially with the number of dates. Furthermore, the expected error cannot be better than the error of the integral discretization obtained at each date, which is of order $1/\sqrt{M}$ in the case of Monte-Carlo sampling. The complexity/error ratio becomes extremely poor as soon as the number of dates N exceeds a few units.

Because we are going to study the continuous time limit $N \to +\infty$, it is necessary to develop suitable tools, which are better adapted to pass to the limit.

What are the important issues to consider?

- As the number of dates N goes to infinity, we need to screen how the error in the approximated calculation of each $u^{(h)}(t_i, x)$ is propagated along the iterations.

- In certain cases, the distribution of $X^{(h)}$ will be more complicated than that of the Euler scheme with Gaussian transition: the density may not exist or not be explicit (if the process X is not discretized or if we include jumps, for example). Thus it would be interesting to have robust error estimates with respect to the distribution of the Markov chain $(X_{t_i}^{(h)})_i$. Thus, in the following, we would rather *see the model as a black box*, i.e. by supposing that we can generate simulations of $X^{(h)}$ without, however, knowing the model equation: a dynamic programming algorithm taking as inputs the data f, g and T and just sequences of independent simulations of $X^{(h)}$ (without other information about the model) will have the advantage of being very flexible and robust with respect to the distribution of the model. Its adjustment will be made a priori in a rather universal and automatic way.

– Deriving non-asymptotic error estimates allows us to better understand how to adjust the convergence parameters, even if sometimes the practical implementation may bring good surprises, i.e. the convergence may take place more rapidly than predicted by the theory.

The non-parametric regression tools that we study in the following, allow us to work out a numerical scheme called an *empirical regression* (or *regression Monte-Carlo* or *empirical least squares*) that has a number of advantages mentioned above.

The curse of dimensionality. When the marginal distributions of $X^{(h)}$ are sampled for calculating the nested functions in (8.0.2), the estimation of these functions becomes more and more delicate as the dimension d of the space increases. Indeed, it is increasingly difficult to satisfactorily fill the d-dimensional space (d large) with a sample of a given size (even large). This is called *the curse of dimensionality*, illustrated by the following example, taken from [69, Chapter 2].

Let $X, X_1, \ldots, X_M$ be independent random variables uniformly distributed on $[0, 1]^d$ and set

$$d_\infty(d, M) = \mathbb{E}\left[\min_{i=1,\ldots,M} |X - X_i|_\infty\right]$$

where for $x = (x^1, \ldots, x^d) \in \mathbb{R}^d$ we set $|x|_\infty = \max_{j=i,\ldots,d} |x^j|$.

The average distance $d_\infty(d, M)$ measures how much a new random point X is close to the grid $(X_1, \ldots, X_M)$, or in other words, how well

244 ■ Simulation by empirical regression

$d\backslash M$	10^2	10^3	10^4	10^5	10^6	10^7	10^8
1	0.002	0.000	0.000	0.000	0.000	0.000	0.000
3	0.081	0.038	0.017	0.008	0.004	0.002	0.001
5	0.166	0.105	0.066	0.042	0.026	0.017	0.010
10	0.287	0.228	0.181	0.144	0.114	0.091	0.072
15	0.345	0.296	0.254	0.218	0.187	0.160	0.137
20	0.378	0.337	0.300	0.268	0.239	0.213	0.190

Table 8.1 Several values of the lower bound of $d_\infty(d, M)$.

the points $(X_1, \ldots, X_M)$ fill the cube $[0,1]^d$. From the independence we deduce (for $t \geq 0$)

$$\mathbb{P}\left(\min_{i=1,\ldots,M} |x - X_i|_\infty \leq t\right) \leq M\mathbb{P}(|x - X_1|_\infty \leq t)$$
$$= M\left[\mathbb{P}(|x^1 - X_1^1| \leq t)\right]^d \leq M(2t)^d,$$

then

$$d_\infty(d, M) = \int_0^\infty \left(1 - \mathbb{P}\left(\min_{i=1,\ldots,M} |X - X_i|_\infty \leq t\right)\right) dt$$
$$\geq \int_0^\infty (1 - M(2t)^d)_+ dt = \frac{d}{2(d+1)} \frac{1}{M^{1/d}}.$$

As we can see in Table 8.1, the influence of the dimension is very significant for d exceeding ten. The decrease of the order $1/M^{1/d}$ is characteristic in this type of problem, namely in the calculation of d-dimensional functions. By additionally supposing the unknown function to be regular, we can reduce the effect of the dimension. All these heuristics will be reflected in the subsequent estimates.

8.2 APPROXIMATION OF CONDITIONAL EXPECTATIONS BY LEAST SQUARES METHODS

Before we tackle the most difficult case of the dynamic programming in the form (8.0.1), let us study a simpler problem with a single date, which will give us an opportunity to introduce adequate terminology and notation.

8.2.1 Empirical regression

Definition 8.2.1 *Consider two random variables O and R, with values in $\mathbb{R}^d$ and $\mathbb{R}$, respectively. We suppose that R is square integrable and we denote $\mathfrak{M}(.)$ as the regression function defined by*

$$\mathbb{E}(R|O) = \mathfrak{M}(O) \quad \text{a.s.} . \tag{8.2.1}$$

In other terms, $\mathfrak{M}(.)$ is the conditional expectation function.

We adapt the statistical terminology, calling R a response *and* O an observation: $\mathfrak{M}(O)$ *is the best estimation of the response R in L_2 sense given an observation O (see (8.2.3))*.

Given a sample $(O^{(m)}, R^{(m)})_{1 \leq m \leq M}$ of size M, an empirical regression procedure aims to produce a function $\widetilde{\mathfrak{M}}_M(.)$ — built from the sample — that approximates (in a certain sense) the unknown regression function $\mathfrak{M}(.)$.

In general, the couples $(O^{(m)}, R^{(m)})_m$ are independent, but this does not happen systematically as we will see in the numerical calculation of the solution for the dynamic programming equation.

Many empirical regression methods exist and we refer the reader to the book [69]: here we follow the method of empirical ordinary least squares, consisting of calculating $\widetilde{\mathfrak{M}}_M(\cdot)$ on a finite-dimensional linear subspace Φ, generated by K basis functions $\{\phi_k(.) : 1 \leq k \leq K\}$:

$$\Phi = \text{Span.}(\phi_1, \ldots, \phi_K)$$
$$= \Big\{ \varphi : \mathbb{R}^d \mapsto \mathbb{R} \text{ such that } \exists \, \alpha \in \mathbb{R}^K$$
$$\text{and } \varphi(.) := \sum_{k=1}^{K} \alpha_k \phi_k(.) := \alpha \cdot \phi(.) \Big\}. \tag{8.2.2}$$

We insist on the latter notation $\varphi = \alpha \cdot \phi$, which permits us to have a concise notation in what follows. Without specifying this later, we will always suppose that all the functions $\varphi \in \Phi$ are such that $\mathbb{E}|\varphi(O)|^2 < +\infty$, or equivalently $\mathbb{E}|\phi_k(O)|^2 < +\infty$ for any k.

IMPORTANT. We do not suppose that the basis functions $(\phi_k)_k$ are orthonormal with respect to the scalar product induced by $L_2(\mathbb{P} \circ O^{-1})$. Indeed, in general, although the Gram-Schmidt orthonormalization is theoretically possible, it is not always possible numerically because the distribution of O is not known or is insufficiently tractable.

There are two straightforward heuristics: the greater are K and the space Φ, the better is the approximation; the larger the sample, the better the estimation of the coefficients in Φ. This will be quantified more precisely in Theorem 8.2.4, with an optimal balance between K and M. In what follows, we always suppose $M \geq K$ to avoid the problem of over-fitting.

Because $\mathfrak{M}(O)$ minimizes

$$\mathbb{E}(R - \mathfrak{M}_O)^2 = \mathbb{E}(R - \mathbb{E}(R|O))^2 + \mathbb{E}(\mathbb{E}(R|O) - \mathfrak{M}_O)^2 \qquad (8.2.3)$$

over all the square integrable and $\sigma(O)$-measurable random variables $\mathfrak{M}_O$, it is natural to take for $\widetilde{\mathfrak{M}}_M(\cdot)$ the function $\varphi \in \Phi$ minimizing the empirical quadratic criteria

$$\frac{1}{M} \sum_{m=1}^{M} (R^{(m)} - \varphi(O^{(m)}))^2.$$

In fact, this means that we choose basis coefficients that are equal to

$$\alpha^M := \arg\min_{\alpha \in \mathbb{R}^K} \frac{1}{M} \sum_{m=1}^{M} (R^{(m)} - \alpha \cdot \phi(O^{(m)}))^2, \qquad (8.2.4)$$

and then to set

$$\widetilde{\mathfrak{M}}_M(.) = \alpha^M \cdot \phi(\cdot). \qquad (8.2.5)$$

8.2.2 SVD method

The minimization (8.2.4) is a simple least squares problem, which is, in addition, linear (no constraints on the coefficients). But because of possible collinearities between the basis functions, there may exist several optimal coefficients that attain the minimum in (8.2.4): this is an embarrassing situation that is generally associated with instability problems.

To choose a numerical solution, there exist several methods: among others, Singular Value Decomposition and Normal Equations, see [64, Chapter 5] for the details. These various approaches differ in their performance concerning the rounding errors (matrix conditioning), the required memory size, the computational cost, and the conclusions depend on the relative values of K and M. The SVD (Singular Value

Decomposition) method is known to be slightly longer in computational time[1] but a little more stable: we explain this method and how to deduce a minimizer (8.2.4); see [64, Theorems 2.5.2 and 5.5.1].

The Singular Value Decomposition of a matrix

$$P = (\phi_i(O^{(m)}))_{1 \leq m \leq M, 1 \leq i \leq K}$$

of size $M \times K$ (with $M \geq K$) is written as

$$P = UP'V^\top \quad \text{with} \quad P' = \begin{pmatrix} \sigma_1 & & 0 \\ & \ddots & \\ 0 & & \sigma_K \\ 0 & \cdots & 0 \end{pmatrix}$$

where U and V are two orthogonal matrices of size $M \times M$ and $K \times K$ respectively, and $\sigma_1 \geq \cdots \geq \sigma_K \geq 0$. So, the quantity to minimize can be written as

$$\sum_{m=1}^{M} (\alpha \cdot \phi(O^{(m)}) - R^{(m)})^2 = |UP'V^\top \alpha - R|^2_{\mathbb{R}^M}$$

$$= |P'V^\top \alpha - U^\top R|^2_{\mathbb{R}^M} \quad \text{(as } U \text{ is an orthogonal matrix)}$$

$$= \sum_{i=1}^{K} |\sigma_i(V^\top \alpha)_i - (U^\top R)_i|^2 + \sum_{i>K} |(U^\top R)_i|^2.$$

The set of minimizing coefficients α of (8.2.4) is:

$$\mathcal{A} = \left\{ \alpha = V \begin{pmatrix} \cdots \\ \mathbf{1}_{\sigma_i>0} \frac{(U^\top R)_i}{\sigma_i} + \mathbf{1}_{\sigma_i=0} \chi_i \\ \cdots \end{pmatrix}, \chi_i \in \mathbb{R} \right\}.$$

We easily verify that for all the coefficients $\alpha \in \mathcal{A}$, the vector $(\alpha \cdot \phi(O^{(m)}))_{1 \leq m \leq M}$ remains the same: thus we have a *unique*[2] *minimizing function*

$$\widetilde{\mathfrak{M}}(\cdot) := \arg\min_{\varphi = \alpha \cdot \phi \in \Phi} \frac{1}{M} \sum_{m=1}^{M} (R^{(m)} - \varphi(O^{(m)}))^2. \tag{8.2.6}$$

On the contrary, its representation in the space Φ via its coefficients

---

[1] Its computational cost is $2MK^2 + 11K^3$.
[2] Along the $(O^{(m)})_{1 \leq m \leq M}$.

can be non-unique. If P is of full rank ($\sigma_K > 0$), then $\mathcal{A}$ contains a single element and the coefficient that solves (8.2.4) is unique. If $\mathrm{rank}(P) < K$, the optimal solution in the SVD sense is associated with the choice $\chi_i = 0$ for any i, in other words,

$$\alpha^M = V \begin{pmatrix} \vdots \\ \mathbf{1}_{\sigma_i>0} \frac{(U^\top R)_i}{\sigma_i} \\ \vdots \end{pmatrix}. \tag{8.2.7}$$

Proposition 8.2.2 *The optimal solution in the SVD sense is, among all the minimizers of (8.2.4), the one with a minimal norm. We call it the* SVD-optimal coefficient.

PROOF:
We have observed that a minimizer is of the form

$$\alpha = V \begin{pmatrix} \vdots \\ \mathbf{1}_{\sigma_i>0} \frac{(U^\top R)_i}{\sigma_i} + \mathbf{1}_{\sigma_i=0} \chi_i \\ \vdots \end{pmatrix}.$$

As V is orthogonal, it preserves the norm, thus we have

$$|\alpha|^2_{\mathbb{R}^K} = \sum_{i=1}^K \left| \mathbf{1}_{\sigma_i>0} \frac{(U^\top R)_i}{\sigma_i} + \mathbf{1}_{\sigma_i=0} \chi_i \right|^2$$

$$= \sum_{i=1}^K \mathbf{1}_{\sigma_i>0} \left| \frac{(U^\top R)_i}{\sigma_i} \right|^2 + \sum_{i=1}^K \mathbf{1}_{\sigma_i=0} |\chi_i|^2,$$

from which the proof easily follows. $\square$

We give several simple properties of the SVD solution to the least squares problem: the first two properties are standard and of deterministic nature; the last one is of probabilistic nature.

Proposition 8.2.3 *Let α^M be the SVD-optimal coefficient associated with the sample $(O^{(m)}, R^{(m)})_{1 \leq m \leq M}$ and to the space Φ. Then we have the following properties.*

i) Linearity: the map $(R^{(m)})_{1 \leq m \leq M} \mapsto \alpha^M$ is linear.

ii) Non-expanding property: $\dfrac{1}{M} \displaystyle\sum_{m=1}^M (\alpha^M \cdot \phi(O^{(m)}))^2 \leq \dfrac{1}{M} \displaystyle\sum_{m=1}^M (R^{(m)})^2.$

Approximation of conditional expectations by least squares methods ■ 249

iii) Solution as conditional expectation: suppose that $(\phi(O^{(m)}))_{m=1,\ldots,M}$ is measurable with respect to the sigma-algebra $\mathcal{Q}$. Then, the SVD-optimal coefficient associated with the response $\mathbb{E}(R|\mathcal{Q}) = (\mathbb{E}(R^{(m)}|\mathcal{Q}))_{m=1,\ldots,M}$ is given by $\mathbb{E}(\alpha^M|\mathcal{Q})$.

PROOF:
The linearity *i)* obviously follows from (8.2.7), because $U, V, (\sigma_i)_i$ remain unchanged when the observation $(O^{(m)})_{1 \le m \le M}$ is the same. Using the same equation and observing that $U, V, (\sigma_i)_i$ are $\mathcal{Q}$-measurable, we prove *iii)*. To prove *ii)*, let us use the Pythagoras decomposition: for any $\alpha \in \mathbb{R}^K$, we have $\frac{1}{M}\sum_{m=1}^{M}(\alpha \cdot \phi(O^{(m)}) - R^{(m)})^2 = \frac{1}{M}\sum_{m=1}^{M}(\alpha^M \cdot \phi(O^{(m)}) - R^{(m)})^2 + \frac{1}{M}\sum_{m=1}^{M}((\alpha^M - \alpha) \cdot \phi(O^{(m)}))^2$. Taking $\alpha = 0$, we obtain the stated inequality. □

8.2.3 Example of approximation space: the local polynomials

In this example, the functions are defined locally, smooth on disjoint subsets. This local property leads to an easier numerical implementation.

Definition. For the space Φ, take the *local polynomials* of degree k, defined locally on cubes $\mathcal{C}_{i_1,\ldots,i_d} = [i_1 \Delta, (i_1+1)\Delta[\times \cdots \times [i_d \Delta, (i_d+1)\Delta[$ for[3] $i_j \in \{-H/\Delta, \ldots, H/\Delta - 1\}$, $1 \le j \le d$. Outside the cube $[-H, H]^d$, the functions of Φ are thus equal to zero (truncation effect at *infinity*). The number of cubes $\mathcal{C}_{i_1,\ldots,i_d}$ is $(2H/\Delta)^d$ and on each of them, we take all the polynomials of d variables for which the maximal degree in each variable is k, thus $(k+1)^d$ polynomials in total. So, we have

$$\dim(\Phi) = K = (2H/\Delta)^d \times (k+1)^d. \quad (8.2.8)$$

A function $\varphi \in \Phi$ may possibly have discontinuities on the boundary of the cubes; see Figure 8.1.

Implementation. Because the cubes have disjoint supports, the calculation of the solution to the least squares problem is reduced to a cube-by-cube resolution, which considerably reduces the computational cost. Finally, on a cube we need to perform a polynomial regression whose parametric dimension remains small (equal to $(k+1)^d$).

---

[3] We do not take into account the rounding effects supposing that H is a multiple of Δ.

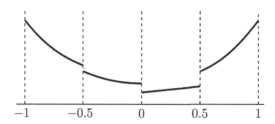

Figure 8.1 A local polynomial function ($d = 1$, $k = 2$, $H = 1$, $\Delta = 0.5$).

As for the evaluation of the value $\varphi(x) = \alpha^M \cdot \phi(x)$ at a point x, its evaluation cost is not K as it might be thought in view of the K-dimensional representation of Φ, but it is in fact equal to the number of polynomials on a cube, i.e. $(k+1)^d$: indeed, because of the tensored grid, we have a *unitary cost* to determine[4] the cube containing x, then a cost $(k+1)^d$ to evaluate the polynomial associated with the cube. In general, the number of cubes is large (i.e., $K \gg (k+1)^d$) and this algorithmic simplification is really important.

8.2.4 Error estimations, robust with respect to the model

We give a first error estimate in the calculation of the empirical regression function. The reader may notice how weak the hypotheses are regarding the joint distribution of (O, R) or their marginal distributions: indeed, we suppose only the condition of a bounded conditional expectation. In this sense, *estimates are robust w.r.t. the model*. In particular, it is enough for the response R to be a bounded random variable in order to satisfy the hypothesis $v)$. Furthermore, we do not assume any hypotheses on the basis functions $(\phi_k)_k$.

Theorem 8.2.4 (regression error estimates) *Consider the following notation and hypotheses:*

i) *For a random couple $(O, R) \in \mathbb{R}^d \times \mathbb{R}$ where R is square integrable, set $\mathfrak{M}(.)$ to be the $\sigma(O)$-measurable variable defined by $\mathfrak{M}(o) := \mathbb{E}(R|O = o)$.*

ii) *Denote by μ the marginal distribution of O and $|\cdot|_{L_2(\mu)}$ the asso-*

---
[4]The indices $i_1, \ldots, i_d$ of the cube are given by $i_j = \lfloor \frac{x_j}{\Delta} \rfloor$.

Approximation of conditional expectations by least squares methods ■ 251

ciated L_2 norm, i.e., for a function φ

$$|\varphi|^2_{L_2(\mu)} := \int_{\mathbb{R}^d} \varphi^2(o)\mu(do).$$

iii) $(O^{(1)}, R^{(1)}), \cdots, (O^{(M)}, R^{(M)})$ are independent copies of the couple (O, R);

iv) Denote by $|\varphi|_{L_2(\mu^M)}$ the L_2 norm associated with the empirical measure of the sample $(O^{(1)}, \ldots, O^{(M)})$:

$$|\varphi|^2_{L_2(\mu^M)} := \frac{1}{M}\sum_{m=1}^{M} \varphi^2(O^{(m)}).$$

v) The conditional variance of R is bounded: $\sigma^2 := \sup_{o \in \mathbb{R}^d} \text{Var}(R|O = o) < +\infty$.

vi) $\Phi = \text{Span}.(\phi_1, \ldots \phi_K)$ is a linear subspace with a dimension of less than K.

vii) The empirical regression function is defined as the function in Φ attaining the minimum

$$\widetilde{\mathfrak{M}}_M(.) := \arg\min_{\varphi \in \Phi} \frac{1}{M}\sum_{m=1}^{M} |R^{(m)} - \varphi(O^{(m)})|^2. \qquad (8.2.9)$$

Then, the averaged quadratic error is upper bounded as follows:

$$\mathbb{E}\left(|\widetilde{\mathfrak{M}}_M - \mathfrak{M}|^2_{L_2(\mu^M)}\right) \leq \sigma^2 \frac{K}{M} + \min_{\varphi \in \Phi} |\varphi - \mathfrak{M}|^2_{L_2(\mu)}. \qquad (8.2.10)$$

Before we prove this result in the next section, let us make some comments.
▷ First, this is a non-asymptotic result, which provides control of the error for a fixed computational cost (it does not require K and M to tend to infinity).
▷ The quadratic error $|\widetilde{\mathfrak{M}}_M - \mathfrak{M}|^2_{L_2(\mu^M)}$ is stochastic for two reasons: on the one hand, because the estimated function $\widetilde{\mathfrak{M}}_M$ is random, as estimated using a sample; on the other hand, because the norm of the

error is taken with respect to the empirical (random) measure. From this, it is possible to deduce upper bounds for $\mathbb{E}(|\widetilde{\mathfrak{M}}_M - \mathfrak{M}|^2_{L_2(\mu)})$ at the cost of some mathematical complications; see the proof of Proposition 8.3.3. Another possible extension consists of allowing the space Φ to be adapted to the observations (*data-driven function-set*), and the mathematical analysis is very similar.

▷ The first term $\sigma^2 \frac{K}{M}$ of the right-hand side of (8.2.10) is traditionally called the *variance term* and corresponds to the *statistical error* incurred while calculating the coefficients α^M using a finite sample: it converges to 0 as the size M of the sample increases.

The second term is due to the *approximation error* in the space Φ and is written as the square of the approximation bias (square of the $L_2(\mu)$ distance from $\mathfrak{M}$ to Φ): by making the space of the approximation richer ($K \to +\infty$), we can expect that this term will converge to 0. The decomposition (8.2.10) is called the *bias/variance decomposition*.

▷ The essential point is that the dimension K has two opposite effects: increasing K reduces the approximation error, but increases the statistical error (over-parameterization problem); conversely, decreasing K makes the statistical estimation of the approximation coefficients easier, but leads to a bad approximation in Φ. Thus there is an optimal adjustment between K and M to be determined; see Figure 8.2. Without this, the approximation by empirical regression may not con-

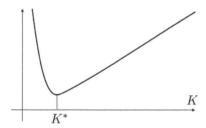

Figure 8.2 Empirical regression error for a given M as a function of K.

verge at all. We give an example of parameter tuning when we have additional information on $\mathfrak{M}$.

8.2.5 Adjustment of the parameters in the case of local polynomials

We consider the space of the local polynomials described in Section 8.2.3 and we suppose that $\mathfrak{M}$ is a bounded function of class

Approximation of conditional expectations by least squares methods ■ 253

$\mathcal{C}^{k+\alpha}(\mathbb{R}^d,\mathbb{R})$, with $k \in \mathbb{N}$ and $\alpha \in (0,1]$, having bounded derivatives and α-Hölder continuous k-th order derivatives.

Keeping the notation of Section 8.2.3, the variance term equals $\sigma^2 \frac{K}{M} = \sigma^2 \frac{(2H/\Delta)^d \times (k+1)^d}{M}$. Now let us evaluate the squared bias term. Using that the cubes have disjoint support and taking on each cube the approximation polynomial of degree k obtained by a Taylor formula, we obtain a simple upper bound on the error

$$\min_{\varphi \in \Phi} |\varphi - \mathfrak{M}|^2_{L_2(\mu)}$$
$$= \mathbb{E}\big(\mathfrak{M}^2(O) \mathbf{1}_{O \notin \mathcal{C}_{i_1,\ldots,i_d}: -H/\Delta \leq i_1,\ldots,i_d \leq H/\Delta - 1}\big)$$
$$+ \sum_{-H/\Delta \leq i_1,\ldots,i_d \leq H/\Delta - 1} \min_{\text{Pol.}\mathcal{P}:\,\deg(\mathcal{P}) \leq k} \mathbb{E}\big(|\mathfrak{M}(O) - \mathcal{P}(O)|^2 \mathbf{1}_{O \in \mathcal{C}_{i_1,\ldots,i_d}}\big)$$
$$\leq C_\mathfrak{M} \Big(\mathbb{P}(|O|_\infty \geq H) + \sum_{-H/\Delta \leq i_1,\ldots,i_d \leq H/\Delta - 1} \Delta^{2(k+\alpha)} \mathbb{P}\big(O \in \mathcal{C}_{i_1,\ldots,i_d}\big)\Big)$$
$$\leq C_\mathfrak{M} \big(\mathbb{P}(|O|_\infty \geq H) + \Delta^{2(k+\alpha)}\big)$$

for a constant $C_\mathfrak{M}$ depending only on $\mathfrak{M}(\cdot)$ and its derivatives.

Suppose moreover that the random variable O has exponential moments: $\mathbb{P}(|O|_\infty \geq H) \leq \lambda \exp(-H/\lambda)$ for any $H \geq 0$, for a certain $\lambda > 0$. Then taking $H = c\log(1/\Delta)$ for a large enough constant c — as soon as $\Delta < 1$ — the term $\mathbb{P}(|O|_\infty \geq H)$ becomes negligible compared to $\Delta^{2(k+\alpha)}$ as $\Delta \to 0$ (this justifies the truncation at *infinity*).

Finally, we can re-write the upper bound (8.2.10) (without detailing the constants):

$$\mathbb{E}\Big(|\widetilde{\mathfrak{M}}_M - \mathfrak{M}|^2_{L_2(\mu^M)}\Big) \leq C \frac{[\log(1/\Delta)\Delta^{-1}]^d}{M} + C\Delta^{2(k+\alpha)}.$$

To achieve the magnitude of the error approximation, it is enough to take $M \underset{\Delta \to 0}{\sim} \Delta^{-d-2(k+\alpha)}[\log(1/\Delta)]^d$.

To better understand the effect of the dimension d, let us express the variables Δ, K, and M as functions of the global error variable ε^2 representing the expected accuracy of the average quadratic error $\mathbb{E}\Big(|\widetilde{\mathfrak{M}}_M - \mathfrak{M}|^2_{L_2(\mu^M)}\Big)$. Neglecting the logarithmic terms and the constants, we obtain the following heuristics for the order of magnitude w.r.t. ε:

$$\begin{cases} \Delta \underset{\varepsilon \to 0}{\sim} \varepsilon^{\frac{1}{k+\alpha}}, \\ K \underset{\varepsilon \to 0}{\sim} \varepsilon^{-\frac{d}{k+\alpha}}, \\ M \underset{\varepsilon \to 0}{\sim} \varepsilon^{-2-\frac{d}{k+\alpha}}, \end{cases} \implies \mathbb{E}\Big(|\widetilde{\mathfrak{M}}_M - \mathfrak{M}|^2_{L_2(\mu^M)}\Big) \underset{\varepsilon \to 0}{\leq} \varepsilon^2. \qquad (8.2.11)$$

We observe a subtle competition between the dimension d and the regularity $k + \alpha$ of the unknown function $\mathfrak{M}$. For a given regularity $k + \alpha$, the larger is the number d of variables on which depends the estimated function, the more significant is the computational cost for achieving a given accuracy: this is the famous *curse of dimensionality*; see our discussion at the beginning of the chapter.

In practice, knowledge of the parameter $k + \alpha$ often comes from a preliminary analysis, which allows us to identify the regularity of the problem and its qualitative properties. Once the regularity is identified, the adjustment of other parameters is easy, at least if we focus on the rate and not on the constants.

In terms of computational cost we get

$$\Big[\mathbb{E}\Big(|\widetilde{\mathfrak{M}}_M - \mathfrak{M}|^2_{L_2(\mu^M)}\Big)\Big]^{\frac{1}{2}} \underset{M \to +\infty}{\leq} CM^{-\frac{(k+\alpha)}{2(k+\alpha)+d}}.$$

If the regularity is large ($k + \alpha \to +\infty$), we get the convergence rate $\sqrt{M}$ as for the central limit theorem. Otherwise, the rate is less and strongly depends on the dimension and the regularity of $\mathfrak{M}(\cdot)$.

8.2.6 Proof of the error estimations

PROOF:

We prove Theorem 8.2.4, following [69, Theorem 11.2]. To simplify, denote $O^{(1:M)} := (O^{(m)})_{1 \leq m \leq M}$ and $\mathbb{E}^{(1:M)}(\cdot) = \mathbb{E}(\cdot | O^{(1:M)})$. Suppose we already have

$$\mathbb{E}\Big(|\widetilde{\mathfrak{M}}_M - \mathfrak{M}|^2_{L_2(\mu^M)} \big| O^{(1:M)}\Big) \leq \sigma^2 \frac{K}{M} + \min_{\varphi \in \Phi} |\varphi - \mathfrak{M}|^2_{L_2(\mu^M)}. \qquad (8.2.12)$$

Then the advertised upper bound (8.2.10) follows directly, by taking the expectation of (8.2.12) and observing that

$$\mathbb{E}\big(\min_{\varphi \in \Phi} |\varphi - \mathfrak{M}|^2_{L_2(\mu^M)}\big)$$

$$\leq \min_{\varphi \in \Phi} \mathbb{E}(|\varphi - \mathfrak{M}|^2_{L_2(\mu^M)}) = \min_{\varphi \in \Phi} \mathbb{E}\Big(\frac{1}{M} \sum_{m=1}^{M} |\varphi(O^{(m)}) - \mathfrak{M}(O^{(m)})|^2\Big)$$

$$= \min_{\varphi \in \Phi} \mathbb{E}\Big(|\varphi(O) - \mathfrak{M}(O)|^2\Big) = \min_{\varphi \in \Phi} |\varphi - \mathfrak{M}|^2_{L_2(\mu)}.$$

Now we show (8.2.12). Because the computations are made conditionally on $O^{(1:M)}$, we can suppose without loss of generality that $(\phi_1, ... \phi_{K_M})$ is an orthonormal family[5] in $L_2(\mu^M)$, possibly with $K_M \leq K$:

$$\frac{1}{M}\sum_{m=1}^{M} \phi_k(O^{(m)})\phi_l(O^{(m)}) = \delta_{k,l}, \quad 1 \leq k,l \leq K_M.$$

Consequently, the solution $\arg\min_{\varphi \in \Phi} \dfrac{1}{M}\sum_{m=1}^{M} |\varphi(O^{(m)}) - R^{(m)}|^2$ is given by

$$\widetilde{\mathfrak{M}}_M(.) = \sum_{j=1}^{K_M} \alpha_j^M \phi_j(.) \quad \text{with} \quad \alpha_j^M = \frac{1}{M}\sum_{m=1}^{M}\phi_j(O^{(m)})R^{(m)}$$

(α^M is the SVD-optimal coefficient). Note that $\mathbb{E}^{(1:M)}(\widetilde{\mathfrak{M}}_M(.)) = \mathbb{E}(\widetilde{\mathfrak{M}}_M(.)|O^{(1:M)})$ is a solution to the least squares problem

$$\min_{\varphi \in \Phi} \frac{1}{M}\sum_{m=1}^{M} |\varphi(O^{(m)}) - \mathfrak{M}(O^{(m)})|^2 = \min_{\varphi \in \Phi} |\varphi - \mathfrak{M}|^2_{L_2(\mu^M)}.$$

Indeed, by Proposition 8.2.3 applied with $\mathcal{Q} = \sigma(O^{(1:M)})$, the conditional expectation $\mathbb{E}(\cdot|O^{(1:M)})$ of the coefficient α^M is the SVD-optimal coefficient associated with the response $(\mathbb{E}(R^{(m)}|O^{(1:M)}))_{1\leq m \leq M}$: but the independence of the sample data implies that $\mathbb{E}(R^{(m)}|O^{(1:M)}) = \mathbb{E}(R^{(m)}|O^{(m)})$; moreover, $\mathbb{E}(R^{(m)}|O^{(m)}) = \mathfrak{M}(O^{(m)})$ by the definition of the regression function.

Then, using the Pythagoras theorem, we deduce that

$$|\widetilde{\mathfrak{M}}_M - \mathfrak{M}|^2_{L_2(\mu^M)}$$
$$= |\widetilde{\mathfrak{M}}_M - \mathbb{E}^{(1:M)}(\widetilde{\mathfrak{M}}_M)|^2_{L_2(\mu^M)} + |\mathbb{E}^{(1:M)}(\widetilde{\mathfrak{M}}_M) - \mathfrak{M}|^2_{L_2(\mu^M)},$$
$$= |\widetilde{\mathfrak{M}}_M - \mathbb{E}^{(1:M)}(\widetilde{\mathfrak{M}}_M)|^2_{L_2(\mu^M)} + \min_{\varphi \in \Phi}|\varphi - \mathfrak{M}|^2_{L_2(\mu^M)}. \qquad (8.2.13)$$

Because $(\phi_j)_j$ are orthonormal in $L_2(\mu^M)$, we have

$$|\widetilde{\mathfrak{M}}_M - \mathbb{E}^{(1:M)}(\widetilde{\mathfrak{M}}_M)|^2_{L_2(\mu^M)} = \sum_{j=1}^{K_M} |\alpha_j^M - \mathbb{E}^{(1:M)}(\alpha_j^M)|^2.$$

---

[5] Removing the elements, collinear in $L_2(\mu^M)$, and performing the Gram-Schmidt procedure for the others.

From $\alpha_j^M - \mathbb{E}^{(1:M)}(\alpha_j^M) = \frac{1}{M}\sum_{m=1}^{M}\phi_j(O^{(m)})(R^{(m)} - \mathfrak{M}(O^{(m)}))$, it follows that

$$\mathbb{E}^{(1:M)}\left(|\widetilde{\mathfrak{M}}_M - \mathbb{E}^{(1:M)}(\widetilde{\mathfrak{M}}_M)|^2_{L_2(\mu^M)}\right)$$
$$= \sum_{j=1}^{K_M} \frac{1}{M^2} \mathbb{E}^{(1:M)}\Bigg(\sum_{m,m'=1}^{M} \phi_j(O^{(m)})\phi_j(O^{(m')})$$
$$\times (R^{(m)} - \mathfrak{M}(O^{(m)}))(R^{(m')} - \mathfrak{M}(O^{(m')}))\Bigg).$$

Then, by the *independence of the data*, we get remarkable simplifications:

$$\mathbb{E}^{(1:M)}\left(\phi_j(O^{(m)})\phi_j(O^{(m')})(R^{(m)} - \mathfrak{M}(O^{(m)}))(R^{(m')} - \mathfrak{M}(O^{(m')}))\right)$$
$$= \phi_j(O^{(m)})\phi_j(O^{(m')})\mathbb{E}^{(1:M)}\left((R^{(m)} - \mathfrak{M}(O^{(m)}))(R^{(m')} - \mathfrak{M}(O^{(m')}))\right)$$
$$= \phi_j(O^{(m)})\phi_j(O^{(m')})$$
$$\times \mathbb{E}\left((R^{(m)} - \mathfrak{M}(O^{(m)}))(R^{(m')} - \mathfrak{M}(O^{(m')}))|O^{(m)}, O^{(m')}\right)$$
$$= \begin{cases} \phi_j^2(O^{(m)})\mathrm{Var}(R^{(m)}|O^{(m)}) & \text{if } m = m', \\ 0 & \text{otherwise.} \end{cases}$$

Thus the hypothesis of uniform bound on the conditional variance leads to

$$\mathbb{E}^{(1:M)}\left(|\widetilde{\mathfrak{M}}_M - \mathbb{E}^{(1:M)}(\widetilde{\mathfrak{M}}_M)|^2_{L_2(\mu^M)}\right)$$
$$\leq \sigma^2 \sum_{j=1}^{K_M} \frac{1}{M^2}\sum_{m=1}^{M}\phi_j^2(O^{(m)}) = \sigma^2 \frac{K_M}{M} \leq \sigma^2 \frac{K}{M},$$

the above equality coming from the unitary norm of ϕ_j in $L_2(\mu^M)$. Gathering this with (8.2.13) for which we take the conditional expectation $\mathbb{E}(\cdot|O^{(1:M)})$, we obtain the inequality (8.2.12). $\square$

We finish this proof with several remarks. Note that the error concerns the regression function and not the coefficients: this is of course the relevant target. But there is a deeper reason; we cannot expect equally good estimates on the error for the coefficients, because of multiple representations of the function $\widetilde{\mathfrak{M}}_M$ in Φ when there are collinearities (which may take place with positive probability).

The reader could easily adapt the previous proof to the case where the functions $(\phi_j)_j$ are random and chosen adaptively with respect to the data $(O^{(1)}, \ldots, O^{(M)})$. For example, in the case of local polynomials, by partitioning $\mathbb{R}^d$ into "cubes" (non-regular) each containing the same number of data, we can better automatically adapt to the distribution of O, ensuring that a minimum of information will be available to calculate the local regression functions.

8.3 APPLICATION TO THE RESOLUTION OF THE DYNAMIC PROGRAMMING EQUATION BY EMPIRICAL REGRESSION

We assume hypotheses from Theorem 7.3.1 concerning the discretization of the backward stochastic differential equation, leading to the simulation of the dynamic programming equation (8.0.1).

In this framework, the unknown functions $u^{(h)}$ are bounded because f and g are so: this follows from (7.4.1), and the bound is explicit and given by

$$\sup_{0 \leq i \leq N} \sup_{x \in \mathbb{R}^d} |u^{(h)}(t_i, x)| \leq |f|_\infty + T|g|_\infty := L. \tag{8.3.1}$$

To *force* the numerical solution to satisfy this bound, we *clip* the empirical regression functions, obtained in what follows, at the level L (see Figure 2.4).

8.3.1 Learning sample and approximation space

We will empirically approximate the regression functions $(u^{(h)}(t_i, .) : 0 \leq i \leq N - 1)$ using the following:

1. A *learning sample* consisting of M simulated independent trajectories $((X_{t_i}^{(h,m)})_{0 \leq i \leq N} : 1 \leq m \leq M)$ having the same distribution as $(X_{t_i}^{(h)})_{0 \leq i \leq N}$.

2. At each date the approximation space is denoted by Φ. To simplify the presentation (but this is not a mathematical necessity), we take the same Φ (dictionary) for all the dates, namely we take a vector space

$$\Phi := \text{Span.}(\phi_1, \ldots, \phi_K)$$

of a finite dimension K. The reader could generalize without difficulty the result for $\Phi^{(i)}$ depending on t_i, which may be meaningful numerically.

Finally, the approximated solutions are denoted by $u^{(h,M,\Phi)}$.

8.3.2 Calculation of the empirical regression functions

To initialize the resolution of the dynamic programming equation, we set
$$u^{(h,M,\Phi)}(T,\cdot) = f(\cdot). \tag{8.3.2}$$

Suppose that an approximation $u^{(h,M,\Phi)}(t_{i+1},\cdot)$ is calculated and now let us construct $u^{(h,M,\Phi)}(t_i,\cdot)$. Set $\alpha^{(M,i)}$, the SVD-optimal coefficient defined by

$$\alpha^{(M,i)} := \arg\min_{\alpha \in \mathbb{R}^K} \frac{1}{M} \sum_{m=1}^M \Big(u^{(h,M,\Phi)}(t_{i+1}, X_{t_{i+1}}^{(h,m)}) \\ + hg\big(t_i, X_{t_i}^{(h,m)}, u^{(h,M,\Phi)}(t_{i+1}, X_{t_{i+1}}^{(h,m)})\big) \\ - \alpha \cdot \phi(X_{t_i}^{(h,m)}) \Big)^2, \tag{8.3.3}$$

which allows us to define the clipped function

$$u^{(h,M,\Phi)}(t_i,\cdot) := \mathcal{C}_L[\alpha^{(M,i)} \cdot \phi](\cdot) = -L \vee [\alpha^{(M,i)} \cdot \phi(\cdot)] \wedge L. \tag{8.3.4}$$

This procedure is then iterated backward from $i = N-1$ to 0; see Algorithm 8.1.

To measure the error on the empirical regression functions, we will (as in Section 8.2.4) use the L_2 empirical norms associated with the simulations: for a function $\varphi : \mathbb{R}^d \mapsto \mathbb{R}$ depending on the processes at a t_i, we define

$$|\varphi|_{L_2(\mu^M)} := \Big(\frac{1}{M} \sum_{m=1}^M \varphi^2(X_{t_i}^{(h,m)}) \Big)^{\frac{1}{2}}.$$

More generally, if the function $\varphi : \mathbb{R}^d \times \mathbb{R}^d \mapsto \mathbb{R}$ depends on the processes at times t_i and t_{i+1}, we set

$$|\varphi|_{L_2(\mu^M)} := \Big(\frac{1}{M} \sum_{m=1}^M \varphi^2(X_{t_i}^{(h,m)}, X_{t_{i+1}}^{(h,m)}) \Big)^{\frac{1}{2}}.$$

In general, depending on the context of use, there will not be any possible confusion between the two conventions.

1 M, N, K, m, i: **int** ; /* numbers of simulations, time steps and basis functions, variable indices */
2 $\mathcal{P}$: **vector of paths** ;   /* set of M trajectories $(X^{(h,m)})_{1 \leq m \leq M}$: the type **path** is a vector of size $N+1$ with the components filled by a simulated trajectory of $(X_{t_i}^{(h)})_{0 \leq i \leq N}$ */
3 α: **vector of coeff** ;   /* set of coefficients $(\alpha^{(M,i)})_{0 \leq i \leq N-1}$: the type **coeff** is a vector of size K */
4 L, T, h: **double** ; /* bound on solution, terminal time and time step */
5 **for** $m = 1$ **to** M **do**
6 | **for** $i = 0$ **to** N **do**
7 | | $\mathcal{P}[m][i] \leftarrow X_{t_i}^{(h,m)}$; /* generation of Euler scheme */
8 /* Initialization of the regressions   */
9 $\alpha[N-1] \leftarrow$ SVD coefficient ;   /* associated with the response functions $f(\cdot) + hg(t_{N-1}, \cdot, f(\cdot))$, to the data $(\mathcal{P}[., N-1], \mathcal{P}[., N])$ and the space Φ */
10 /* Iteration of the regressions   */
11 **for** $i = N-2$ **to** 0 **do**
12 | $\alpha[i] \leftarrow$ SVD coefficient ;   /* associated with the response function $\mathcal{C}_L[\alpha[i+1] \cdot \phi(\cdot)] + hg(t_i, \cdot, \mathcal{C}_L[\alpha[i+1] \cdot \phi(\cdot)])$, to the data $(\mathcal{P}[., i], \mathcal{P}[., i+1])$ and to the space Φ */
13 **Return** α ;   /* all the coefficients representing all the unknown functions $u^{(h,M,\Phi)}(t_i, \cdot) = \mathcal{C}_L[\alpha[i] \cdot \phi(\cdot)]$ */

Algorithm 8.1: Resolution of the dynamic programming equation by empirical regression.

260 ■ Simulation by empirical regression

8.3.3 Equation of the error propagation

We establish an equation between the average empirical quadratic errors at times t_i and t_{i+1}, describing how the errors are locally propagating. The notation is the following.

- In addition to the coefficient $\alpha^{(M,i)}$ calculated using the erroneous function $u^{(h,M,\Phi)}(t_{i+1},\cdot)$, we define the coefficient calculated with the exact function $u^{(h)}(t_{i+1},\cdot)$:

$$\overline{\alpha}^{(M,i)} := \arg\min_{\alpha \in \mathbb{R}^K} \frac{1}{M} \sum_{m=1}^{M} \Big(u^{(h)}\big(t_{i+1}, X_{t_{i+1}}^{(h,m)}\big)$$
$$+ hg\big(t_i, X_{t_i}^{(h,m)}, u^{(h)}(t_{i+1}, X_{t_{i+1}}^{(h,m)})\big)$$
$$- \alpha \cdot \phi(X_{t_i}^{(h,m)}) \Big)^2$$
$$= \arg\min_{\alpha \in \mathbb{R}^K} \Big| u^{(h)}(t_{i+1},\cdot) + hg\big(t_i, \cdot, u^{(h)}(t_{i+1},\cdot)\big) - \alpha \cdot \phi(\cdot) \Big|^2_{L_2(\mu^M)}.$$
(8.3.5)

- We denote by $\mu_{X_{t_i}^{(h)}}$ the distribution of $X_{ih}^{(h)}$.

We will justify that the error propagation satisfies the following iterative equation.

Proposition 8.3.1 (local error propagation) *Under the previous hypotheses and notation, we have for $0 \leq i \leq N-1$*

$$\mathbb{E}\Big| u^{(h,M,\Phi)}(t_i,\cdot) - u^{(h)}(t_i,\cdot) \Big|^2_{L_2(\mu^M)}$$
$$\leq (1+N^{-1})(1+hL_g)^2 \mathbb{E}\Big| u^{(h,M,\Phi)}(t_{i+1},\cdot) - u^{(h)}(t_{i+1},\cdot) \Big|^2_{L_2(\mu^M)}$$
$$+ (1+N)4L^2\frac{K}{M} + \min_{\varphi \in \Phi}\Big| \varphi - u^{(h)}(t_i,\cdot) \Big|^2_{L_2(\mu_{X_{t_i}^{(h)}})}. \quad (8.3.6)$$

PROOF:
Because $u^{(h)}$ is bounded by L and because the clipping is a 1-Lipschitz operator, we have

$$\Big| u^{(h,M,\Phi)}(t_i,\cdot) - u^{(h)}(t_i,\cdot) \Big|^2_{L_2(\mu^M)} \leq \Big| \alpha^{(M,i)} \cdot \phi(\cdot) - u^{(h)}(t_i,\cdot) \Big|^2_{L_2(\mu^M)}.$$
(8.3.7)

Application to the resolution of the dynamic programming equation ■ 261

▷ **Decomposition into three contributions.** To simplify the presentation, write $X_{t_i}^{(h,1:M)} = \{X_{t_i}^{(h,m)} : 1 \leq m \leq M\}$. By Proposition 8.2.3-iii), $\mathbb{E}(\overline{\alpha}^{(M,i)}|X_{t_i}^{(h,1:M)})$ is the SVD-optimal coefficient associated with the observations $X_{t_i}^{(h,1:M)}$ and with the responses

$$\mathbb{E}\left(u^{(h)}(t_{i+1}, X_{t_{i+1}}^{(h,m)}) + hg(t_i, X_{t_i}^{(h,m)}, u^{(h)}(t_{i+1}, X_{t_{i+1}}^{(h,m)}))\big|X_{t_i}^{(h,1:M)}\right)$$
$$= \mathbb{E}\left(u^{(h)}(t_{i+1}, X_{t_{i+1}}^{(h,m)}) + hg(t_i, X_{t_i}^{(h,m)}, u^{(h)}(t_{i+1}, X_{t_{i+1}}^{(h,m)}))\big|X_{t_i}^{(h,m)}\right)$$
$$= u^{(h)}(t_i, X_{t_i}^{(h,m)}).$$

So, the Pythagoras decomposition provides the equality

$$\left|\alpha^{(M,i)} \cdot \phi(\cdot) - u^{(h)}(t_i, \cdot)\right|^2_{L_2(\mu^M)}$$
$$= \inf_{\varphi \in \Phi}\left|\varphi(\cdot) - u^{(h)}(t_i, \cdot)\right|^2_{L_2(\mu^M)} \tag{8.3.8}$$
$$+ \left|(\alpha^{(M,i)} - \mathbb{E}(\overline{\alpha}^{(M,i)}|X_{t_i}^{(h,1:M)})) \cdot \phi(\cdot)\right|^2_{L_2(\mu^M)}.$$

Then, using the triangular inequality for the $L_2(\mu^M)$ norm and the inequality $(a+b)^2 \leq (1+N)a^2 + (1+N^{-1})b^2$, we obtain (combined with (8.3.7))

$$\left|u^{(h,M,\Phi)}(t_i, \cdot) - u^{(h)}(t_i, \cdot)\right|^2_{L_2(\mu^M)}$$
$$\leq \inf_{\varphi \in \Phi}\left|\varphi(\cdot) - u^{(h)}(t_i, \cdot)\right|^2_{L_2(\mu^M)}$$
$$+ (1+N)\left|(\overline{\alpha}^{(M,i)} - \mathbb{E}(\overline{\alpha}^{(M,i)}|X_{t_i}^{(h,1:M)})) \cdot \phi(\cdot)\right|^2_{L_2(\mu^M)}$$
$$+ (1+N^{-1})\left|(\alpha^{(M,i)} - \overline{\alpha}^{(M,i)}) \cdot \phi(\cdot)\right|^2_{L_2(\mu^M)}. \tag{8.3.9}$$

▷ **Estimation of the different terms.** The first term is connected to the *approximation* of $u^{(h)}(t_i, \cdot)$ in the space Φ:

$$\mathbb{E}\left(\inf_{\varphi \in \Phi}\left|\varphi(\cdot) - u^{(h)}(t_i, \cdot)\right|^2_{L_2(\mu^M)}\right)$$
$$\leq \inf_{\varphi \in \Phi}\mathbb{E}\left(\frac{1}{M}\sum_{m=1}^{M}|\varphi(X_{t_i}^{(h,m)}) - u^{(h)}(t_i, X_{t_i}^{(h,m)})|^2\right)$$
$$= \inf_{\varphi \in \Phi}\left|\varphi(\cdot) - u^{(h)}(t_i, \cdot)\right|^2_{L_2(\mu_{X_{t_i}^{(h)}})}. \tag{8.3.10}$$

In the second term of (8.3.9), we retrieve the *estimation error* in the calculation of the regression function $u^{(h)}(t_i, \cdot)$ given the responses

$(u^{(h)}(t_{i+1}, X^{(h,m)}_{t_{i+1}}) + hg(t_i, X^{(h,m)}_{t_i}, u^{(h)}(t_{i+1}, X^{(h,m)}_{t_{i+1}})))_{1 \leq m \leq M}$ and the observations $(X^{(h,m)}_{t_i})_{1 \leq m \leq M}$. As in the proof of Theorem 8.2.4 (first term on the right of the equality (8.2.13)), we can bound this term from above in expectation:

$$\mathbb{E}\left(\left|(\overline{\alpha}^{(M,i)} - \mathbb{E}(\overline{\alpha}^{(M,i)}|X^{(h,1:M)}_{t_i})) \cdot \phi(\cdot)\right|^2_{L_2(\mu^M)}\right) \leq (L + h|g|_\infty)^2 \frac{K}{M} \tag{8.3.11}$$

where $(L + h|g|_\infty)^2$ is an upper bound, uniform in x, of the variance of $u^{(h)}(t_{i+1}, X^{(h)}_{t_{i+1}}) + hg(t_i, X^{(h)}_{t_i}, u^{(h)}(t_{i+1}, X^{(h)}_{t_{i+1}}))$ conditionally on $X^{(h)}_{t_i} = x$.

Finally, concerning the last term of (8.3.9), we first invoke the non-expanding property of the empirical projection, then the Lipschitz regularity of g:

$$\left|(\alpha^{(M,i)} - \overline{\alpha}^{(M,i)}) \cdot \phi(\cdot)\right|^2_{L_2(\mu^M)}$$
$$\leq \left|[u^{(h,M,\Phi)}(t_{i+1}, \cdot) + hg(t_i, \cdot, u^{(h,M,\Phi)}(t_{i+1}, \cdot))]\right.$$
$$\left. - [u^{(h)}(t_{i+1}, \cdot) + hg(t_i, \cdot, u^{(h)}(t_{i+1}, \cdot))]\right|^2_{L_2(\mu^M)}$$
$$\leq (1 + hL_g)^2 \left|u^{(h,M,\Phi)}(t_{i+1}, \cdot) - u^{(h)}(t_{i+1}, \cdot)\right|^2_{L_2(\mu^M)}. \tag{8.3.12}$$

This upper bound highlights the effects of the error propagation along the dynamic programming equation.

▷ **End of the proof.** To sum up, combining the inequalities (8.3.9)-(8.3.10)-(8.3.11)-(8.3.12) and taking the expectation yields

$$\mathbb{E}\left|u^{(h,M,\Phi)}(t_i, \cdot) - u^{(h)}(t_i, \cdot)\right|^2_{L_2(\mu^M)}$$
$$\leq \inf_{\varphi \in \Phi} \left|\varphi(\cdot) - u^{(h)}(t_i, \cdot)\right|^2_{L_2(\mu_{X^{(h)}_{t_i}})} + (1+N)(L + h|g|_\infty)^2 \frac{K}{M}$$
$$+ (1+N^{-1})(1+hL_g)^2 \mathbb{E}\left|u^{(h,M,\Phi)}(t_{i+1}, \cdot) - u^{(h)}(t_{i+1}, \cdot)\right|^2_{L_2(\mu^M)}.$$

We conclude the result using a rough bound $(L + h|g|_\infty)^2 \leq 4L^2$. □

The previous result provides an iterative relation for the error $\mathbb{E}\left|u^{(h,M,\Phi)}(t_i, \cdot) - u^{(h)}(t_i, \cdot)\right|^2_{L_2(\mu^M)}$, which becomes zero for $i = N$. Using $(1+hL_g)^2(1+N^{-1}) \leq 1 + \frac{1}{N}(1+2L_gT)^2$ and the same calculation as for (7.3.7), we easily get the following.

Theorem 8.3.2 (global error measured in empirical norm)
With the previous notation and hypotheses, we have for any $0 \leq i \leq N-1$

$$\mathbb{E}\left|u^{(h,M,\Phi)}(t_i,\cdot) - u^{(h)}(t_i,\cdot)\right|^2_{L_2(\mu^M)}$$
$$\leq e^{(1+2L_gT)^2} \sum_{j=i}^{N-1} \left[(1+N)4L^2\frac{K}{M} + \min_{\varphi\in\Phi}\left|\varphi - u^{(h)}(t_j,\cdot)\right|^2_{L_2(\mu_{X_{t_j}^{(h)}})}\right].$$
(8.3.13)

To complete the analysis and to be coherent with the norm used in the quantification of the discretization error from the previous chapter, we would like to measure the error in norm $|\cdot|_{L_2(\mu_{X_{t_i}^{(h)}})}$ instead of $|\cdot|_{L_2(\mu^M)}$. We will use a variant of Theorem 2.4.10 from Chapter 2, which we will prove at the end.

Proposition 8.3.3 (deviation between empirical mean and exact mean) With the previous notation and hypotheses, we have for $0 \leq i \leq N-1$

$$\mathbb{E}\left|\left|u^{(h,M,\Phi)}(t_i,\cdot) - u^{(h)}(t_i,\cdot)\right|^2_{L_2(\mu^M)} - \left|u^{(h,M,\Phi)}(t_i,\cdot) - u^{(h)}(t_i,\cdot)\right|^2_{L_2(\mu_{X_{t_i}^{(h)}})}\right|$$
$$\leq 101L^2\sqrt{\frac{(K+1)\log(6M)}{M}}.$$

By combining Theorem 8.3.2 and Proposition 8.3.3, we deduce the main result quantifying the error of the empirical regression scheme for solving the dynamic programming equation.

Theorem 8.3.4 (global error of the empirical regression scheme) With the previous notation and hypotheses, we have for any $0 \leq i \leq N-1$

$$\mathbb{E}\left|u^{(h,M,\Phi)}(t_i,\cdot) - u^{(h)}(t_i,\cdot)\right|^2_{L_2(\mu_{X_{t_i}^{(h)}})}$$

$$\leq e^{(1+2L_gT)^2} \sum_{j=i}^{N-1} \left[(1+N)4L^2\frac{K}{M} + \min_{\varphi\in\Phi}\left|\varphi - u^{(h)}(t_j,\cdot)\right|^2_{L_2(\mu_{X^{(h)}_{t_j}})}\right]$$

$$+ 101L^2\sqrt{\frac{(K+1)\log(6M)}{M}}. \qquad (8.3.14)$$

Two contributions of different nature appear: the first term corresponds to the superposition of the empirical regression errors incurred at each date after t_i; we can observe that with this approach, they are unavoidable. There is a slight deterioration of the error control due to the factor $(1+N)$. This comes from the structure of the dynamic programming which involves multiple regression dates and nested regression problems. The second term with $\log(\cdot)$ interprets uniform fluctuations between the errors measured with respect to the empirical and the true distributions.

When $M = +\infty$, the upper bound for the error becomes just (up to a factor) $\sum_{j=i}^{N-1} \min_{\varphi\in\Phi}\left|\varphi - u^{(h)}(t_j,\cdot)\right|^2_{L_2(\mu_{X^{(h)}_{t_j}})}$, i.e., the accumulation of the approximation errors in the spaces Φ at each date, which is rather intuitive.

Proof of Proposition 8.3.3. Let us introduce the class of functions $\mathcal{G}_i = \{(\mathcal{C}_L\varphi - u^{(h)}(t_i,\cdot))^2 : \varphi \in \Phi\}$ and denote by Z the random variable for which we calculate the expectation

$$Z = \left|\left|u^{(h,M,\Phi)}(t_i,\cdot) - u^{(h)}(t_i,\cdot)\right|\right|^2_{L_2(\mu^M)}$$
$$- \left|u^{(h,M,\Phi)}(t_i,\cdot) - u^{(h)}(t_i,\cdot)\right|^2_{L_2(\mu_{X^{(h)}_{t_i}})}\right|.$$

As $u^{(h,M,\Phi)}(t_i,\cdot) \in \mathcal{C}_L\Phi$, we can write

$$\mathbb{E}(Z) = \int_0^{+\infty} \mathbb{P}(Z > \varepsilon)d\varepsilon$$
$$\leq \int_0^{+\infty} \mathbb{P}\left(\exists \psi \in \mathcal{G}_i : \left|\frac{1}{M}\sum_{m=1}^M \psi(X^{(h,m)}_{t_i}) - \int_{\mathbb{R}^d}\psi(o)\mu_{X^{(h)}_{t_i}}(do)\right| > \varepsilon\right)d\varepsilon.$$

Let us verify that $\mathcal{G}_i$ can be covered in L_1 for any probability measure μ. Let $\varepsilon' > 0$; from Theorem 2.4.6, there exists $\{\mathcal{C}_L\varphi_j : 1 \leq j \leq n\}$ a

Application to the resolution of the dynamic programming equation

$\frac{\varepsilon'}{4L}$-covering of $\mathcal{C}_L\Phi$ and for any $(\mathcal{C}_L\varphi - u^{(h)}(t_i,\cdot))^2 \in \mathcal{G}_i$ there exists j such that

$$\left|(\mathcal{C}_L\varphi - u^{(h)}(t_i,\cdot))^2 - (\mathcal{C}_L\varphi_j - u^{(h)}(t_i,\cdot))^2\right|_{L_1(\mu)}$$
$$= \left|([\mathcal{C}_L\varphi - u^{(h)}(t_i,\cdot)] - [\mathcal{C}_L\varphi_j - u^{(h)}(t_i,\cdot)])\right.$$
$$\left.\times ([\mathcal{C}_L\varphi - u^{(h)}(t_i,\cdot)] + [\mathcal{C}_L\varphi_j - u^{(h)}(t_i,\cdot)])\right|_{L_1(\mu)}$$
$$\leq |\mathcal{C}_L\varphi - \mathcal{C}_L\varphi_j|_{L_1(\mu)} 4L \leq \varepsilon'.$$

We have shown the first inequality below:

$$\mathcal{N}_1(\varepsilon', \mathcal{G}_i, \mu) \leq \mathcal{N}_1\left(\frac{\varepsilon'}{4L}, \mathcal{C}_L\Phi, \mu\right) \leq \left(\frac{24L^2}{\varepsilon'}\right)^{2(K+1)}, \qquad (8.3.15)$$

while the second is just the one from Theorem 2.4.6, valid if $0 < \varepsilon' \leq 2L^2$. Plugging this upper bound into the one from Theorem 2.4.7 (with $B = 4L^2$), we obtain

$$\mathbb{P}\left(\exists \psi \in \mathcal{G}_i : \left|\frac{1}{M}\sum_{m=1}^M \psi(X_{t_i}^{(h,m)}) - \int_{\mathbb{R}^d} \psi(o)\mu_{X_{t_i}^{(h)}}(\mathrm{d}o)\right| > \varepsilon\right)$$
$$\leq 8\left(\frac{192L^2}{\varepsilon}\right)^{2(K+1)} \exp\left(-\frac{\varepsilon^2 M}{2048L^4}\right),$$

under the condition $\varepsilon \leq 16L^2$; if this condition on ε is not satisfied, the upper bound still remains true since the probability to bound is then equal to 0 in view of the ψ-bounds equal to $4L^2$. Plugging this inequality into that for $\mathbb{E}(Z)$, we obtain for any $\varepsilon_0 \geq 0$:

$$\mathbb{E}(Z) \leq \varepsilon_0 + 8\int_{\varepsilon_0}^{+\infty} \left(\frac{192L^2}{\varepsilon}\right)^{2(K+1)} \exp\left(-\frac{\varepsilon^2 M}{2048L^4}\right) \mathrm{d}\varepsilon.$$

For $\varepsilon_0 \geq \frac{192L^2}{\sqrt{6M}}$, we deduce

$$\mathbb{E}(Z) \leq \varepsilon_0 + 8(6M)^{(K+1)} \int_{\varepsilon_0}^{+\infty} \frac{\sqrt{6M}\varepsilon}{192L^2} \exp\left(-\frac{\varepsilon^2 M}{2048L^4}\right) \mathrm{d}\varepsilon$$
$$= \varepsilon_0 + (6M)^{(K+1)} \frac{128\sqrt{6}L^2}{3\sqrt{M}} \exp\left(-\frac{\varepsilon_0^2 M}{2048L^4}\right).$$

The choice $\varepsilon_0 = \frac{L^2}{\sqrt{M}}\sqrt{2048(K+1)\log(6M)}$ verifies $(6M)^{(K+1)}\exp\left(-\frac{\varepsilon_0^2 M}{2048 L^4}\right) = 1$ and $\varepsilon_0 \geq \frac{192 L^2}{\sqrt{6M}}$. The estimate

$$\mathbb{E}(Z) \leq \frac{L^2}{\sqrt{M}}\sqrt{(K+1)\log(6M)}\left(\sqrt{2048} + \frac{128\sqrt{6}}{3\sqrt{(K+1)\log(6M)}}\right)$$

readily follows, as well as the one stated in the theorem after some simple bounds. □

8.3.4 Optimal adjustment of the convergence parameters in the case of local polynomials

We aim at getting an error of order $1/\sqrt{N}$ — i.e., $\mathbb{E}\left|u^{(h,M,\Phi)}(t_i,\cdot) - u^{(h)}(t_i,\cdot)\right|^2_{L_2(\mu_{X_{t_i}^{(h)}})} = O(N^{-1})$ — as the discretization error from Theorem 7.3.1. As in Section 8.2.5, our attention is focused on the rate of M, K as a function of N and not on the constants, and we suppose that the unknown functions $u^{(h)}(t_i, \cdot)$ are $C^{k+\alpha}(\mathbb{R}^d, \mathbb{R})$ with regularity constants uniform in i. The approximation space is that of the local polynomials of degree k, on cubes with an edge Δ, with a truncation[6] beyond the threshold $c\log(1/\Delta)$ for c large enough.

In this case, $K \sim [\log(1/\Delta)\Delta^{-1}]^d$ and the computations of Section 8.2.5 lead to

$$\sum_{j=i}^{N-1}\left[(1+N)4L^2\frac{K}{M} + \min_{\varphi \in \Phi}\left|\varphi - u^{(h)}(t_j,\cdot)\right|^2_{L_2(\mu_{X_{t_j}^{(h)}})}\right]$$
$$\leq CN\left[N\frac{[\log(1/\Delta)\Delta^{-1}]^d}{M} + \Delta^{2(k+\alpha)}\right].$$

To ensure this term is of order $1/N$, it is enough to choose $\Delta \underset{N\to+\infty}{\sim} N^{-\frac{1}{(k+\alpha)}}$ up to a constant. As for the number of simulations that allow us to have the statistical error of the same order, we must choose

$$M \underset{N\to+\infty}{\gtrsim} N^3(\log(1/\Delta)\Delta^{-1})^d \sim (\log(N))^d N^{\frac{d}{(k+\alpha)}+3}.$$

We hold the sufficient rule $M \sim N^{\frac{d}{(k+\alpha)}+3+\eta}$ with $\eta > 0$, which we

---
[6]Valid choice because the Euler scheme has exponential moments.

denote simply as $M \sim N^{\frac{d}{(k+\alpha)}+3^+}$. With this choice, the contribution with the logarithmic term in (8.3.14) is negligible with respect to the sought order N^{-1}: indeed, we have $\sqrt{\frac{K}{M}} \lesssim \frac{1}{N^{\frac{3}{2}}}$.

To sum up, we get:

$$M \underset{N \to +\infty}{\sim} N^{\frac{d}{(k+\alpha)}+3^+}. \tag{8.3.16}$$

Let us count the elementary operations of the algorithm, to analyze its complexity (up to constants):

- Cost $M \times N$ to simulate M trajectories of the Euler scheme with step T/N.

- Cost $N \times M$ to calculate the N regression coefficients (taking advantage of the algorithmic simplification about the computation of regressions on local polynomials; see discussion in Section 8.2.3).

Consequently, for a global error $\mathcal{E} \underset{N \to +\infty}{\sim} N^{-\frac{1}{2}}$, the complexity is

$$\mathcal{C} \underset{N \to +\infty}{\sim} MN \sim N^{\frac{d}{(k+\alpha)}+4^+} \sim \mathcal{E}^{-2\frac{d}{(k+\alpha)}-8^+}. \tag{8.3.17}$$

Without surprise, the algorithm suffers from the curse of dimensionality. The order of convergence is obviously worse than that of a simple Monte-Carlo evaluation of the expectation ($\mathcal{C} \sim \mathcal{E}^{-2}$), but on the other hand, the presented algorithm gives much more information because it provides N functions in the whole space $\mathbb{R}^d$, whose dependencies are non-linear. Here again, the rate (8.3.17) shows a competition between the dimension and the regularity.

The interested reader can find more efficient versions of this type of algorithm in [63] (by exploiting the dynamic programming (7.4.1) to get a to better complexity $\mathcal{E}^{-\frac{d}{(k+\alpha)}-8^+}$ improving the dependence with respect to dimension), in [13] (adding control variate) or in [62] (including importance sampling).

8.4 EXERCISES

Exercise 8.1 (Are polynomials dense in L_2?) *In Subsection 8.2.5, we have discussed how to tune parameters of local polynomial basis to* obtain a convergent approximation in L_2. When we consider a global polynomial basis $(1, x, x^2, \dots)$, the convergence is a delicate question, especially when the support of the distribution is infinite.

i) Consider a log-normal model, i.e. $O = \exp(W_1)$ where W is a Brownian motion. Show that
$$\mathbb{E}\left(O^k \sin(2\pi \log(O))\right) = 0, \quad \forall k \geq 0.$$

ii) Deduce that the best approximation (in $L_2(\mathbb{P} \circ O^{-1})$) of $\sin(2\pi \log(O))$ on a global polynomial basis is 0.

Comment: it shows that we cannot expect a square integrable function (even bounded as above) of log-normal variables to be well approximated on a global polynomial basis.

iii) A sufficient condition for polynomials to be dense in $L_2(\mathbb{P} \circ O^{-1})$ is that the random variable O has exponential moments (this is related to the Stieltjes moment problem). In the case of log-normal variable O, propose a basis for well approximating $\sin(2\pi \log(O))$.

Exercise 8.2 (regression Monte-Carlo versus nested Monte-Carlo) *A company wants to evaluate the opportunity to start at time 1 a new project which will give at time 2 a profit $f_2(X_2)$ where (X_1, X_2) is a Markov chain that describes the evolution of the uncertain economy. The Markov chain starts from a fixed known value $X_0 = x_0$. The expected profit at time 1, i.e. $u(x) = \mathbb{E}(f_2(X_2) \mid X_1 = x)$, is compared to a benchmark profit $f_1(x)$; the investment opportunity is then evaluated by measuring the mean profit excess*
$$C = \mathbb{E}\Big(\big(u(X_1) - f_1(X_1)\big)_+\Big).$$
We suppose that f_1 and f_2 are bounded.

i) **Naive Monte-Carlo approach (a.k.a. Monte-Carlo in the Monte-Carlo).** We have M i.i.d. simulations of X_1, denoted by $(X_1^m)_{1 \leq m \leq M}$, and for each X_1^m we have M i.i.d. simulations of X_2 conditionally to X_1^m, denoted by $(X_2^{m,m'})_{1 \leq m' \leq M}$.

(a) We set
$$\hat{u}(X_1^m) := \frac{1}{M} \sum_{m'=1}^{M} f_2(X_2^{m,m'}),$$

$$\hat{C} := \frac{1}{M} \sum_{m=1}^{M} (\hat{u}(X_1^m) - f_1(X_1^m))_+.$$

Explain heuristically why $\hat{C}$ is a convergent estimator of C as $M \to +\infty$. What is the computational cost of this method?

(b) Prove the bound on the quadratic error:

$$\mathbb{E}\Big((\hat{C} - C)^2\Big) \leq 2\frac{|u - f_1|_\infty^2}{M}$$
$$+ 2\mathbb{E}\left(\frac{1}{M} \sum_{m=1}^{M} (\hat{u}(X_1^m) - u(X_1^m))^2\right).$$

(c) Deduce that $\mathbb{E}\big((\hat{C} - C)^2\big) = O(M^{-1})$.

ii) **Regression Monte-Carlo.** We study how the computation of $\hat{u}$ by empirical regression can significantly improve the convergence when the function u is smooth enough. For this, consider an i.i.d. sample $(X_1^m, X_2^m)_{1 \leq m \leq M}$ with distribution (X_1, X_2) and consider a linear approximation space $\mathcal{F}$ spanned by K basis functions. Set

$$\tilde{u}(.) := \arg\inf_{\varphi \in \mathcal{F}} \frac{1}{M} \sum_{m=1}^{M} |\varphi(X_1^m) - f_2(X_2^m)|^2,$$

$$\tilde{C} := \frac{1}{M} \sum_{m=1}^{M} (\tilde{u}(X_1^m) - f_1(X_1^m))_+.$$

(a) As in i), show that (for some constant $c > 0$)

$$\mathbb{E}\Big((\tilde{C} - C)^2\Big) \leq c \left(\frac{K}{M} + \inf_{\varphi \in \mathcal{F}} \mathbb{E}|u(X_1) - \varphi(X_1)|^2\right).$$

(b) Assume that X_1 takes values in $D = [-H, H]^d$, that $u \in \mathcal{C}_b^k(D, \mathbb{R})$, and that $\mathcal{F}$ is spanned by local polynomials of degree $k-1$ on cubes with edge δ covering D. Give a bound of the quadratic error $\mathbb{E}(\tilde{C} - C)^2$.

(c) Optimize δ as a function of M, then deduce a bound on the quadratic error as a function of M.

(d) Compare the two methods and conclude according to the values of k and d.

Exercise 8.3 (exponential deviation bound and expectation)
The inequalities below are useful to bound the cost of switching from L_2-norms in empirical measure to those in true measure, and vice versa, like in Proposition 8.3.3.

i) *Assume that $(Z_M)_{M\geq 1}$ is a sequence of non-negative random variables satisfying*

$$\mathbb{P}(Z_M > \varepsilon) \leq c\left(\frac{a}{\varepsilon}\right)^K \exp(-bM\varepsilon), \quad \forall \varepsilon > 0, \quad \forall M \geq 1,$$

for some constants $K \geq 1, a > 0, b > 0, c > 0$. Prove the following bound:

$$\mathbb{E}(Z_M) \leq \frac{1}{bM}\Big(1 + [\log(c)]_+ + K\log[(1+ab)M]\Big).$$

Hint: reduce first to $c \geq 1$. Then write $\mathbb{E}(Z_M) \leq \varepsilon_0 + \int_{\varepsilon_0}^{+\infty} \mathbb{P}(Z_M > \varepsilon)d\varepsilon$ with the choice $\varepsilon_0 = \frac{1}{bM}\log\left(c((1+ab)M)^K\right) \geq \frac{a}{M(1+ab)}$ (the last inequality is proved using $\log(1+x) \geq \frac{x}{1+x}$ for $x \geq 0$).

ii) *Give a similar bound on $\mathbb{E}(Z_M)$ where*

$$\mathbb{P}(Z_M > \varepsilon) \leq c\left(\frac{a}{\varepsilon}\right)^K \exp(-bM\varepsilon^2), \quad \forall \varepsilon > 0, \quad \forall M \geq 1,$$

and $K \geq 2, a > 0, b > 0, c > 0$.

Exercise 8.4 (regression using local averaging) *We study a few bias properties of the local averaging estimator. Let $(\mathcal{C}_i : 1 \leq i \leq n)$ be n cells forming a partition of $\mathbb{R}^d$. For a point $x \in \mathbb{R}^d$, we write $\mathcal{C}(x)$ for the cell containing x.*

Consider an M-sample of the couple (Response, Observation) $(R^{(m)}, O^{(m)})_{1 \leq m \leq M}$, considered as independent random variables with the same distribution; we suppose that the variable $R^{(m)}$ is square integrable. We define the weight

$$\omega_m(x) = \frac{\mathbf{1}_{O^{(m)} \in \mathcal{C}(x)}}{\sum_{j=1}^{M} \mathbf{1}_{O^{(j)} \in \mathcal{C}(x)}}$$

(with the convention $0/0 = 0$): it is non-zero if and only if the observation $O^{(m)}$ is in the same cell as x, and in that case, its inverse is equal to the number of observations in this cell. The local averaging estimator

of $\mathfrak{M}(x) := \mathbb{E}(R \mid O = x)$ is defined by the average of responses over the observations in the neighborhood of x:

$$\mathfrak{M}_M(x) = \sum_{m=1}^{M} \omega_m(x) R^{(m)}.$$

i) Let $X \stackrel{d}{=} \mathcal{B}in(N,p)$. Prove that $\mathbb{E}\left(\frac{1}{1+X}\right) \leq \frac{1}{(N+1)p}$.

ii) Show that $\mathbb{E}(\mathfrak{M}_M(x)|O^{(1:M)}) = \sum_{m=1}^{M} \omega_m(x)\mathfrak{M}(O^{(m)})$.

iii) Suppose that $\mathfrak{M}(.) \geq 0$. Establish the following inequalities:

(a) $\mathbb{E}(\mathfrak{M}_M(x)) \leq \mathbb{E}\left(\frac{\mathbf{1}_{O \in \mathcal{C}(x)}}{\mathbb{P}(O \in \mathcal{C}(x))} \mathfrak{M}(O)\right)$.

(b) For any $x \in \mathbb{R}^d$, $\sup_{y \in \mathcal{C}(x)} \mathbb{E}(\mathfrak{M}_M(y)) \leq \sup_{y \in \mathcal{C}(x)} \mathfrak{M}(y)$.

(c) If O' is an independent copy of O and independent of the sample, then
$$\mathbb{E}(\mathfrak{M}_M(O')) \leq \mathbb{E}(\mathfrak{M}(O)).$$

CHAPTER 9

Interacting particles and non-linear equations in the McKean sense

In this last chapter, we study non-linear diffusion processes, introduced in the 1960s by McKean [111], and we tackle their simulation: processes of this type correspond to stochastic differential equations in which the distribution of the diffusion process is itself an unknown of the equation; we talk about *non-linear diffusions in the McKean sense*.

9.1 HEURISTICS

9.1.1 Macroscopic scale versus microscopic scale

For a d-dimensional process, the form of the equation is

$$\begin{cases} X_t = X_0 + \int_0^t \overline{b}^{\mu_s}(X_s)\mathrm{d}s + \int_0^t \overline{\sigma}^{\mu_s}(X_s)\mathrm{d}W_s, & t \geq 0, \\ \mu_t = \text{distribution of } X_t, \end{cases} \quad (9.1.1)$$

where we use the notation

$$\overline{\varphi}^\nu(x) := \int_{\mathbb{R}^d} \varphi(x,y)\nu(\mathrm{d}y) \quad (9.1.2)$$

for a given probability measure ν on $\mathbb{R}^d$ and a measurable function $\varphi : \mathbb{R}^d \times \mathbb{R}^d$. Here the coefficients of X are described by two functions $b, \sigma : (x,y) \in \mathbb{R}^d \times \mathbb{R}^d \mapsto b(x,y), \sigma(x,y)$. In Part **B**, drift and diffusion coefficients do not depend on the variable y, leading to a standard diffusion process ($\bar{b}^\mu = b$ and $\bar{\sigma}^\mu = \sigma$): here the coefficients depend on the distribution of X and not only on its position. In (9.1.1), the initial condition X_0 is a random variable independent of W.

Taking $\sigma(x,y) = \sigma(x)$ and applying the Itô formula (4.4.4) in expectation[1] to the function $f \in \mathcal{C}^2$ with compact support, we obtain that $\mathbb{E}(f(X_t)) - \mathbb{E}(f(X_0))$ is equal to

$$\int_{\mathbb{R}^d} f(x)\mu_t(\mathrm{d}x) - \int_{\mathbb{R}^d} f(x)\mu_0(\mathrm{d}x)$$

$$= \int_0^t \Bigl(\sum_{i=1}^d \int_{\mathbb{R}^d} \partial_{x_i} f(x) \bar{b}_i^{\mu_s}(x) \mu_s(\mathrm{d}x)$$

$$+ \frac{1}{2} \sum_{i,j=1}^d \int_{\mathbb{R}^d} \partial_{x_i,x_j}^2 f(x) [\sigma\sigma^\mathsf{T}]_{i,j}(x) \mu_s(\mathrm{d}x)\Bigr) \mathrm{d}s$$

$$= \int_0^t \int_{\mathbb{R}^d} f(x) \Bigl[-\sum_{i=1}^d \partial_{x_i}\bigl(\mu_s \bar{b}_i^{\mu_s}(\cdot)\bigr)$$

$$+ \frac{1}{2} \sum_{i,j=1}^d \partial_{x_i,x_j}^2 \bigl(\mu_s [\sigma\sigma^\mathsf{T}]_{i,j}(\cdot)\bigr) \Bigr] (\mathrm{d}x)\,\mathrm{d}s$$

where the last equality holds in the sense of distributions. So, the solution process (9.1.1) gives a probabilistic interpretation to the non-linear partial differential equation written for the probability measures $(\mu_t)_t$:[2]

$$\begin{cases} \partial_t \mu_t = \dfrac{1}{2} \sum_{i,j=1}^d \partial_{x_i,x_j}^2 \bigl(\mu_t [\sigma\sigma^\mathsf{T}]_{i,j}(x)\bigr) \\ \qquad - \sum_{i=1}^d \partial_{x_i}\Bigl(\mu_t \int_{\mathbb{R}^d} b_i(x,y)\mu_t(\mathrm{d}y)\Bigr), \quad t \geq 0,\, x \in \mathbb{R}^d, \\ \mu_0 \quad \text{given probability measure (distribution of } X_0). \end{cases}$$

(9.1.3)

A particular case of this equation (when $\sigma \equiv 0$) is the Vlasov (1908–1975) kinetic equation, fundamental in the physics of plasmas: it describes the time evolution of the distribution of charged particles in

---

[1] Supposing that the conditions letting the stochastic integral contribution disappear are satisfied.

[2] The PDE is in the sense of distributions, i.e., calculated by integrating with respect to smooth test functions f.

a plasma. More generally, these equations are called *McKean-Vlasov equations*.

Essentially, the equation (9.1.1) can be seen, under certain conditions, as a mean-field limit[3] of a large number N of stochastic differential equations:

$$\begin{cases} X_t^{(i,N)} = X_0^{(i,N)} + \dfrac{1}{N}\sum_{j=1}^{N}\int_0^t b(X_s^{(i,N)}, X_s^{(j,N)})\mathrm{d}s \\ \qquad\qquad + \dfrac{1}{N}\sum_{j=1}^{N}\int_0^t \sigma(X_s^{(i,N)}, X_s^{(j,N)})\mathrm{d}W_s^i, \quad t\geq 0, \quad 1\leq i \leq N. \end{cases}$$
(9.1.4)

Here, each stochastic differential equation $X^{(i,N)}$ is driven by a Brownian motion W^i independent of the others, and the N stochastic differential equations interact through their coefficients averaged over all the processes. The system (9.1.4) is interpreted as a description at the microscopic scale of *interacting physical particles* in the same environment, while the initial model (9.1.1) gives a macroscopic point of view for a representative particle.

The different points of view (given by deterministic or stochastic equations (9.1.1-9.1.3-9.1.4)) are useful and complement each other in applications: either for the numerical simulation by a Monte-Carlo method of the probability measure μ_t, the solution of (9.1.3), or to understand how the microscopic averaging operates at a macroscopic level.

In this chapter, we assume regular interaction hypotheses, which largely simplify the results of existence and limit proofs. We follow the approach of Sznitman [137]. The irregular (or singular) case is much more difficult. Furthermore, the dependence of the averaged coefficients $\bar{b}^\mu$ and $\bar{\sigma}^\mu$ could be non-linear in μ. For extensions and more complete references, we refer the reader to [137], [12], [65], [83].

9.1.2 Examples and applications

We give several illustrations of the interacting particle phenomena.

▷ **Aggregation phenomena in population dynamics or economics.** The aggregation phenomena in the populations or social networks offer a

---

[3] In Theorem 9.3.1, we prove this limit and give an estimation of the speed of convergence.

natural setting for interaction models. The interactions of the McKean-Vlasov type are studied in [117]; for example, in [16], it is described how *Polyergus rufescens* ant-soldiers — each of their N positions corresponds to a stochastic differential equation $X^{(i,N)}$ — organize themselves and gather in colonies for looting. For applications in economics, see [40].

▷ **Interbank system stability and systemic risk.** Consider a financial system with N banks; the treasuries of the banks depend on each other because when they need short-term funding, they can borrow money from each other. Denote by $X^{(i,N)}$ the logarithm of the monetary reserve of the bank i. Simply modeling the interbank lending system (see [42]), we get the following dynamics

$$X_t^{(i,N)} = X_0^{(i,N)} + \int_0^t \frac{\alpha}{N} \sum_{j=1}^N (X_s^{(j,N)} - X_s^{(i,N)}) ds$$
$$+ \sigma W_t^i, \quad t \geq 0, \quad 1 \leq i \leq N,$$

the factor $\alpha \geq 0$ is interpreted as an authorized interbank lending rate. This parameter α can be adjusted by the regulation authorities to guarantee better stability of the system. The equation above is of the form (9.1.4).

▷ **Random matrices.** Given an integer $N \geq 1$ and standard independent Brownian motions $(W^{i,j}, 1 \leq i < j \leq N)$ and $(W^i, 1 \leq i \leq N)$, let us define a symmetric random matrix A of size N:

$$A_{i,i}(t) = W_t^i / \sqrt{N/2} \quad \text{and} \quad A_{i,j}(t) = A_{j,i}(t) = W_t^{i,j} / \sqrt{N}.$$

Its N real-valued increasing eigenvalues $\lambda_1(t) \leq \cdots \leq \lambda_N(t)$ are interacting stochastic processes, because of the re-ordering; by Dyson [34], they satisfy the dynamics

$$d\lambda_i(t) = \frac{1}{N} \sum_{j \neq i} \frac{1}{\lambda_i(t) - \lambda_j(t)} dt + \sqrt{\frac{2}{N}} d\widetilde{W}_t^i, \quad 1 \leq i \leq N,$$

for certain independent Brownian motions $(\widetilde{W}^i)_{1 \leq i \leq N}$. We retrieve an equation of the form (9.1.4). The corresponding solution is called *Dyson*

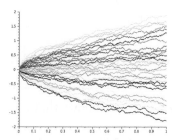 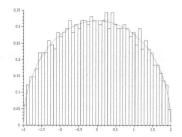

Figure 9.1 On the left, the trajectories of $N = 30$ eigenvalues associated with the Dyson Brownian motion. On the right, the histogram of the eigenvalues for $N = 1000$ at time 1 with the semicircle distribution.

Brownian motion. The drift term interprets the repulsion between the eigenvalues (see Figure 9.1 on the left). On the right of the figure, we represent the empirical measure of the eigenvalues (at time 1), which is known to converge as $N \to +\infty$ to the semicircle distribution (Wigner distribution) with density $\frac{1}{2\pi}\sqrt{4-x^2}\mathbf{1}_{|x|\leq 2}$. For more details and references, see [3, Chapter 4].

▷ **Burgers equation.** This is a partial differential equation important in fluid mechanics, introduced by Burgers [23]; it can be applied to the models of gas dynamics, to road traffic, and this is a simple model of turbulence (considered as a one-dimensional toy model of the Navier-Stokes equations). The equation takes the form

$$\partial_t v = \frac{1}{2}\sigma^2 \partial_{xx}^2 v - v\partial_x v, \tag{9.1.5}$$

where v represents the speed and $\nu = \frac{1}{2}\sigma^2$ is the viscosity coefficient. Then v is interpreted as the c.d.f. of the measure μ associated with the non-linear diffusion with $d = 1$, $\sigma(x,y) = \sigma$ and $b(x,y) = \mathbf{1}_{x>y}$ (Heaviside function). Indeed, setting $u(t,x) = \int_{-\infty}^{x} \mu_t(dy)$ and starting from (9.1.3), the reader can verify (formally) that u solves (9.1.5).

In fact, the Burgers equation admits an explicit solution and this can serve as a test for the numerical solution, before handling more complicated equations. The solution is obtained thanks to the Cole-Hopf transformation. Starting from the solution w of the heat equation

$\partial_t w = \frac{1}{2}\sigma^2 \partial_{xx}^2 w$, which we suppose to be non-zero for any t, we verify that the function $u = -\sigma^2 \frac{\partial_x w}{w}$ satisfies (9.1.5).

9.2 EXISTENCE AND UNIQUENESS OF NON-LINEAR DIFFUSIONS

To simplify the presentation, we suppose in what follows that $\sigma(x,y) = \mathrm{Id}$.

As in Part **B**, we work with a filtered probability space, from which we also require to make the initial condition X_0 (with distribution μ_0) $\mathcal{F}_0$-measurable. Under the assumption that the coefficients are Lipschitz and bounded, we show the existence and the uniqueness of (9.1.1).

Theorem 9.2.1 (existence-uniqueness) *Suppose that the coefficient $b : \mathbb{R}^d \times \mathbb{R}^d \mapsto \mathbb{R}^d$ is bounded and globally Lipschitz, and that $\sigma(x,y) = \mathrm{Id}$. Then there exists a unique solution to the non-linear diffusion equation* (9.1.1).

PROOF:

As for Theorem 4.3.1, we construct the solution using a fixed point argument. For this, we introduce the Wasserstein distance $D_T(\cdot,\cdot)$, defined on the set $\mathcal{M}(\mathcal{C}_T)$ of the probability measures on the continuous trajectories $\mathcal{C}_T := \mathcal{C}([0,T], \mathbb{R}^d)$:

$$D_T(\nu_1, \nu_2) := \inf_{\nu \in \mathcal{M}(\mathcal{C}_T \times \mathcal{C}_T) \text{ with marginals } \nu_1 \text{ and } \nu_2}$$

$$\left\{ \int (\sup_{s \leq T} |\omega_{1,s} - \omega_{2,s}| \wedge 1)\nu(\mathrm{d}\omega_1, \mathrm{d}\omega_2) \right\}.$$

This defines a complete metric on $\mathcal{M}(\mathcal{C}_T)$, whose topology is that of weak convergence. For ν in $\mathcal{M}(\mathcal{C}_T)$, let us associate $\Phi(\nu)$ the distribution of the solution defined as

$$X_t = X_0 + \int_0^t \bar{b}^{\nu_s}(X_s) \mathrm{d}s + W_t.$$

Note that this stochastic differential equation is well defined (Theorem 4.3.1) because $x \mapsto \bar{b}^{\nu_s}(x) = \int_{\mathbb{R}^d} b(x,y)\nu_s(\mathrm{d}y) = \int_{\mathcal{C}_T} b(x, \omega_s)\nu(\mathrm{d}\omega)$ is uniformly Lipschitz in x. Given two probability measures ν_1 and ν_2 in $\mathcal{M}(\mathcal{C}_T)$, compare the two stochastic differential equations corresponding to X_1 and X_2: for any coupling ν on $\mathcal{C}_t$ with marginal distributions ν_1 and ν_2, we have

$$\sup_{s \leq t} |X_{1,s} - X_{2,s}|$$

$$\leq \int_0^t \Big| \int_{\mathcal{C}_t} b(X_{1,s}, \omega_{1,s}) \nu_1(d\omega_1) - \int_{\mathcal{C}_t} b(X_{2,s}, \omega_{2,s}) \nu_2(d\omega_2) \Big| ds$$

$$\leq \int_0^t \Big[L_b |X_{1,s} - X_{2,s}|$$

$$+ \int_{\mathcal{C}_t \times \mathcal{C}_t} \big([L_b |\omega_{1,s} - \omega_{2,s}|] \wedge [2|b|_\infty] \big) \nu(d\omega_1, d\omega_2) \Big] ds,$$

where L_b is the Lipschitz constant of b.

Set $K = \max(L_b, 2|b|_\infty)$; then, taking the infimum over the couplings ν and using the Gronwall lemma, we get

$$\sup_{s \leq t} |X_{1,s} - X_{2,s}| \leq K \int_0^t \big[|X_{1,s} - X_{2,s}| + D_s(\nu_1, \nu_2) \big] ds$$

$$\leq K e^{KT} \int_0^t D_s(\nu_1, \nu_2) ds.$$

From this we deduce a precise control of $D_t(\Phi(\nu_1), \Phi(\nu_2)) \leq K e^{KT} \int_0^t D_s(\nu_1, \nu_2) ds$. We can easily conclude the result: we take $\nu \in \mathcal{M}(\mathcal{C}_T)$ and iterating this procedure, we get $D_T(\Phi^k(\nu), \Phi^{k+1}(\nu)) \leq \frac{(KTe^{KT})^k}{k!} D_T(\nu, \Phi(\nu))$, which allows us to prove easily that $(\Phi^k(\nu))_k$ is a Cauchy sequence, converging to the fixed point of Φ. This implies existence and uniqueness. Finally, we verify that the solutions constructed on the intervals $[0, T]$, with arbitrary T, are consistent in time. □

9.3 CONVERGENCE OF THE SYSTEM OF INTERACTING DIFFUSIONS, PROPAGATION OF CHAOS AND SIMULATION

Under the same hypotheses, we show that each stochastic differential equation $X^{(i,N)}$ of the particles system (9.1.4) converges to the stochastic differential equation where the empirical measure is replaced by the distribution of the process. This is the passage from the microscopic scale to the macroscopic scale mentioned at the beginning, where one particle in a much populated environment approximately behaves like in a mean-field interaction (due to the normalization with the factor $1/N$ in the equations).

Theorem 9.3.1 (convergence in L_1) *Consider the hypotheses of Theorem 9.2.1 and denote μ_t the common distribution of the $(X_t^{(i)})_i$,* which are the solutions of

$$X_t^{(i)} = X_0^{(i)} + \int_0^t \overline{b}^{\mu_s}(X_s^{(i)})ds + W_t^i, \quad t \geq 0, \quad 1 \leq i \leq N, \quad (9.3.1)$$

where $(X_0^{(i)})_i$ are independent random variables with a given distribution μ_0 and $(W^i)_i$ are independent Brownian motions.
For $t \geq 0$ and $1 \leq i \leq N$, set

$$X_t^{(i,N)} = X_0^{(i)} + \frac{1}{N}\sum_{j=1}^N \int_0^t b(X_s^{(i,N)}, X_s^{(j,N)})ds + W_t^i. \quad (9.3.2)$$

Then, for any T, we have

$$\sup_{1 \leq i \leq N} \mathbb{E}\left(\sup_{t \leq T} |X_t^{(i,N)} - X_t^{(i)}|\right) = O(N^{-1/2}).$$

PROOF:
The rate $\sqrt{N}$ is not surprising because this is the size of the fluctuations of empirical means associated with independent random variables (Chapter 2). Here, this is a bit more delicate because all the diffusions interact. By the triangular inequality, we have

$$|X^{(i,N)} - X^{(i)}|_T^* := \sup_{t \leq T} |X_t^{(i,N)} - X_t^{(i)}|$$

$$\leq \int_0^T \frac{1}{N}\sum_{j=1}^N |b(X_s^{(i,N)}, X_s^{(j,N)}) - b(X_s^{(i)}, X_s^{(j,N)})|ds$$

$$+ \int_0^T \frac{1}{N}\sum_{j=1}^N |b(X_s^{(i)}, X_s^{(j,N)}) - b(X_s^{(i)}, X_s^{(j)})|ds$$

$$+ \int_0^T |\frac{1}{N}\sum_{j=1}^N b(X_s^{(i)}, X_s^{(j)}) - \overline{b}^{\mu_s}(X_s^{(i)})|ds$$

$$\leq L_b \int_0^T |X_s^{(i,N)} - X_s^{(i)}|ds + L_b \int_0^T \frac{1}{N}\sum_{j=1}^N |X_s^{(j,N)} - X_s^{(j)}|ds$$

$$+ \int_0^T |\frac{1}{N}\sum_{j=1}^N b(X_s^{(i)}, X_s^{(j)}) - \overline{b}^{\mu_s}(X_s^{(i)})|ds.$$

By symmetry of the distributions, $\mathbb{E}|X^{(i,N)} - X^{(i)}|_T^*$ does not depend on i, and applying the Gronwall lemma, we deduce that

$$\mathbb{E}|X^{(1,N)} - X^{(1)}|_T^* = \frac{1}{N}\sum_{i=1}^N \mathbb{E}|X^{(i,N)} - X^{(i)}|_T^*$$

$$\leq 2L_b \int_0^T \frac{1}{N} \sum_{j=1}^N \mathbb{E}|X_s^{(j,N)} - X_s^{(j)}|ds$$

$$+ \int_0^T \mathbb{E}|\frac{1}{N} \sum_{j=1}^N b(X_s^{(1)}, X_s^{(j)}) - \overline{b}^{\mu_s}(X_s^{(1)})|ds$$

$$\leq e^{2L_b T} \int_0^T \mathbb{E}|\frac{1}{N} \sum_{j=1}^N b(X_s^{(1)}, X_s^{(j)}) - \overline{b}^{\mu_s}(X_s^{(1)})|ds.$$

But, by the Cauchy-Schwarz inequality,

$$\left(\mathbb{E}|\frac{1}{N} \sum_{j=1}^N b(X_s^{(1)}, X_s^{(j)}) - \overline{b}^{\mu_s}(X_s^{(1)})|\right)^2$$

$$\leq \frac{1}{N^2} \mathbb{E}\left(\sum_{j,k=1}^N (b(X_s^{(1)}, X_s^{(j)}) - \overline{b}^{\mu_s}(X_s^{(1)}))^\mathsf{T} (b(X_s^{(1)}, X_s^{(k)}) - \overline{b}^{\mu_s}(X_s^{(1)}))\right).$$

Remark that the terms for $j \neq k$ and $j \neq 1$ are zero in expectation, because on the one hand the common distribution of $X_s^{(j)}$ is μ_s and $\overline{b}^{\mu_s}$ is the average of b with respect to the second variable (i.e. $\mathbb{E}(b(x, X_s^{(j)})) = \overline{b}^{\mu_s}(x)$), and on the other hand $X^{(1)}$ and $X^{(j)}$ are independent. The terms for $j \neq k$ and $k \neq 1$ equal zero for the same reason. Thus, the only N remaining terms are for $j = k$; owing to b bounded, we deduce

$$\frac{1}{N^2} \mathbb{E}\left(\sum_{j,k=1}^N (b(X_s^{(1)}, X_s^{(j)}) - \overline{b}^{\mu_s}(X_s^{(1)}))^\mathsf{T} (b(X_s^{(1)}, X_s^{(k)}) - \overline{b}^{\mu_s}(X_s^{(1)}))\right)$$

$$\leq \frac{4|b|_\infty^2}{N}$$

and we conclude

$$\mathbb{E}|X^{(i,N)} - X^{(i)}|_T^* \leq T e^{2L_b T} \frac{2|b|_\infty}{\sqrt{N}}.$$

$\square$

Actually, beyond the behavior of a specific particle, we can describe the asymptotic dependence of a group of n particles when the total number N of particles of the environment goes to infinity. Though being in interaction, at the limit, the diffusions behave independently.

Theorem 9.3.2 (propagation of chaos) *Under the hypotheses of Theorem 9.3.1, the distribution of $(X^{(i_1,N)}, \ldots, X^{(i_n,N)})$ (for any n-tuple $1 \leq i_1 < \cdots < i_n \leq N$) is asymptotically (as $N \to +\infty$) that of an n-tuple of independent particles.*

We admit this result whose proof is beyond the scope of this book, and we refer the reader to [137].

The terminology *propagation of chaos* was introduced by Kac [85] to describe the equations of statistical physics starting from microscopic equations.

Here, the result is in agreement with the intuition that we may have when the interactions are regular and bounded: as $N \to +\infty$, the influence of a particle on another one decreases rapidly, although without disappearing completely. When the interaction potentials are irregular (Burgers equation with the Heaviside function) or singular (very localized interactions in turbulence models), the limit may become significantly more delicate to obtain. In the simulation of (9.1.1), we may also be interested in the effects of the time discretization of the process, and of finite approximation of the initial distribution; see [17].

▷ **Simulation algorithm.** The simulation of the solution of a non-linear diffusion in the McKean sense

$$\begin{cases} X_t = X_0 + \int_0^t \bar{b}^{\mu_s}(X_s)\mathrm{d}s + W_t, & t \geq 0, \\ \mu_t = \text{distribution of } X_t, \end{cases}$$

can be performed by simulating a system of N interacting particles, given by (9.3.2): this provides the empirical measure

$$\mu_t^N(\mathrm{d}x) = \frac{1}{N} \sum_{i=1}^N \delta_{X_t^{(i,N)}}(\mathrm{d}x) \qquad (9.3.3)$$

which will approximate the unknown distribution μ_t, as $N \to +\infty$. If necessary, we can then generate simulations of X by simulating a standard diffusion (as in Chapter 5) given by

$$X_t^N = X_0 + \int_0^t \bar{b}^{\mu_s^N}(X_s^N)\mathrm{d}s + W_t, \quad t \geq 0. \qquad (9.3.4)$$

Indeed, the previous results give us the convergence of μ_t^N to μ_t.

Theorem 9.3.3 (convergence of the empirical measure) *Under the hypotheses and notation of Theorem 9.3.1, for any t and any continuous function f with compact support, we have*

$$\int_{\mathbb{R}^d} f(x)\mu_t^N(\mathrm{d}x) \xrightarrow[N \to +\infty]{L_1} \int_{\mathbb{R}^d} f(x)\mu_t(\mathrm{d}x).$$

PROOF:

For any $\varepsilon > 0$, there exists a Lipschitz function f_ε such that $|f - f_\varepsilon|_\infty \leq \varepsilon/3$. Thus, because μ_t is the common distribution of the $(X^{(i)})_i$, we have

$$\mathbb{E}\left|\int_{\mathbb{R}^d} f(x)\mu_t^N(\mathrm{d}x) - \int_{\mathbb{R}^d} f(x)\mu_t(\mathrm{d}x)\right|$$

$$\leq \frac{2\varepsilon}{3} + \mathbb{E}\left|\int_{\mathbb{R}^d} f_\varepsilon(x)\mu_t^N(\mathrm{d}x) - \int_{\mathbb{R}^d} f_\varepsilon(x)\mu_t(\mathrm{d}x)\right|$$

$$\leq \frac{2\varepsilon}{3} + \mathbb{E}\left|\frac{1}{N}\sum_{i=1}^N \left(f_\varepsilon(X_t^{(i,N)}) - f_\varepsilon(X_t^{(i)})\right)\right|$$

$$+ \mathbb{E}\left|\frac{1}{N}\sum_{i=1}^N \left(f_\varepsilon(X_t^{(i)}) - \mathbb{E}(f_\varepsilon(X_t^{(i)}))\right)\right|$$

$$\leq \frac{2\varepsilon}{3} + L_{f_\varepsilon} \sup_{1\leq i \leq N} \mathbb{E}\left|X_t^{(i,N)} - X_t^{(i)}\right| + \frac{|f_\varepsilon|_\infty}{\sqrt{N}}$$

where for the last term, we have used an upper-bound on the variance as in the proof of Theorem 9.3.1. Taking N large enough, the upper-bound becomes smaller than ε, which finishes the proof.

The reader may notice that the previous arguments can be applied to the trajectories and not only to their marginals. □

APPENDIX A

Reminders and complementary results

In this appendix we gather several results throughout the book.

A.1 ABOUT CONVERGENCES

A.1.1 Convergence a.s., in probability and in L_1

The Borel-Cantelli "lemma" is a powerful tool for a.s. asymptotic analysis of events.

Theorem A.1.1 (Borel-Cantelli lemma [80, Chapter 10]) *Let $(A_n)_{n \geq 1}$ be a sequence of events in Ω.*

a) If $\sum_{n \geq 1} \mathbb{P}(A_n) < +\infty$, then $\mathbb{P}(\limsup_{n \to +\infty} A_n) = 0$, i.e. a.s. at most a finite number of $(A_n)_n$ occurs.

*b) Suppose that the events $(A_n)_n$ are independent.
If $\mathbb{P}(\limsup_{n \to +\infty} A_n) = 0$, then $\sum_{n \geq 1} \mathbb{P}(A_n) < +\infty$.*

The result below is useful for deducing the L_1 convergence from the convergence in probability, under uniform integrability conditions.

Theorem A.1.2 ([80, Chapter 27]) *Let $(X_n)_{n \geq 1}$ be a sequence of real-valued random variables. Suppose that*

a) X_n converges in probability to X, and

b) $\sup_{n \geq 1} \mathbb{E}(|X_n|^p) < +\infty$ for a certain $p > 1$.

Then X_n converges to X in L_1 and in particular $\mathbb{E}(X_n) \to \mathbb{E}(X)$.

A.1.2 Convergence in distribution

We recall that if X is a random variable with values in $\mathbb{R}^d$, the characteristic function $u \in \mathbb{R}^d \mapsto \Phi_X(u) = \mathbb{E}(e^{iu \cdot X})$ uniquely characterizes the distribution of X. For example, $\Phi_X(u) = e^{ium - \frac{1}{2}u \cdot Ku}$, with $m \in \mathbb{R}^d$ and K a symmetric non-negative matrix of size d, is the characteristic function of a d-dimensional Gaussian random variable $X \stackrel{\mathrm{d}}{=} \mathcal{N}(m, K)$ (Gaussian vector).

Theorem A.1.3 (Lévy theorem, [80, Chapter 19]) *Let $(X_n)_n$ be random variables with values in $\mathbb{R}^d$.*

a) If X_n converges in distribution to X as $n \to +\infty$, then $\Phi_{X_n}(u)$ converges to $\Phi_X(u)$ for any $u \in \mathbb{R}^d$.

b) Conversely, if for any $u \in \mathbb{R}^d$ $\Phi_{X_n}(u)$ converges to $\Phi(u)$ as $n \to +\infty$ and if Φ is continuous at 0, then this is the characteristic function of a random variable X ($\Phi = \Phi_X$) and X_n converges in distribution to X.

It is remarkable that for Gaussian variables, the convergence in distribution is equivalent to the convergence of mean and variance parameters.

Lemma A.1.4 (convergence of Gaussian random variables) *Consider a sequence $(X_n)_n$ of real-valued random variables such that $X_n \stackrel{\mathrm{d}}{=} \mathcal{N}(\mu_n, \sigma_n^2)$. Then X_n converges in distribution as $n \to +\infty$ if and only if the sequence of the parameters $(\mu_n, \sigma_n^2)_n$ converges.*

In the case of convergence, the limit distribution of X_n is Gaussian, with parameters $(\lim_n \mu_n, \lim_n \sigma_n^2)$.

PROOF:
$\Rightarrow$) Part a) of Theorem A.1.3 implies that the characteristic function of X_n, $\Phi_{X_n}(u) = \exp(iu\mu_n - \frac{1}{2}u^2\sigma_n^2)$, converges for any $u \in \mathbb{R}$: it remains to deduce from this the convergence of μ_n and σ_n^2.

The absolute value of $\Phi_{X_n}(u)$ equals $\exp(-\frac{1}{2}u^2\sigma_n^2)$ and converges: taking $u \neq 0$, we deduce the convergence of σ_n^2. It follows that $\exp(iu\mu_n)$ converges for any u, as well as

$$\int_0^\infty \exp(iu\mu_n)\exp(-u)du = \frac{1}{1 - i\mu_n}$$

by an application of the dominated convergence theorem. The convergence of μ_n is proved.

⇐) If $(\mu_n, \sigma_n^2) \to (\mu, \sigma^2)$,

$$\Phi_{X_n}(u) = e^{iu\mu_n - \frac{1}{2}u^2\sigma_n^2} \to e^{iu\mu - \frac{1}{2}u^2\sigma^2} := \Phi(u).$$

Therefore, Φ is continuous at $u = 0$, and this is the characteristic function of a random variable with the distribution $\mathcal{N}(\mu, \sigma^2)$. We conclude by b) of Theorem A.1.3. □

Theorem A.1.5 (of Slutsky [80, Chapter 18]) *Let $(X_n)_n$ be a sequence of d-dimensional random variables such that X_n converges in distribution to X.*

a) *If Y_n is such that $|X_n - Y_n|$ converges to 0 in probability, then Y_n converges in distribution to X.*

b) *If Z_n converges in probability to a constant matrix c (with d columns), then $Z_n X_n$ converges in distribution to cX.*

A.2 SEVERAL USEFUL INEQUALITIES

A.2.1 Inequalities for moments

We start by very standard inequalities, see [80, Chapter 23] for example.

- **Jensen inequality**: if X is a d-dimensional random variable and $\varphi : \mathbb{R}^d \mapsto \mathbb{R}$ is a continuous convex function, if X and $\varphi(X)$ are integrable, then
$$\varphi(\mathbb{E}(X)) \leq \mathbb{E}(\varphi(X)). \tag{A.2.1}$$

- **Hölder inequality**: if X and Y are (multidimensional) random variables having moments of order $p > 1$ and $q > 1$, respectively, for some conjugate exponents p and q (i.e. $1/p + 1/q = 1$), then
$$\mathbb{E}(|X||Y|) \leq \mathbb{E}(|X|^p)^{1/p} \mathbb{E}(|Y|^q)^{1/q}. \tag{A.2.2}$$

The most famous case is the Cauchy-Schwarz inequality for $p = q = 2$.

- **Minkowski inequality**: if X and Y are d-dimensional random variables having finite moments of order $p \geq 1$, then
$$\mathbb{E}(|X + Y|^p)^{1/p} \leq \mathbb{E}(|X|^p)^{1/p} + \mathbb{E}(|Y|^p)^{1/p}. \tag{A.2.3}$$

- **Growth of L_p norms** ($p \geq 1$): we denote by $|X|_p := \mathbb{E}(|X|^p)^{1/p}$ the L_p norm of a random variable X (possibly multidimensional), so that (A.2.3) can be seen as the triangular inequality for this norm. If $1 \leq q \leq p$, then

$$|X|_q \leq |X|_p. \tag{A.2.4}$$

This follows from the Jensen inequality, applied to the random variable $|X|^q$ and to the convex function $\varphi(x) = x^{p/q}$. When $q = 1$, this is also written in the form

$$(\mathbb{E}|X|)^p \leq \mathbb{E}(|X|^p),$$

proving the non-expanding property of expectation.

It is known that $\mathbb{E}(X)$ is the best quadratic approximation of X, meaning that $\mathbb{E}|X - \mathbb{E}(X)|^2 = \inf_x \mathbb{E}|X - x|^2$. The next result (apparently not very well known) shows that this is also true in L_p, up to a factor 2.

Proposition A.2.1 (the best approximation in L_p) *Let X be a d-dimensional random variable which is in L_p ($p \geq 1$). Then*

$$\inf_{x \in \mathbb{R}^d} |X - x|_p \leq |X - \mathbb{E}(X)|_p \leq 2 \inf_{x \in \mathbb{R}^d} |X - x|_p. \tag{A.2.5}$$

PROOF:
The left-hand side inequality is obvious. For the right-hand side, let us use (A.2.3) and (A.2.4) to write, that for any x:

$$|X - \mathbb{E}(X)|_p \leq |X - x|_p + |\mathbb{E}(x - X)|_p \leq |X - x|_p + |x - X|_p = 2|X - x|_p.$$

We can then minimize the above inequality in x and conclude. $\square$

Proposition A.2.2 (L_p upper bound for a bounded real-valued random variable) *Let X be a real-valued random variable such that $\mathbb{P}(X \in [a, b]) = 1$ for $-\infty < a < b < +\infty$. Then, for $p \geq 1$,*

$$\inf_{x \in \mathbb{R}^d} |X - x|_p \leq \frac{b - a}{2}. \tag{A.2.6}$$

In particular, the standard deviation of X is upper bounded by $\frac{b-a}{2}$.

PROOF:
It is enough to take $x = \frac{a+b}{2}$ and observe that $|X - x| \leq \frac{b-a}{2}$. In the case $p = 2$, the left-hand side term is $\sqrt{\text{Var}(X)}$, which allows us to conclude the result. □

Proposition A.2.3 (upper bound for exponential moments of a real-valued bounded random variable) *With the same notation and hypotheses as in Proposition A.2.2, for any $s \geq 0$ we have*
$$\mathbb{E}(e^{s(X-\mathbb{E}(X))}) \leq e^{s^2(b-a)^2/8}. \qquad (A.2.7)$$

PROOF:
We suppose $\mathbb{E}(X) = 0$, otherwise we can recenter X.
The convexity of the exponential function gives $e^{sX} \leq \frac{X-a}{b-a}e^{sb} + \frac{b-X}{b-a}e^{sa}$ because $X \in [a,b]$ a.s. Taking the expectation and setting $p = -a/(b-a)$ and $u = s(b-a)$, we obtain

$$\mathbb{E}(e^{sX}) \leq \frac{-a}{b-a}e^{sb} + \frac{b}{b-a}e^{sa} = (1 - p + pe^u)e^{-pu} := \exp(\Phi(u)).$$

A direct calculation proves that

$$\Phi'(u) = -p + \frac{p}{p + (1-p)e^{-u}}, \qquad \Phi''(u) = \frac{p(1-p)e^{-u}}{(p + (1-p)e^{-u})^2} \leq \frac{1}{4},$$

the upper bound coming from the inequality $\alpha\beta \leq \frac{1}{4}(\alpha + \beta)^2$ for any positive α, β. Finally, observing that $\Phi(0) = \Phi'(0) = 0$, we deduce (for a certain $\tilde{u} \in [0, u]$)

$$\Phi(u) = \Phi(0) + u\Phi'(0) + \frac{u^2}{2}\Phi''(\tilde{u}) \leq \frac{u^2}{8} = \frac{s^2(b-a)^2}{8}.$$

□

The following result gives a universal estimate of the p-th moment of a sum of independent random variables, as a function of the moments of each term in the sum (for a proof, see [125, Theorem 2.9]).

Theorem A.2.4 (Rosenthal inequality) *Let $(X_1, \ldots, X_M)$ be a sequence of independent real-valued random variables, having finite moments of order $p \geq 2$, with $\mathbb{E}(X_m) = 0$. Then there exists a certain universal constant c_p, such that*

$$\mathbb{E}\Big(\Big|\sum_{m=1}^M X_m\Big|^p\Big) \leq c_p\Big(\sum_{m=1}^M \mathbb{E}(|X_m|^p) + \Big(\sum_{m=1}^M \mathbb{E}(X_m^2)\Big)^{p/2}\Big).$$

A.2.2 Inequalities in the deviation probabilities

To quantify the upper deviations of a real-valued random variable X, the simplest way is to use the Markov inequality, valid if $X \geq 0$ a.s.:

$$\mathbb{P}(X \geq x) \leq \mathbb{E}\left(\frac{X}{x} 1_{X \geq x}\right) \leq \frac{\mathbb{E} X}{x}, \quad x > 0. \quad (\text{A.2.8})$$

For a random variable X that is not sign-constant, we may get a result using a non-negative increasing function $\varphi : \mathbb{R} \mapsto \mathbb{R}^+$ and the inclusion

$$\{X \geq x\} \subset \{\varphi(X) \geq \varphi(x)\},$$

which leads to

$$\mathbb{P}(X \geq x) \leq \frac{\mathbb{E}(\varphi(X))}{\varphi(x)}.$$

Taking $\varphi(x) = x^q$ ($q > 0$) and the random variable $|X - \mathbb{E}(X)|$, we obtain an estimation of the tail distribution as a function of the moments of X:

$$\mathbb{P}(|X - \mathbb{E} X| \geq x) \leq \frac{\mathbb{E}(|X - \mathbb{E} X|^q)}{x^q}, \quad x > 0. \quad (\text{A.2.9})$$

The case $q = 2$ is the well-known Chebychev inequality.

The Chernov inequality (or Chebychev exponential inequality) is rather based on exponential functions $\varphi(x) = e^{sx}$ for $s \geq 0$. Here, we get

$$\mathbb{P}(X - \mathbb{E} X \geq x) \leq e^{-sx} \mathbb{E}(e^{s(X - \mathbb{E} X)}), \quad x \geq 0, s \geq 0. \quad (\text{A.2.10})$$

These upper bounds and their different versions have multiple consequences; we present only one of them which we use a lot. To learn more, see [18].

Theorem A.2.5 (Hoeffding inequality) *Let $(X_1, \ldots, X_M)$ be a sequence of independent bounded real-valued random variables such that $\mathbb{P}(X_m \in [a_m, b_m]) = 1$. Then for any $\varepsilon > 0$ we have*

$$\mathbb{P}\left(\left|\frac{1}{M} \sum_{m=1}^{M}(X_m - \mathbb{E}(X_m))\right| > \varepsilon\right) \leq 2 \exp\left(-\frac{2M\varepsilon^2}{\frac{1}{M}\sum_{m=1}^{M}|b_m - a_m|^2}\right).$$

PROOF:

Let us use the inequality (A.2.10), the independence of $(X_m)_m$, and Proposition A.2.3:

$$\mathbb{P}\left(\frac{1}{M}\sum_{m=1}^{M}(X_m - \mathbb{E}(X_m)) > \varepsilon\right)$$

$$\leq e^{-s\varepsilon}\mathbb{E}\left(\exp\left(s\sum_{m=1}^{M}(X_m - \mathbb{E}(X_m))/M\right)\right)$$

$$= e^{-s\varepsilon}\prod_{m=1}^{M}\mathbb{E}\left(\exp\left(s(X_m - \mathbb{E}(X_m))/M\right)\right)$$

$$\leq e^{-s\varepsilon}\prod_{m=1}^{M}\exp\left(s^2\frac{(b_m - a_m)^2}{8M^2}\right), \quad s \geq 0.$$

Taking the smallest upper bound, i.e. for $s = \frac{4M^2\varepsilon}{\sum_{m=1}^{M}(b_m-a_m)^2}$, we get

$$\mathbb{P}\left(\frac{1}{M}\sum_{m=1}^{M}(X_m - \mathbb{E}(X_m)) > \varepsilon\right) \leq \exp\left(-\frac{2M\varepsilon^2}{\frac{1}{M}\sum_{m=1}^{M}|b_m - a_m|^2}\right).$$

We bound $\mathbb{P}\left(\frac{1}{M}\sum_{m=1}^{M}(X_m - \mathbb{E}(X_m)) < -\varepsilon\right)$ in the same way, working with $-X_m$ instead of X_m. This finishes the proof. □

Bibliography

[1] Y. Achdou and O. Pironneau. *Computational Methods for Option Pricing.* SIAM series, Frontiers in Applied Mathematics, Philadelphia, 2005.

[2] G. Allaire. *Numerical Analysis and Optimization.* Oxford University Press, 2007.

[3] G.W. Anderson, A. Guionnet, and O. Zeitouni. *An Introduction to Random Matrices.* Cambridge University Press, 2009.

[4] C. Ané, S. Blachère, D. Chafaï, P. Fougères, I. Gentil, F. Malrieu, C. Roberto, and G. Scheffer. *Sur les inégalités de Sobolev logarithmiques*, volume 10 of *Panoramas et Synthèses*. Société Mathématique de France, Paris, 2000.

[5] S. Asmussen and P.W. Glynn. *Stochastic Simulation: Algorithms and Analysis.* Stochastic Modelling and Applied Probability 57. New York, NY: Springer, 2007.

[6] S. Asmussen, P.W. Glynn, and J. Pitman. Discretization error in simulation of one-dimensional reflecting Brownian motion. *Ann. Appl. Probab.*, 5(4):875–896, 1995.

[7] R. Avikainen. On irregular functionals of SDEs and the Euler scheme. *Finance and Stochastics*, 13:381–401, 2009.

[8] F. Bach and E. Moulines. Non-asymptotic analysis of stochastic approximation algorithms for machine learning. *Advances in Neural Information Processing Systems (NIPS)*, 2011.

[9] V. Bally and D. Talay. The law of the Euler scheme for stochastic differential equations: I. Convergence rate of the distribution function. *Probab. Theory Related Fields*, 104-1:43–60, 1996.

[10] N. Bartoli and P. Del Moral. *Simulation et algorithmes stochastiques.* Cépaduès-Editions, 2001.

[11] W.F. Bauer. The Monte-Carlo method. *J. Soc. Indust. Appl. Math.*, 6(4):438–451, 1958.

[12] S. Benachour, B. Roynette, D., Talay, and P. Vallois. Nonlinear self-stabilizing processes. I: Existence, invariant probability, propagation of chaos. *Stochastic Processes Appl.*, 75(2):173–201, 1998.

[13] C. Bender and J. Steiner. Least-squares Monte-Carlo for BSDEs. In R. Carmona, P. Del Moral, P. Hu, and N. Oudjane, editors, *Numerical Methods in Finance*, pages 257–289. Series: Springer Proceedings in Mathematics, Vol. 12, 2012.

[14] A. Benveniste, M. Metivier, and P. Priouret. *Adaptive Algorithms and Stochastic Approximations*. Springer-Verlag, New York, 1990.

[15] C. Bernardi and Y. Maday. Spectral methods. In *Handbook of Numerical Analysis, Vol. V*, Handb. Numer. Anal., V, pages 209–485. North-Holland, Amsterdam, 1997.

[16] S. Boi, V. Capasso, and D. Morale. Modeling the aggregative behavior of ants of the species *Polyergus rufescens*. *Nonlinear Anal., Real World Appl.*, 1(1):163–176, 2000.

[17] M. Bossy and D. Talay. A stochastic particle method for the McKean-Vlasov and the Burgers equation. *Math. Comp.*, 66(217):157–192, 1997.

[18] S. Boucheron, G. Lugosi, and P. Massart. *Concentration Inequalities. A Nonasymptotic Theory of Independence*. Clarendon Press, Oxford, 2013.

[19] L. Breiman. *Probability*. Society for Industrial and Applied Mathematics (SIAM), Philadelphia, PA, 1992. Corrected reprint of the 1968 original.

[20] P. Bremaud. *Markov Chains: Gibbs Fields, Monte-Carlo Simulation, and Queues*, volume 31. Springer Science & Business Media, 1999.

[21] M. Broadie, P. Glasserman, and S. Kou. A continuity correction for discrete barrier options. *Mathematical Finance*, 7:325–349, 1997.

[22] M. Broadie and O. Kaya. Exact simulation of stochastic volatility and other affine jump diffusion processes. *Oper. Res.*, 54(2):217–231, 2006.

[23] J.M. Burgers. A mathematical model illustrating the theory of turbulence. *Adv. in Appl. Mech.*, 1:171–199, 1948.

[24] R.E. Caflisch. Monte-Carlo and quasi-Monte-Carlo methods. In *Acta Numerica, 1998*, volume 7 of *Acta Numer.*, pages 1–49. Cambridge Univ. Press, Cambridge, 1998.

[25] J. W. Cahn and S. M. Allen. A macroscopic theory for antiphase boundary motion and its application to antiphase domain coarsening. *Acta Metall.*, 27:1084–1095, 1979.

[26] O. Cappé, E. Moulines, and T. Rydén. *Inference in Hidden Markov Models*. Springer Series in Statistics. Springer, New York, 2005.

[27] M. Cessenat, R. Dautray, G. Ledanois, P.L. Lions, E. Pardoux, and R. Sentis. *Méthodes probabilistes pour les équations de la physique*. Collection CEA, Eyrolles, 1989.

[28] K.L. Chung. *A Course in Probability Theory*. Academic Press Inc., San Diego, CA, third edition, 2001.

[29] B.A. Cipra. The Best of the 20th Century: Editors Name Top 10 Algorithms. *SIAM News* http://www.siam.org/news/news.php?id=637, 33(4), 2000.

[30] J.C. Cox, J.E. Ingersoll, and S.A. Ross. A theory of the term structure of interest rates. *Econometrica*, 53(2):385–407, 1985.

[31] L. Devroye. *Nonuniform Random Variate Generation*. Springer-Verlag, New York, 1986.

[32] M. Duflo. *Random Iterative Models*, volume 34 of *Applications of Mathematics*. Springer-Verlag, Berlin, 1997. Translated from the 1990 French original by Stephen S. Wilson and revised by the author.

[33] R. Durrett. *Brownian Motion and Martingales in Analysis*. Wadsworth Mathematics Series. Wadsworth International Group, Belmont, CA, 1984.

[34] F. J. Dyson. A Brownian-motion model of the eigenvalues of a random matrix. *J. Mathematical Phys.*, 3:1191–1198, 1962.

[35] R. Eckhardt. Stam Ulam, John Von Neumann and the Monte-Carlo method. *Los Alamos Science*, Special Issue:131–143, 1987.

[36] B. Efron. *The Jackknife, the Bootstrap and Other Resampling Plans*, volume 38 of *CBMS-NSF Regional Conference Series in Applied Mathematics*. Society for Industrial and Applied Mathematics (SIAM), Philadelphia, Pa., 1982.

[37] N. El Karoui, S. Hamadène, and A. Matoussi. Backward stochastic differential equations and applications. In R. Carmona, editor, *Indifference Pricing: Theory and Applications*, Chapter 8, pages 267–320. Springer-Verlag, 2008.

[38] N. El Karoui, S.G. Peng, and M.C. Quenez. Backward stochastic differential equations in finance. *Math. Finance*, 7(1):1–71, 1997.

[39] W.J. Ewens. *Mathematical Population Genetics. I: Theoretical Introduction.* Interdisciplinary Mathematics 27. New York, NY: Springer, second edition, 2004.

[40] W. Finnoff. Law of large numbers for a general system of stochastic differential equations with global interaction. *Stochastic Processes and Their Applications*, 46(1):153–182, 1993.

[41] R.A. Fisher. The advance of advantageous genes. *Ann. Eugenics*, 7:335–369, 1937.

[42] J-P. Fouque and L.H. Sun. Systemic risk illustrated. In J-P. Fouque and J. Langsam, editors, *Handbook on Systemic Risk*. Cambridge University Press, 2013.

[43] M.I. Freidlin. *Functional Integration and Partial Differential Equations*. Annals of Mathematics Studies – Princeton University Press, 1985.

[44] A. Friedman. *Partial Differential Equations of Parabolic Type*. Prentice-Hall, 1964.

[45] A. Friedman. *Stochastic Differential Equations and Applications. Vol. 1.* New York, San Francisco, London: Academic Press, a subsidiary of Harcourt Brace Jovanovich, Publishers. XIII, 1975.

[46] A. Friedman. *Stochastic Differential Equations and Applications. Vol. 2.* New York, San Francisco, London: Academic Press, a subsidiary of Harcourt Brace Jovanovich, Publishers. XIII, 1976.

[47] N. Frikha and S. Menozzi. Concentration bounds for stochastic approximations. *Electronic Communications in Probability*, 17:1–15, 2012.

[48] D. Funaro. *Polynomial Approximation of Differential Equations*, volume 8 of *Lecture Notes in Physics. New Series: Monographs.* Springer-Verlag, Berlin, 1992.

[49] J. Galambos. *The Asymptotic Theory of Extreme Order Statistics.* R.E. Kreiger, Malabar, FL, 1987.

[50] G.R. Gavalas. *Nonlinear Differential Equations of Chemically Reacting Systems.* Springer Tracts in Natural Philosophy, Vol. 17. Springer-Verlag New York Inc., New York, 1968.

[51] D. Geman. Random fields and inverse problems in imaging. In *École d'été de Probabilités de Saint-Flour XVIII—1988*, volume 1427 of *Lecture Notes in Math.*, pages 113–193. Springer, Berlin, 1990.

[52] S. Geman and D. Geman. Stochastic relaxation, Gibbs distributions, and the Bayesian restoration of images. *IEEE Trans. Pattern Anal. Mach. Intell.*, 6:721–741, 1984.

[53] D. Gilbarg and N.S. Trudinger. *Elliptic Partial Differential Equations of Second Order.* Springer Verlag, second edition, 1983.

[54] M.B. Giles. Multilevel Monte-Carlo path simulation. *Operation Research*, 56:607–617, 2008.

[55] P. Glasserman, P. Heidelberger, and P. Shahabuddin. Asymptotically optimal importance sampling and stratification for pricing path-dependent options. *Math. Finance*, 9(2):117–152, 1999.

[56] P. Glasserman and D.D. Yao. Some guidelines and guarantees for common random numbers. *Management Science*, 38(6):884–908, 1992.

[57] P.W. Glynn and W. Whitt. The asymptotic validity of sequential stopping rules for stochastic simulations. *Ann. Appl. Probab.*, 2(1):180–198, 1992.

[58] E. Gobet. Euler schemes and half-space approximation for the simulation of diffusions in a domain. *ESAIM: Probability and Statistics*, 5:261–297, 2001.

[59] E. Gobet. Advanced Monte-Carlo methods for barrier and related exotic options. In P.G. Ciarlet, A. Bensoussan, and Q. Zhang, editors, *Handbook of Numerical Analysis, Vol. XV, Special Volume: Mathematical Modeling and Numerical Methods in Finance*, pages 497–528. Elsevier, Netherlands: North-Holland, 2009.

[60] E. Gobet and S. Menozzi. Stopped diffusion processes: boundary corrections and overshoot. *Stochastic Processes and Their Applications*, 120:130–162, 2010.

[61] E. Gobet and R. Munos. Sensitivity analysis using Itô-Malliavin calculus and martingales. Application to stochastic control problem. *SIAM Journal of Control and Optimization*, 43:5:1676–1713, 2005.

[62] E. Gobet and P. Turkedjiev. Adaptive importance sampling in least-squares Monte-Carlo algorithms for backward stochastic differential equations. In revision for *Stochastic Processes and Their applications*, 2015.

[63] E. Gobet and P. Turkedjiev. Linear regression MDP scheme for discrete backward stochastic differential equations under general conditions. *Math. Comp.*, 85(299):1359–1391, 2016.

[64] G. Golub and C.F. Van Loan. *Matrix Computations*, 3rd ed. Baltimore, MD: The Johns Hopkins Univ. Press. xxvii, 694 p.

[65] C. Graham and S. Méléard. Probabilistic tools and Monte-Carlo approximations for some Boltzmann equations. *ESAIM, Proc.*, 10:77–126, 2001.

[66] L. Gross. Logarithmic Sobolev inequalities. *Amer. J. Math.*, 97(4):1061–1083, 1975.

[67] J. Guyon and P. Henry-Labordère. *Nonlinear Pricing*. CRC Financial Mathematics. Chapman and Hall, 2014.

[68] I. Gyöngy and M. Rásonyi. A note on Euler approximations for SDEs with Hölder continuous diffusion coefficients. *Stochastic Processes Appl.*, 121(10):2189–2200, 2011.

[69] L. Gyorfi, M. Kohler, A. Krzyzak, and H. Walk. *A Distribution-Free Theory of Nonparametric Regression.* Springer Series in Statistics, 2002.

[70] M. Hairer, M. Hutzenthaler, and A. Jentzen. Loss of regularity for Kolmogorov equations. *Annals of Probability*, 43(2):468–527, 2015.

[71] B. Hajek. Cooling schedules for optimal annealing. *Math. Oper. Res.*, 13(2):311–329, 1988.

[72] F.H. Harlow and N. Metropolis. Computing and computers: Weapons simulation leads to the computer era. *Los Alamos Science*, Winter/Spring:132–141, 1983.

[73] T. Hastie, R. Tibshirani, and J. Friedman. *The Elements of Statistical Learning: Data Mining, Inference, and Prediction.* Springer Series in Statistics. New York, NY: Springer, second edition, 2009.

[74] W.K. Hastings. Monte-Carlo sampling methods using Markov chains and their applications. *Biometrika*, 57:97–109, 1970.

[75] S. Heinrich. Multilevel Monte-Carlo Methods. In *LSSC '01 Proceedings of the Third International Conference on Large-Scale Scientific Computing*, volume 2179 of *Lecture Notes in Computer Science*, pages 58–67. Springer-Verlag, 2001.

[76] D. Henry. *Geometric Theory of Semilinear Parabolic Equations*, volume 840 of *Lecture Notes in Mathematics*. Springer-Verlag, Berlin, 1981.

[77] S.L. Heston. A closed-form solution for options with stochastic volatility with applications to bond and currency options. *The review of Financial Studies*, 6(2):327–343, 1993.

[78] A.L. Hodgkin, A.F. Huxley, and B. Katz. Measurement of current-voltage relations in the membrane of the giant axon of Loligo. *Journal of Physiology*, available at http://www.sfn.org/skins/main/pdf/HistoryofNeuroscience/hodgkin1.pdf, 116:424–448, 1952.

[79] K. Itô and H.P. McKean. *Diffusion Processes and Their Sample Paths.* Berlin-Heidelberg-New York: Springer-Verlag, 1965.

[80] J. Jacod and P. Protter. *Probability Essentials*. Springer, second edition, 2003.

[81] A. Jentzen, T. Müller-Gronbach, and L. Yaroslavtseva. On stochastic differential equations with arbitrary slow convergence rates for strong approximation. arXiv:1506.02828, 2015.

[82] B. Jourdain and J. Lelong. Robust adaptive importance sampling for normal random vectors. *Ann. Appl. Probab.*, 19(5):1687–1718, 2009.

[83] B. Jourdain, S. Méléard, and W.A. Woyczynski. Nonlinear SDEs driven by Lévy processes and related PDEs. *ALEA, Lat. Am. J. Probab. Math. Stat.*, 4:1–29, 2008.

[84] M. Kac. On distributions of certain Wiener functionals. *Trans. Amer. Math. Soc.*, 65:1–13, 1949.

[85] M. Kac. Foundations of kinetic theory. *Proc. 3rd Berkeley Sympos. Math. Statist. Probability* 3, 171–197, 1956.

[86] M. Kac. *Integration in Function Spaces and Some of Its Applications*. Accademia Nazionale dei Lincei, Pisa, 1980. Lezioni Fermiane.

[87] J. Kaipio and E. Somersalo. *Statistical and Computational Inverse Problems*, volume 160. Springer Science & Business Media, 2006.

[88] I. Karatzas and S.E. Shreve. *Brownian Motion and Stochastic Calculus*. Springer Verlag, second edition, 1991.

[89] A. Kebaier. Romberg extrapolation: A new variance reduction method and applications to option pricing. *Ann. Appl. Probab.*, 15(4):2681–2705, 2005.

[90] A.G.Z. Kemna and A.C.F. Vorst. A pricing method for options based on average asset values. *J. Banking Finan.*, 14:113–129, 1990.

[91] A.J. Kinderman and J.F. Monahan. Computer generation of random variables using the ratio of uniform deviates. *ACM Transactions on Mathematical Software (TOMS)*, 3(3):257–260, 1977.

[92] P.E. Kloeden and E. Platen. *Numerical Solution of Stochastic Differential Equations.* Springer Verlag, 1995.

[93] A. Kolmogoroff. Über die Summen durch den Zufall bestimmter unabhängiger Größen. *Math. Ann.*, 99(1):309–319, 1928.

[94] A.N. Kolmogorov, I.G. Petrovsky, and N.S. Piskunov. Etude de l'équation de la diffusion avec croissance de la quantité de matière et son application à un problème biologique. *Bulletin Université d'état à Moscou, Série internationale A 1*, pages 1–26, 1937.

[95] H.A. Kramers. Brownian motion in a field of force and the diffusion model of chemical reactions. *Physica*, 7, 1940.

[96] P. Krée and C. Soize. *Mathematics of Random Phenomena*, volume 32 of *Mathematics and its Applications*. D. Reidel Publishing Co., Dordrecht, 1986.

[97] H. Kunita. *Stochastic Flows and Stochastic Differential Equations.* Cambridge Studies in Advanced Mathematics. 24. Cambridge: Cambridge University Press, 1997.

[98] H.J. Kushner and G.G. Yin. *Stochastic Approximation and Recursive Algorithms and Applications*, volume 35 of *Applications of Mathematics (New York)*. Springer-Verlag, New York, second edition, 2003. Stochastic Modelling and Applied Probability.

[99] M.T. Lacey and W. Philipp. A note on the almost sure central limit theorem. *Statist. Probab. Lett.*, 9(3):201–205, 1990.

[100] O.A. Ladyzenskaja, V.A. Solonnikov, and N.N. Ural'ceva. *Linear and Quasi-Linear Equations of Parabolic Type.* Vol. 23 of Translations of Mathematical Monographs, American Mathematical Society, Providence, 1968.

[101] T. Lagache and D. Holcman. Effective motion of a virus trafficking inside a biological cell. *SIAM J. Appl. Math.*, 68(4):1146–1167, 2008.

[102] P. L'Ecuyer and C. Lemieux. Recent advances in randomized Quasi-Monte-Carlo methods. In *Modeling Uncertainty: An Examination of Stochastic Theory, Methods, and Applications*, M. Dror, P. L'Ecuyer, and F. Szidarovszki, eds., pages 419–474. Kluwer Academic Publishers, 2002.

[103] M. Ledoux. Concentration of measure and logarithmic Sobolev inequalities. In *Séminaire de Probabilités, XXXIII*, volume 1709 of *Lecture Notes in Math.*, pages 120–216. Springer, Berlin, 1999.

[104] V. Lemaire and G. Pages. Unconstrained recursive importance sampling. *Annals of Applied Probability*, 20(3):1029–1067, 2010.

[105] G.M. Lieberman. *Second Order Parabolic Differential Equations*. World Scientific Publishing Co. Inc., River Edge, NJ, 1996.

[106] J.S. Liu. *Monte-Carlo Strategies in Scientific Computing*. Springer Series in Statistics. Springer-Verlag, New York, 2001.

[107] F. Longstaff and E.S. Schwartz. Valuing American options by simulation: A simple least squares approach. *The Review of Financial Studies*, 14:113–147, 2001.

[108] J. Ma and J. Yong. *Forward-Backward Stochastic Differential Equations*. Lecture Notes in Mathematics, 1702, Springer-Verlag, 1999. A course on stochastic processes.

[109] A.W. Marshall and I. Olkin. Families of multivariate distributions. *J. Amer. Statist. Assoc.*, 83(403):834–841, 1988.

[110] G. Maruyama. Continuous Markov processes and stochastic equations. *Rendiconti del Circolo Matematico di Palermo, Serie II*, 4:48–90, 1955.

[111] H.P. McKean. A class of Markov processes associated with nonlinear parabolic equations. *Proc. Natl. Acad. Sci. USA*, 56:1907–1911, 1966.

[112] H.P. McKean. Application of Brownian motion to the equation of Kolmogorov-Petrovskii-Piskunov. *Comm. Pure Appl. Math.*, 28(3):323–331, 1975.

[113] A.J. McNeil, R. Frey, and P. Embrechts. *Quantitative Risk Management*. Princeton series in finance. Princeton University Press, 2005.

[114] N. Metropolis. The beginning of the Monte-Carlo method. *Los Alamos Science*, Special Issue:125–130, 1987.

[115] N. Metropolis, A.W. Rosenbluth, M.N. Rosenbluth, A.H. Teller, and E. Teller. Equations of state calculations by fast computing machines. *Journal of Chemical Physics*, 21(6):1087–1092, 1953.

[116] N. Metropolis and S. Ulam. The Monte-Carlo method. *Journal of the American Statistical Association*, 44:335–341, 1949.

[117] D. Morale, V. Capasso, and K. Oelschläger. An interacting particle system modelling aggregation behavior: From individuals to populations. *J. Math. Biol.*, 50(1):49–66, 2005.

[118] B. Moro. The full Monte. *Risk*, 8:57–58, Feb. 1995.

[119] M. Musiela and M. Rutkowski. *Martingale Methods in Financial Modelling*. Springer Verlag, second edition, 2005.

[120] J. Neveu. *Discrete-Parameter Martingales*, volume 10. Elsevier, 1975.

[121] N.J. Newton. Variance reduction for simulated diffusions. *SIAM Journal on Applied Mathematics*, 54(6):1780–1805, 1994.

[122] H. Niederreiter. *Random Number Generation and Quasi-Monte-Carlo Methods*, volume 63 of *CBMS-NSF Regional Conference Series in Applied Mathematics*. Society for Industrial and Applied Mathematics (SIAM), Philadelphia, PA, 1992.

[123] E. Pardoux. BSDEs, weak convergence and homogenization of semilinear PDEs. In *Nonlinear Analysis, Differential Equations and Control (Montreal, QC, 1998)*, volume 528 of *NATO Sci. Ser. C Math. Phys. Sci.*, pages 503–549. Kluwer Acad. Publ., Dordrecht, 1999.

[124] C.J. Pérez, J. Martín, C. Rojano, and F.J. Girón. Efficient generation of random vectors by using the ratio-of-uniforms method with ellipsoidal envelopes. *Statistics and Computing*, 18(2):209–217, 2008.

[125] V.V. Petrov. *Limit Theorems of Probability Theory*, volume 4 of *Oxford Studies in Probability*. The Clarendon Press Oxford University Press, New York, 1995.

[126] A. Rasulov, G. Raimova, and M. Mascagni. Monte-Carlo solution of Cauchy problem for a nonlinear parabolic equation. *Math. Comput. Simulation*, 80(6):1118–1123, 2010.

[127] D. Revuz and M. Yor. *Continuous Martingales and Brownian Motion*. Comprehensive Studies in Mathematics. Berlin: Springer, third edition, 1999.

[128] C. Rhee and P.W. Glynn. A new approach to unbiased estimation for SDEs. In C. Laroque, J. Himmelspach, R. Pasupathy, O. Rose, and A. M. Uhrmacher, editors, *Proceedings of the 2012 Winter Simulation Conference*, pages 495–503, 2012.

[129] C.P. Robert and G. Casella. *Monte-Carlo Statistical Methods*. Springer Texts in Statistics. Springer-Verlag, New York, second edition, 2004.

[130] G. Rubino and B. Tuffin. *Rare Event Simulation Using Monte-Carlo Methods*. Wiley, 2009.

[131] P.A. Samuelson. Proof that properly anticipated prices fluctuate randomly. *Industrial Management Review*, 6:42–49, 1965.

[132] M. Schmidt, N. Le Roux, and F. Bach. Minimizing finite sums with the Stochastic Average Gradient. *hal-00860051*, 2013.

[133] N. Shigesada and K. Kawasaki, editors. *Biological Invasions: Theory and Practice*. Oxford Series in Ecology and Evolution, Oxford: Oxford University Press, 1997.

[134] A. Sklar. Fonctions de répartition à n dimensions et leurs marges. *Publ. Inst. Statist. Univ. Paris*, 8:229–231, 1959.

[135] A.V. Skorokhod. Branching diffusion processes. *Theor. Probab. Appl.*, 9:445–449, 1965.

[136] J. Smoller. *Shock Waves and Reaction-Diffusion Equations*, volume 258 of *Fundamental Principles of Mathematical Sciences*. Springer-Verlag, New York, second edition, 1994.

[137] A.S. Sznitman. Topics in propagation of chaos. In *Saint Flour Probability Summer School—1989*, volume 1464 of *Lecture Notes in Math.*, pages 164–251. Springer, Berlin, 1991.

[138] S. Meyn and R.L. Tweedie. *Markov Chains and Stochastic Stability*. Cambridge University Press, Cambridge, second edition, 2009.

[139] D. Talay and L. Tubaro. Expansion of the global error for numerical schemes solving stochastic differential equations. *Stochastic Analysis and Applications*, 8-4:94–120, 1990.

[140] S. Ulam. Random processes and transformations. *Proceedings of the International Congress of Mathematicians*, 2:264–275, 1950.

[141] O. Vasicek. An equilibrium characterisation of the term of structure. *Journal of Financial Economics*, 5:177–188, 1977.

[142] A. Visintin. *Models of Phase Transitions*. Progress in Nonlinear Differential Equations and their Applications, 28. Birkhäuser Boston Inc., Boston, MA, 1996.

[143] S. Wasserman. *All of Statistics: A Concise Course in Statistical Inference*. Springer Texts in Statistics. Springer, 2004.

[144] S. Wasserman and P. Pattison. Logit models and logistic regressions for social networks: I. An introduction to Markov random graphs and p*. *Psychometrika*, 61(3):401–426, 1996.

[145] P. Wesseling. *An Introduction to Multigrid Methods*. Pure and Applied Mathematics (New York). John Wiley & Sons Ltd., Chichester, 1992.

Index

ε-cover
    covering number, 72, 73, 265
    of a set, 72

Brownian motion
    Brownian bridge, 125
    Dyson equation, 276
    finite differences method for PDEs, 130
    Gaussian process, 120
    heat equation, 128
    Itô formula, 132
    Levy construction, 127
    properties, 124
    quadratic variation, 132
    random walk, 119
    Wiener approximation, 122

central limit theorem
    confidence intervals, 55
    sectioning method, 57
    substitution method, 56

clipping, 74

concentration
    confidence intervals, 68, 194
    empirical measure versus true measure, 263
    exponential bounds, 67
    Gaussian, 77, 84
    logarithmic Sobolev inequality, 79
    uniform
        on a class of functions, 70
        on a clipped vector space, 76
        uniform deviation in L_1, 77

confidence intervals
    central limit theorem, 55
    confidence intervals, 194
    multi-level method, 200
    non asymptotic, 68
    Student distribution, 57

convergence
    almost sure central limit theorem, 66
    approximation by local polynomials, 253
    Berry-Essen bounds, 65
    central limit theorem, 52
    confidence intervals, 194
    dynamic programming equation, 260
    Edgeworth expansion, 66
    empirical regression, 251
    Gaussian random variables, 286
    law of iterated logarithm, 66
    law of large numbers, 49
    propagation of chaos, 281
    random walk, 119
    uniform law of large numbers, 76

curse of dimensionality, 11, 243

diffusion process, 138

entropy, 78

Euler scheme, 164
    backward equation and

dynamic programming
equation, 228
exit time, 178
simulation, 166
strong convergence, 170
weak convergence, 173

Feynman-Kac formula and PDE
Cauchy condition, 144
Cauchy-Dirichlet condition, 149
comparison with Monte-Carlo method, 131
diffusion process, 142
Dirichlet condition, 153
elliptic equation, 148
heat equation and Brownian motion, 128
KPP equation and branching process, 235
semi-linear PDE and backward equations, 221

Gaussian process, 120
Gaussian vector, 43, 286
generation of random variable, 32
acceptance-rejection method, 38
Archimedean copula, 46
Beta distribution, 39
Cauchy distribution, 34
conditional distribution, 37
discrete distribution, 33
exponential distribution, 33
Gamma distribution, 40
Gaussian distribution, 35, 36, 43
geometric distribution, 34
inversion method, 33
Pareto distribution, 35
Rayleigh distribution, 34
Student distribution, 42
Weibull distribution, 35
with dependency, 44
generation of random variables
Gamma distribution, 48
Gaussian distribution, 48

inequality
exponential, 78, 289
Hoeffding, 69, 290
logarithmic Sobolev, 78
most used probabilistic inequalities, 287
infinitesimal generator, 143
Itô formula
Brownian motion, 132
diffusion process, 145
Itô process, 137

Milshtein scheme, 187
modeling
aggregation model, 275
Allen-Cahn equation in materials physics, 217, 223
arithmetic Brownian motion, 139
Burgers equation, 277
diffusion process, 138
equilibrium model, 142
Fisher-Wright, 141
geometric Brownian motion, 140
Hodgkin-Huxley model in neuroscience, 216, 223
KPP equation in ecology, 215, 223
McKean-Vlasov diffusion equation, 273

Ornstein-Uhlenbeck process, 140
portfolio optimization in finance, 218, 224
random matrices, 276
reaction/diffusion equation, 215
square root process, 141
stochastic control, 219, 224
stochastic differential equation, 138
systemic bank risk, 276

random number generator, 31
Quasi Monte-Carlo, 12

sectioning method, 57
sensitivity, 61
   Bismut-Elworthy-Li formula, 155
   diffusion process, 153
   Euler scheme, 178
   Gaussian distribution, 63, 65
   likelihood method, 63, 155
   Ornstein-Uhlenbeck process, 160
   pathwise differentiation method, 63, 154
   Poisson distribution, 64
   resimulation method, 62
simulation of process
   branching process, 234
   Brownian bridge, 125
   Brownian motion, 123, 124
   dynamic programming equation and backward stochastic equations, 257
   Euler scheme, 166
   exit time, 178
   McKean interaction, 279

Milshtein scheme, 186
multi-level method, 197
randomized multi-level method, 202
sensitivity, 178
stochastic integral
   localization, 146
   properties, 134
   stochastic differential equation, 138
   Wiener integral, 136
substitution method
   bias, 59
   central limit theorem, 56
   convex function, 61

variance reduction
   antithetic sampling, 89
   conditioning, 92
   control variates, 94
      diffusion process, 206
   importance sampling, 97
      adaptive methods, 110
      diffusion process, 207
      Esscher transform, 107
   important sampling
      change of mean and variance, 103
   stratification, 92

For Product Safety Concerns and Information please contact our
EU representative GPSR@taylorandfrancis.com Taylor & Francis
Verlag GmbH, Kaufingerstraße 24, 80331 München, Germany